Integrated Maintenance and Energy Management in the Chemical Industries

Kiran R. Golwalkar

Integrated Maintenance and Energy Management in the Chemical Industries

 Springer

Kiran R. Golwalkar
Consulting Chemical Engineer
Nagpur, Maharashtra, India

ISBN 978-3-030-32525-1   ISBN 978-3-030-32526-8   (eBook)
https://doi.org/10.1007/978-3-030-32526-8

© Springer Nature Switzerland AG 2019
This work is subject to copyright. All rights are reserved by the Publisher, whether the whole or part of the material is concerned, specifically the rights of translation, reprinting, reuse of illustrations, recitation, broadcasting, reproduction on microfilms or in any other physical way, and transmission or information storage and retrieval, electronic adaptation, computer software, or by similar or dissimilar methodology now known or hereafter developed.
The use of general descriptive names, registered names, trademarks, service marks, etc. in this publication does not imply, even in the absence of a specific statement, that such names are exempt from the relevant protective laws and regulations and therefore free for general use.
The publisher, the authors, and the editors are safe to assume that the advice and information in this book are believed to be true and accurate at the date of publication. Neither the publisher nor the authors or the editors give a warranty, express or implied, with respect to the material contained herein or for any errors or omissions that may have been made. The publisher remains neutral with regard to jurisdictional claims in published maps and institutional affiliations.

This Springer imprint is published by the registered company Springer Nature Switzerland AG
The registered company address is: Gewerbestrasse 11, 6330 Cham, Switzerland

Dedicated to those maintenance and operating technicians who work hard in a spirit of cooperation for the safe, pollution free, and efficient running of the chemical industries.

Preface

Chemical industries provide many essential products for meeting the daily requirements of the citizens and also for the economic growth of a country.

Managements run the chemical industries successfully by meeting the product quality and delivery schedules specified by the clients while operating safely and efficiently without causing environmental pollution.

It is well known that the cost of production depends on the yield of raw materials, smooth working of the plant, energy consumption, and recovery. However, it is very difficult to control the cost of raw material which depends on the availability as per specifications and certain external factors like weather conditions, transport bottlenecks, power tariff, etc.

Hence, it is very important to have an efficient integrated maintenance and energy management for ensuring smooth, efficient running of the chemical plant which can reduce the cost of production. *It is also very important to keep in mind that the selling price of products will depend on the competition in market.*

A large number of chemical industries need considerable energy (mainly electrical and thermal) for carrying out various operations like handling and pre-treatment of raw materials, operation of the reactors, further processing (purification) of crude products, and safe disposal of any hazardous waste generated. Senior engineers should look into all these activities carefully for reducing the need for energy.

The operating cost of the process units and machinery can thus be reduced by integrated maintenance and energy management (minimising losses, conservation of energy, and maximising recovery from process streams).

Conflicting Priorities

Important departments of the industry such as purchase and sales, production plants, maintenance section for process units and machinery, stores (inventory control), and utility arrangements (water, power, and fuels) face different situations and have their own priorities in their working areas. Human resources, research and development, logistics, and finance also have their own priorities.

However, priorities of these departments/sections may not be concurrent.

Managements shall prepare guidelines and instructions for these various departments so that all of them understand each other's difficulties and work in close cooperation to meet the common goal of successful running of the organization while ensuring safety in all activities.

Revenue Allocation

The revenue obtained by sale of products shall be judiciously allocated to procure (i) necessary raw materials and other inputs; (ii) spares for maintenance of process units, plant machinery, energy recovery units and facilities for effluents treatment; (iii) wages to employees; (iv) provision for depreciation of equipments; (v) expenses on research and development; (vi) taxes to statutory authorities; and (vii) payment to investors.

Managements of running plants also have to consider and take decisions for:

- Acquiring idle plants (which may be available at low cost) and reviving them
- Expansion of production capacity to meet increased demand
- Diversification to new products for increasing the presence in market
- Modernisation of existing process plants for improving efficiency
- Making certain changes in the plant for safer working and/or better pollution control

Suggestions can be generated in-house after consultation with various departments for all of the above to get improved design or better construction of processing units.

This book gives guidelines for:

Integrated Maintenance: Planning and Implementation
(Very important for a safe, efficient plant which does not cause environmental pollution):

(i) Assistance for preparing purchase orders for process units and machinery (see below), (ii) ensuring safety of human beings and plant assets, (iii) assistance to increase equipment life, (iv) guidance for operating the plant within safe limits as recommended by OEM, (v) regular condition monitoring of equipments and preventive maintenance, (vi) analysis of breakdowns, (vii) arranging timely repairs, (viii) replacement of components by procuring correct spares, (ix) planning for major maintenance jobs and safe procedures for all activities

Procurement activities: classification of process units, materials and machinery
- Capital equipments such as *process units and machinery*

- Process inputs such as *raw materials*, *catalysts*, *stabilisers*, *tower internals*, and *filter aids*
- Equipments related to energy and utilities such as *waste heat recovery boilers*, *steam turbine/gas turbine driven generators*, *electrical motors*, and *refrigeration plants*

Maintenance engineers have the responsibility to maintain the chemical plant so that it can work in a safe, smooth, pollution free, and efficient manner.

Their jobs involve not only looking after machineries like pumps, compressors, conveyors, and electrical equipments but also working in close cooperation with process operation engineers and plant operators for the cleaning of process units like reactors, converters, evaporators, and absorption towers, thorough inspection of their internals, and repair/replace the faulty units.

This may require entering inside such processing units and remain there for several hours for cleaning, inspection, and repairs till the units are ready for restart.

Assistance to Plant Management

The senior management must involve chemical process engineers along with maintenance engineers when selecting the process technology and machinery when a new plant is set up, an idle plant is revived, or an existing plant is diversified.

These engineers have to apply their knowledge and experience to check whether the process units and machinery will be able to deliver the required performance in terms of rated capacity per day and total production per annum (which depends on total running time per year), while handling the various chemicals in severe operating conditions. The design and construction of the equipments should be carefully checked during the procurement of machinery itself so that deterioration of performance shall not occur (less output, not able to meet quality, development of unsafe conditions) during operation of the plant and heavy costs are not incurred due to frequent breakdowns.

The required performance, necessary features, material of constructions, operating conditions etc shall be frozen and clearly mentioned in the inquiry document for process units and machinery.

Drafting the purchase orders: terms and conditions to be included
For a new plant: safe, efficient process units and machinery with proper conditions for performance guarantees for output and environmental pollution control

Expansion/diversification of existing plant: possibility for improving the plant layout, better process control by the use of appropriate *well-maintained* instrumentation, maximising energy efficiency

(Terms and conditions for performance guarantee for the machinery must be specified in the purchase order and agreed upon by the supplier. Include terms and conditions for annual maintenance contract also if required.)

Maintenance of Plant:
- Process units along with their internals (absorption tower internals, filter cloth, etc.)
- Machinery essential for the safe and smooth operation
- Effluent treatment facilities for controlling environmental pollution
- Important electrical equipments such as transformers, motors, rectifiers
- Instrumentation and civil works

Inventory Control

Keeping adequate stock of essential inputs required for process units, Effluent treatment Plant

Maintaining sufficient spares for machinery in the plant

Energy Management: Maximizing Energy Efficiency
- Reduction of consumption by improving the plant layout, for easier handling of procurement of the right materials
- Minimisation of losses of energy by better refractory lining and insulation
- Evaluation of potential for energy recovery
- Selection, operation, and maintenance of efficient heat recovery equipments
- Cogeneration (power and heat)

Useful tips are also given for (i) increasing life of plant units, (ii) good lubrication of machinery, (iii) reducing vibrations in running units for better performance, and (iv) repairs to protective lining of process units.

These tips are useful as they can minimise the breakdowns and hence can reduce energy losses (which occur during frequent stop and restart of the plant).

The main aim of the book is to acknowledge the responsible hard work put in by the maintenance teams of various disciplines and to create an understanding among production, marketing, finance, and human resources departments to work in cooperation.

The book also aims to highlight the management of (i) maintenance of process units and their internal equipment *and* (ii) maintenance of the machinery used in the plant *for* (iii) safe and smooth operation thereby maximising energy efficiency of the production unit without causing environmental pollution.

The book thus aims to highlight the **Integrated Maintenance and Energy Management** for the successful running of a chemical plant.

This book has therefore been written in a simplified manner for a better appreciation of the practical situations by the readers (who could be plant managers, fresh engineers, and students of chemical, mechanical electrical, and industrial engineering).

Suggestions are welcome from the readers for including more information and improving the book further.

Nagpur, Maharashtra, India   Kiran R. Golwalkar

Acknowledgement

I am grateful to the following executives:

Mr Michael Luby, Senior Publishing Editor, Engineering (Springer Science + Business Media) New York, USA, for the permission to refer the following books published by Springer for information related to waste heat recovery boiler, steam turbines, electrical systems, converters, procurement of equipments, safety matters, cooling towers, history cards, etc.

1. *Process Equipment Procurement in the Chemical and Related Industries*
2. *Production Management of the Chemical Industries*

Mr Brian Halm, Project Coordinator, Springer Science+Business Media, New York, USA, for the support and guidance *in writing the present book.*

I am also especially grateful to **Mr Peter Shyamal N.**, Production Editor (Books), Springer Nature, SPi, Chennai (India), Mr Arun Pandian and Ms Priyanka for their support to complete this book.

I wish to thank the managements of the following organisations for providing the permissions to use the data and information related to their products:

1. SKG Refractories, MIDC Industrial Area, Butibori, India
2. Evergreen Technologies, Mumbai, India
3. V K Engineers, MIDC Tarapur, India
4. Calderys India Refractories Ltd, Mount Road, Sadar, Nagpur, India

Many useful suggestions were received from my colleagues, seniors, vendors of process units, machinery, and energy recovery units and the technicians who operated and maintained them. These have also been used while writing this book. I thank all of them.

I thank my students Chirag Thakur and Ms Shruti Bhoyar, who were helpful in writing this book.

Mr Shripad Paraskar gave sincere support for typing the manuscript and for preparing some drawings.

I have referred to my class notes, standard textbooks on chemical engineering, *Chemical Engineers' Handbook*, technical literature advertised by suppliers, and my own experience of working in India and other countries for the procurement, commissioning, operation, and maintenance of various chemical plants.

I have referred to standard textbooks on chemical process industries, *Chemical Engineers' Handbook*, my own classroom notes, and the practical experience gained while working in various capacities for design, modernising, diversification, erection, commissioning, operation, and maintenance of chemical industries in India (GRASIM Industries, Gopalanand Rasayan, Kesoram Rayon, Tanfac Industries, Parksons Dyestuff, IDI Chemicals Ltd, National Rayon Corporation Ltd, SMS Infrastructure, Vidula Chemicals and Manufacturing Industries, V K Engineers, Evergreen Technologies, SKG Refractories, Calderys India, NEAT Consultancy Ltd, etc.), in Kenya (East Africa Heavy Chemicals), Thailand (Thai Rayon Co.), and Indonesia (PT Indo-Bharat Rayon) for writing on various process units, their construction, plant machinery, and their operation and maintenance in various chapters of the book.

Various advertisements by manufacturers of process units, plant machinery, and heat recovery equipments as well as technical discussions with sales engineers, technical representatives, and experienced operating and maintenance engineers of these units have also been used for the present book.

I am deeply grateful to **Respected Mr Michael Luby,** Senior Publishing Editor, Engineering (Springer Science + Business Media) New York, USA, for the permission to refer to the above books published by Springer for matters related to Waste Heat Recovery Boiler, Steam Turbines, electrical systems, converters, procurement of equipments, safety matters, cooling towers, etc and wish to acknowledge the same.

Kiran Golwalkar

Contents

Chapter 1
Introduction to Some Important Chemical Industries

1.1 Introduction to Some Important Chemical Industries

Process units, machineries and equipments in many chemical industries are operated at high temperatures, high pressures and/or are subjected to corrosive and erosive conditions.

Typical equipments used in chemical plants are:

- Crusher, pulveriser, grinders, calciners, melters, filters for pre-treatment of raw materials
- Conveyors of various types
- Process units operating at high temperatures
- Reactors, condensers, waste heat recovery units
- Water treatment, effluent treatment, air pollution control
- Steam heated (jacketed/traced) process units, pipes, pumps and storages
- Cooling systems
- Electrical equipments

1.2 Importance of Process Units and Machinery

Production engineers and maintenance engineers should carefully understand properties of all materials being handled. Material Safety Data Sheets (**MSDS**) as well as the operating conditions of the process units and machinery in the plants must be carefully looked in to.

Some typical units used for carrying out reactions are process reactors (specially operating at high pressures) while units like filters, melters, calciners are used for pre-treatment of raw materials. Heat exchangers are necessary for heating and cooling of process streams for efficient running of the plant. Condensers are required for condensation of reaction products and recovery of volatile materials; while

© Springer Nature Switzerland AG 2019
K. R. Golwalkar, *Integrated Maintenance and Energy Management in the Chemical Industries*, https://doi.org/10.1007/978-3-030-32526-8_1

absorption towers are necessary for completing the reactions, for absorbing valuable volatile materials and for pollution control.

It is necessary to operate and maintain all of them always in proper working condition for **safe, pollution free and smooth working of the process plant**.

Careful monitoring of process operations and condition of equipments is required to prevent any mishap since many of these materials being handled are corrosive, inflammable/dangerous and high temperature/high pressure units are involved.

Statutory rules must be complied with to produce, store, handle, operate and maintain the production facilities at site.

Certain mechanical equipments in these process plants need considerable energy for their operation and careful attention during operation. They operate almost continuously and are subject to erosion and corrosion. Some typical equipments and machinery required in these plants are highlighted in the section following description of the processes.

The actual process design, machinery required, installation, operation and maintenance efforts required will depend on product mix, scale of operations, local climatic conditions, quality of raw materials available.

1.3   Brief Description of Some Important Chemical Industries

1.3.1   Manufacture of Sulphuric Acid

Raw sulphur is molten by steam coils and filtered by Pressure leaf filters. It is fed to the furnace at a controlled rate to produce SO_2. Temperature of the exit gases is reduced by the WHRB before passing through the multistage converter for oxidising the SO_2 to SO_3. It is then absorbed in Inter Pass and Final absorption towers in stream of circulating sulphuric acid. Product acid is transferred to storage tanks. The drying tower serves to dry the air for the plant.

Main Equipments
- Steam coils in sulphur melter ....subjected to corrosion (free acidity in sulphur)
- Filtration system for liquid sulphur... subjected to corrosion (free acidity in sulphur)
- Sulphur feeding pumps...............subjected to corrosion (free acidity in sulphur) and erosion due to some particulate matter (ash)
- Sulphur burning Furnace............operation at high temperature; corrosive gases are present.
- Air Blower ....required to supply air to run the plant, very important machine
- Water treatment plant...... required. to supply DM water to the WHRB and process units plant

- Waste heat recovery boilers WHRB...... required to cool the furnace exit gases and recover heat as steam. Subjected to corrosive gases at high temperature; covered by statutory regulations.
- Economiser(s) ... required to cool the converter exit gases and recover heat as hot water; subjected to corrosive gases at high temperature; covered by statutory regulations due to high pressure working.
- Boiler feed water pump(s) and piping ... required. to supply DM water to the WHRB, covered by statutory regulations due to high pressure working.
- Converter system and heat exchangers....... required to convert SO_2 to SO_3; subjected to corrosive gases at high temperature
- Acid Circulation Tank, Acid Towers, Pumps, acid coolers ....required to operate the SO_3 absorption system and supply dry air to furnace; subjected to corrosive concentrated acid
- Water Pumps...for producing DM water for boilers, for producing acid, for cooling acid
- Instrumentation subjected to corrosive gases at high temperature, and to concentrated acid

1.3.2   Manufacture of 25%, 65% Oleum, Liquid SO₃ Plants

- These plants are add-on units to a sulphuric acid plant. Gases from exit of converter third pass are passed through cold heat exchanger CHE and economiser (optional). The gases are now passed through 25% oleum tower to absorb SO_3 and thereafter through Inter pass absorption tower for further absorption of SO_3 if sulphuric acid is also to be produced. Exit gases from IPAT are reheated by CHE and hot heat exchanger HHE to conversion temperature again; and passed through fourth and fifth passes of converter. Exit gases from converter are passed through economiser and oleum boiler (if SO_3 vapours are to be produced). The equipments required are:
- 25% oleum circulation tower with internals.
- 25% oleum circulation pumps and cooling system.
- Oleum boiler for generating SO_3 vapours. (it can be heated by process gases or steam).
- Heat exchangers for hot oleum exiting from boiler/incoming oleum to boiler
- 65% oleum tower with internals.
- 65% oleum circulation pumps and cooling system.
- SO_3 condenser of special design
- Cooling tower
- Storage tanks
- Dispatch pumps for 25% and 65% oleum tower.
- Steam tracing of piping for 25%, 65% oleum lines and liquid SO_3 tank.

1.3.3 Manufacture of Papermaker's Alum

Alumina hydrate is a pure raw material. It is reacted with sulphuric acid in a reactor with acid resistant lead and brick lining and equipped with steam coils to initiate the reaction. The reacted material is heated in evaporator to remove excess water and is then taken out for pouring in to moulds. Sulphuric acid to be used for this product should not contain more than 50 ppm iron. Only demineralised water should be used. The steam coils shall be of stainless steel 316 L or of hardened lead (if permitted by client).

- Hoist …required to take up alumina hydrate bags to reactor
- Main reactor with lead and brick lining (MS-LL-BL)… subjected to corrosion due to free acidity
- Boiler with accessories…essential to supply steam to reactor and evaporator
- Fuel tank, Fuel oil pump, Air Blower ….to operate the boiler
- Water treatment plant…to feed the boiler and reactor
- Steam coils…to run the reactor and evaporator; subjected to acidic corrosion
- Acid transfer pump, and acid day tank… to feed the reactor; subjected to corrosive concentrated acid

1.3.4 Manufacture of Alum

Bauxite lumps are crushed and ground to 150 mesh (about 85%) and then fed to reactor lined with acid resistant bricks. 98% sulphuric acid and filtrate (with recovered alum) from filter press is added to the reactor. Live steam or specially lined steam pipe is used for stirring and reaction. The reaction mass is sent to clarifier to settle the unreacted impurities. Clarified solution is evaporated by using steam coils made from stainless steel 316 L or hardened lead. The concentrated solution is poured in moulds, crushed and bagged.

- Bauxite feeder …to feed bauxite lumps to crusher; subject to erosion
- Crushing and Grinding Machines…to produce bauxite powder; subject to erosion
- Bag filter for dust control…for pollution control
- Acid Resistant Brick lined Reactor…for main reaction, subjected to erosion as well as acidic corrosion
- Sulphuric acid main storage tank, transfer pump and day tank…for the reactor; subjected to corrosive concentrated acid
- Boiler with accessories, Fuel storage, Oil firing system, air blower…… essential to supply steam to reactor and evaporator.
- (Steam is available from exit of back pressure type steam turbine in some plants).
- Thickener (settler and clarifier tanks) with acid resistant lining … subjected to acidic corrosion

- Filter press with accessories (air compressor)…to recover alum solution from waste water; subjected to pressurised liquor and acidic corrosion
- Evaporators with SS 316 L or lead steam coils … subjected to acidic corrosion;
- Dispatch section……Weighing and Packing of alum slabs/crushed pieces…

1.3.5   Manufacture of Single Super Phosphate

Rock phosphate is ground to about 100 mesh and mixed with 65–70% sulphuric acid in a mixer with Ni-hard blades. Gases are evolved with fluorine compounds, acid mist, CO_2 and water vapour due to reaction with sulphuric acid (a small quantity of CaF_2 is generally present in rock phosphate).

Some un-reacted particles of rock phosphate also get entrained. The gases are scrubbed in wet cyclone, venturi scrubber, spray tower and polishing alkali scrubber (optional). Generally rubber lined induced draft fan is used to suck the gases from mixers-reactors. The outgoing scrubbing liquor contains silica and H2SiF6 solution. The silica is taken out from bottom as precipitated material and the acidic solution is recycled to the mixer-reactor. Reaction material is dropped to a vessel below and reaction is allowed to continue for a few hours. It is then cured for a few days in the shed before bagging.

Important units and machines are:
- Storage shed for rock phosphate lumps…large shed subject to dust, movement of material handling machinery
- Pulveriser/grinder…subjected to big lumps; and can cause dust pollution during operation (which is controlled by providing cyclone separator, bag filter and suction fan)
- Bucket elevator, screw feeder
- AR brick lined mixer (reactor) with paddles of special alloy…for main reaction; subjected to acidic corrosion, erosion due to particles and higher temperature,
- Den Cutter for reacted mass…for continued reaction; subjected to acidic corrosion, erosion due to un particles of rock phosphate
- Acidic liquor (H2SiF6) recycle system from scrubbers… subjected to acidic corrosion and erosion due to particles of silica
- Dilution of sulphuric acid and sending to feed tank for mixer (reactor)… subjected to acidic corrosion
- Air pollution control units (scrubbing liquor circulation pumps for ventury scrubber, spray tower, polishing tower, induced draft fan)… subjected to acidic corrosion and erosion due to particles of silica and unreacted particles of rock phosphate
- ID fan…(rubber lined)..subject to erosive particles of silica and acidic gass
- Precipitated Silica removal from APC units…essential to control erosion due to particles of silica

- Effluent Treatment Plant…to treat acidic liquor which cannot be reused
- Covered Storage shed is provided with hoists, electrical overhead travelling crane (with operator console). The shed is for curing reacted material and storing finished product.

1.3.6 Manufacture of Caustic Soda and Chlorine by Electrolysis

Common Salt is dissolved in depleted brine (with fresh water if required) and treated to remove impurities. It is filtered and fed at controlled rate to the electrolysis cell with nafion membranes (which allow migration of sodium ions to cathode). Dil. NaOH solution with DM water is fed to cathode chamber. Reaction with OH(−) ions produce NaOH while hydrogen gas is evolved. Chlorine gas (it is generally with water vapour) evolved from anode chamber. DC power for electrolysis is supplied by rectifiers.

NaOH solution exits from cathode chamber with higher concentration.

The wet chlorine gas is dried, compressed, cooled and liquefied. It is then filled in cylinders.

Likewise, hydrogen is also compressed and filled up in cylinders.

Hydrogen and Chlorine are reacted in a graphite furnace to produce HCl gas which is absorbed in DM water to produce HCl acid.

- Salt feeder and dissolver …for maintaining concentration of feed brine to electrolytic cells Brine Purifying units and filters…… subject to corrosive conditions
- Brine Feeding system, Flow meters… subject to corrosive conditions
- Electrolytic cells with accessories (Anodes, and Cathodes assemblies) … subject to corrosive conditions
- Membranes and fixing arrangements … subject to corrosive conditions
- Rectifiers…for supply of DC power to electrolysis cells
- Bus-bars…carry heavy currents; even minor resistance/any loose contact can waste lot of power
- Gas evacuation from Electrolytic cell… subject to corrosive conditions due to wet chlorine; any leak can be dangerous
- Chlorine compressor (very important machine) Dryers, Liquefaction system… Any leak of pressurised chlorine gas is dangerous
- Chlorine Cylinders filling arrangement… … Any leak of pressurised chlorine gas is dangerous
- Hydrogen gas compressor (very important machine) dryers and filling in cylinders… Any leak of pressurised hydrogen gas is dangerous

- Corrosion resistant Graphite Furnace for HCl production…subject to corrosion and heat evolution inside. Leak proof construction and cooling arrangement required.
- HCl gas absorbers and storage tanks… subject to corrosive conditions due to wet HCl
- Demineralised (DM) Water plant…required for NaOH production
- Air pollution control (APC) … subject to corrosive conditions due to wet HCl
- Induced Draught fans for chlorine handling area, near electrolytic cells… subject to corrosive conditions
- Diesel Generator (DG) sets for emergency…required in top working order always

1.3.7   Manufacture of Nitric Acid

Liquid ammonia is evaporated at controlled rate by steam pressure and is sent to catalytic reactor. Filtered, compressed, preheated air (also at a controlled rate) is fed to the reactor. The concentration of NH_3 is carefully adjusted at inlet to the reactor to prevent extremely rapid reaction.

Exit gases are cooled by WHRB to generate by product steam, and through a air pre-heater for compressed air The gases are absorbed in a multistage absorber in DM water with cooling. Exit gases are at a pressure and are used for power recovery.

- Liquid Ammonia storage……high pressure unit storing dangerous NH_3
- Evaporator with PRV-Pressure Regulating Valve for steam…shall never overheat the unit by excess supply of HP steam to ammonia evaporator.
- Flow control valve for ammonia….high pressure accurate (proportion) controller;
- Air compressor unit with accessories (very important machine) ….,high pressure units with filter at inlet and air preheater in discharge line.
- Catalytic NH_3 Oxidiser with platinum based catalyst…operation at high temperature and pressure
- Waste Heat Recovery Boiler WHRB… operation at high temperature and pressure
- Demineralised (DM) Water plant… required for WHRB, absorption of NO_2 to get nitric acid production
- Cooling facilities
- Multi Stage Absorber (stainless steel) … required for absorption of NO_2
- Power recovery system …for energy efficiency by using pressure of exit gases to drive air compressor

1.3.8   Carbon Di-Sulphide (Electric Furnace Process)

- *(this is a special chemical product)*
- Liquid filtered sulphur is fed at controlled rate to the electrically heated refractory lined furnace and reacted with calcined charcoal to produce CS_2 vapours.
- Charcoal is calcined to remove moisture and volatile matter before charging to the furnace.
- Excess sulphur is separated and the vapours are condensed. The CS_2 is now refined by distillation.
- Uncondensed gases can contain vapours of CS_2 and some H_2S which is sent to oil scrubbers to remove remaining CS_2. The gases are now sent to sulphur recovery unit.
- Facilities required are;
- Sulphur melter with steam coils, sulphur transfer pumps, sulphur feed tanks, metering pumps. Charcoal conveyor belt, weighing system,
- Charcoal calciner, rotary screen, railway tracks and charging vessels
- Electrical overhead travelling crane,
- Electric furnace with internal refractory and accessories (charging boxes with accessories, graphite electrodes separator with steam jacket, primary condenser with cooling water supply, secondary condenser with chilled water supply, safety vents),
- Refrigeration system with accessories (chilled water pumps, circulation tanks, cooling tower)
- Oil scrubber units, heater and condenser, distillation units, sulphur separation units, gas holders,
- The product is stored under water in storage tanks and transferred by adding water to the tanks which displaces the CS_2.
- **Sulphur recovery units** with scrubbing system, piping for cooling water, chilled water, gas, steam jacketed piping for liquid sulphur, steam piping, alkali piping. Electrical transformers and switchgear.
- Careful condition monitoring of the units is required to prevent any mishap since almost all materials being handled are inflammable/dangerous.
- Statutory rules must be complied with to store, handle, dispatch all these dangerous materials.

1.3.9   Manufacture of Viscose Rayon (Staple Fiber)

Cellulosic pulp made from bamboo, eucalyptus wood and other materials is soaked in 16–20% NaOH solution for a few hours in a steeping unit. The excess alkali is squeezed out and the alk-cell is matured in a maturing drum at controlled temperature and residence time.

It is then reacted with CS_2 to produce alkali-cellulose xanthate. This is dissolved in more NaOH solution and the viscous solution is subject to vacuum to remove air.

It is supplied to the spinning machines (having special alloy discs with thousands of minute holes) for producing rayon fiber by reaction with an acidic bath containing sulphuric acid and certain added chemicals.

The fibers are to be produced as per specifications of diameter and length.

The fibers from all discs are collected in form of a rope and cut at specified length. The fibers are dried, and packed in bales of 200 kgs generally.

Gases are evolved during reaction with acidic bath and further treatment of fibers. CS_2 is recovered from these gases and Na_2SO_4 is recovered from spent spinning bath.

- Pulp receiving and soaking unit...to produce alk-cellulose, with NaOH recovery
- Alk-cell maturing drum with controlled conditions.
- Xanthation... Xanthators which operate at negative pressure for reaction of matured alk-cell with carbon-di-sulphide. No ingress of air is allowed as it can be dangerous
- Viscose preparation...ripening of viscose at low temperature under vacuum
- Multiple filtering stages for viscose
- Spinning machines with accessories (very accurate metering pumps, spinerettes of special alloy, rayon tow cutter, acid bath arrangement)
- Chemical recovery plants...from gases evolved at spinning machines, from spent spin bath liquor of spinning machine.
- Multiple effect evaporators for Na_2SO_4 recovery, CS_2 recovery... to minimise generation of effluent, and air pollution control for CS_2 and sulphur recovery
- Fiber dryers...drying by steam at some what high temperature
- Packing and weighing
- In a large rayon plant CS_2 NaOH and Sul acid are produced at site.

1.3.10   Manufacture of Refractory Bricks

(*which serve as an important input to many chemical industries*)

The raw materials (minerals with alumina, silica etc.) are ground and mixed in appropriate proportion with binders and moulded as per shapes required. The green bricks are dried by hot air and then baked in oil/coke fired tunnel furnace at high temperature. Heat is recovered from the hot baked (fired) bricks as they come out from the furnace. Hot air is generated as the bricks are cooled.

- Grinder..................internal parts are subject to mechanical erosion during grinding
- Weighing machines
- Mixing of binder chemical .... internal parts are subject to mechanical erosion

- Moulding machine............internals are subject to high pressure
- Drying furnace............ operates with high pressure hydraulic pusher mechanism
- Baking Furnace..............operates with high temperature operation, high pressure hydraulic pusher
- Trolleys and rail tracks ......subject to mechanical force and high temperature
- Oil/coke firing.........burners are subject to high temperature and/or erosion
- Set of Air blowers for heat recovery......subject to high temperature air

1.3.11 Petroleum Industries

Crude petroleum oil is desalted, preheated in a fired heater and then sent to distillation/fractionation columns. The various fractions are separated at different temperatures (LPG, kerosene, LDO, naphtha, lubricants and solid residue of petroleum coke) and stored separately.

The crude oil is also used for producing ethylene, propylene by cracking.

An Alkylation unit in the refinery uses Sulphuric acid for producing additives to petrol. Spent sulphuric acid is produced in this process and is regenerated in a separate plant.

Hydrogen is produced separately and used for desulphurisation of fuels. This generates H_2S gas which is sent to sulphuric acid regeneration plant and/or sulphur recovery unit.

Various petrochemicals like para xylene, purified tere phthalic acid, polyester are produced in petrochemical production units *which need large amounts of energy.*

The refinery complex design depends on quality of crude petroleum and product mix desired.

Power is generated by Gas turbines and steam turbines to run the plant.

The process units are subject to corrosive conditions (due to water, presence of sulphur compounds and hydrogen embrittlement), high temperatures and high pressures and hence special alloy steels of appropriate composition and/or with protective linings are used for the process units. *The knowhow is proprietary and readers may contact concerned refinery management for more information.*

The main units are (but not limited to) given in following indicative list.

- Crude oil heater
- Heat exchangers
- Distillation/Fractionating columns
- Condensers
- Crackers
- Flare towers
- Power recovery units
- Gas and Steam turbines
- Storage tanks

- Hydrogen generation
- Desulphuriser
- Sulphur recovery unit
- Regeneration of spent sulphuric acid

1.3.12   Manufacture of Sugar

Sugar cane is cleaned, cut and crushed to produce juice. It is treated with SO_2, lime and is clarified. The clear juice is evaporated in multiple effect evaporators. The waste bagasse produced after extraction of juice is burnt to generate steam. It is used to generate power and is also supplied to the evaporators for concentrating the clarified juice.

- Cutter, crusher for sugar cane.........mechanical units subject to erosion
- Lime treatment and SO_2 treatment ... subject to corrosion due to SO_2; generates Ca-sulphite precipitates and needs filteration.
- Multiple effect evaporators MEE......subject to increasing vacuum in stages
- Boiler, Water treatment....essential for running steam turbines and MEE
- Ash removal......disposal problem (may be used as additive to fertilisers)
- Sugar crystallizer and centrifuge..............handling large quantities
- bagging and weighing......... handling large quantities and accurate dispatch

1.4   Some Important Equipments in Process Plants Described Above

1.4.1   Some Typical Process Units and Mechanical Equipments

The table below indicates some typical equipments in plants described above which need considerable energy and careful attention during operation.

Name of Equipment	Plant/process
1. Crusher	Single super phosphate, alum.
2. Air blower	Sulphuric acid plant
3. Circulation pumps	Acid towers in sulphuric acid plant
4. Refrigeration system	CS_2 Plant, chlorine liquefaction.
5. Air pollution control system	Sulphuric acid, Single super phosphate plants
6. Electric furnace	CS_2 Plant
7. Charcoal calcinations and conveyor system	CS_2 plant

Name of Equipment	Plant/process
8. Air compressors	Nitric acid plant
9. Evaporators	Rayon plant, sugar plant, alum plant
10. Distillation columns	Petroleum industry
11. Process boilers	Liquid SO_3 production, CS_2 purification
12. Rotary dryers	Sodium sulphate recovery plant, Rice husk ash processing plant.
13. Electrolysis units.	Caustic soda plants.
14. Gas compressor and condenser	Chlorine liquefaction plant.
15. Water treatment plant (boiler feed water and special process plants)	Sulphuric acid, nitric acid, HCl, papermakers alum, rayon plant
16. Effluent Treatment plant	Rayon plants, alum plants, single super phosphate

1.4.2   Typical Essential Features of Equipments Used in Above Industries

- Material of construction should be erosion and corrosion resistant.
- Corrosion resistant paint must be put on lower plates and side walls
- Design and Fabrication should be as per ASME Sec VIII
- There should be inspection windows/small nozzles closed by covers which can be opened for inspecting the internals without dismantling the equipment.
- Should be capable of resisting high temperatures.
- There shall be provision for lubrication for critical components while the machines are running as far as possible.
- Silencers/filters must be provided at inlet of air blower.
- Strainers must be provided at inlet of metering pumps, plate heat exchangers, sensors for pH measurement, conductivity measurement instruments.
- Should be possible to easily monitor and control the speed of important equipments.
- The machines and drive motors should be available as a pre-assembled and aligned unit on common base plate.
- These should be grouted firmly on foundation. (with special arrangement to prevent loosening of foundation bolts)
- There shall be appropriate in-built safety features, internal pressure relief valves, safety vents, rupture discs for concerned equipments.
- Expansion joints in refractory lining must be provided for units operating at high temperatures. Reinforcing pad plates should be provided.
- Lifting lugs must be available for each unit
- Reinforcements and gussets at nozzles for entry, exit and manholes
- Appropriate nozzles must be available to monitor pressure, temperature of critical parts, condition of bearings of these units.

1.4.3   Process and Operation which Have Considerable Energy Changes

- Burning of sulphur to produce SO_2 gas in sulphuric acid plant.
- Conversion of SO_2 to SO_3 in sulphuric acid plant.
- Absorption of SO_3 to produce sulphuric acid.
- Production of pure SO_3 vapours by boiling oleum.
- Electrolysis of salt to produce caustic soda.
- Oxidation of ammonia in nitric acid plant.
- Distillation of crude petroleum to produce various gases and fuels.
- Evaporation of sugar juice to produce sugar.
- Recovery of sulphur from H_2S in petroleum refinery
- Production of CS_2 by electric furnace.
- Evaporation of dilute alum solution to produce solid slabs.
- Destruction of hazardous waste by Plasma Gasification Process.
- Crushing, grinding, pulverising of bauxite, rock phosphate in Alum plant, Super Phosphate plants
- Refrigeration units in CS_2, liquid Chlorine, rayon plants.

Chapter 2
Managing the Organisation

Managing the Chemical Industrial Organisation

The chemical industries make a very useful contribution by producing items required for daily healthcare, fertilizers for agriculture, synthetic fibers for clothing and pharmaceuticals for the treatment and well being of the citizen as well as for the overall national economical progress by petroleum processing and manufacture of paints, paper etc.

However, *though very useful*, these industries can pollute the environment and could be very dangerous if not operated and maintained properly. Governments of all countries have therefore made strict laws to ensure safety of personnel and control of environmental pollution and have agencies to monitor their operations closely.

2.1 Typical Features of Chemical Industries

- Effluents are generated in considerable quantities (toxic, harmful substances) which can require considerable equipments/facilities for careful treatment before disposal
- There could be fire or explosion hazards
- Many chemical industries consume large thermal and electrical energy.
- There is scope in many chemical industries for recovery of energy by co-generation
- Many chemical industries are run by automatic operations. However very careful continuous *manual* monitoring of the process plant and machinery is necessary.
- This generally needs very attentive manpower.
- Many chemical industries are run continuously round the clock though some of them are run in a batch wise manner.

© Springer Nature Switzerland AG 2019

K. R. Golwalkar, *Integrated Maintenance and Energy Management in the Chemical Industries*, https://doi.org/10.1007/978-3-030-32526-8_2

2.2   Guidelines from Board of Directors for Integrated Management

The Board of Directors shall issue guidelines to the Plant management for building up a very good reputation for the organisation by ensuring safe, pollution free, efficient operation to manufacture the products as per quality (specifications) and timely delivery required by the clients at minimum cost.

This can be done by following safe operating and maintenance procedures, compliance with statutory rules for pollution control, maximising the yield of the products with minimum wastage of raw materials, minimising the consumption of energy and utilities (i.e. improving overall efficiency of the manufacturing plant) by efficient maintenance of the process units, plant machinery and infrastructure in the premises.

An integrated management of maintenance and energy is thus very essential *in addition to efficient working of all other departments/facilities in the premises.*

However it is seen that various departments have different interests and priorities which may be at variance with each other even though all of them have to work together for achieving aims of the organization.

Typically, the sales department wants to maximise the sale of products; and expects the production to be maximised and ready for delivery as per schedule.

The production engineers would like to run the plant continuously at maximum capacity (sometimes even above the rated capacity) to meet the targets given to them.

However, the maintenance engineers generally advise them to *run it preferably at a little below the maximum rated capacity* in order to minimise breakdowns and certainly not more than the maximum rated capacity (i.e. overloading). They also want the running units to be released as per their request for carrying out preventive maintenance activities.

The Stores (Inventory control) are desirous of minimising the stock of spares for ease of handling, for reducing the risk of damage or obsolescence during long time of storage and minimising blocking of funds in excess inventory.

Hence the management should make it clear to all concerned about the aims and policies of the organisation; and ask all such department to work towards these aims within the framework of the guidelines **from Board of Directors.**

2.3   Modern Management: Aims, Policies and Priorities

Progressive modern managements have well defined aims and policies. These are reviewed periodically in the light of performance of the organisation (generally at intervals of previous 3–6 months).

2.3.1 Typical Aims

- To have a very safe plant (by appropriate timely maintenance)
- To achieve zero pollution level and zero accident level
- To maximise the market share by maximum client satisfaction
- To maximise the total profit
- To manufacture products which will meet specifications by clients (zero defect in case of pharmaceutical products, chemicals to be used for research and analytical work)
- To minimize the cost of production by efficient operations and appropriate maintenance
- To minimise the dependency on external power supply
- To develop a new product(s) every year/2 years
- To establish the reputation (brand name) in the market for quality products, timely delivery and at reasonable cost
- To consider taking over of idle plants for increasing capacity or diversifying
- To expand the plant capacity by maximum use of existing available infrastructure/removing bottlenecks
- To consider vertical expansion of the organisation (producing some inputs in-house/producing value added downstream items by using own products)
- To develop in house capacity for Research & Development (for improving existing process, plant or machinery)
- To develop facilities for energy recovery and power generation (if possible)
- To develop know how and engineering capacity for export of entire project/ becoming a technology supplier rather than a technology importer

2.4 Typical Policies

- Never violate any statutory rules
- Always meet the delivery schedules as specified by the clients
- Maximise the use of existing assets which are underutilised (by taking up job work for external parties/supply partially processed material to external parties) *such as (a) evaporation of dilute solutions; (b) supply of steam/hot air or (c) supply of filtered sulphur to nearby industries from existing units*
- Maximise the use of cheap local input raw materials, packing material, indigenous catalysts, stabilisers, skills available with local people, local infrastructure if suitable for the technology being used.)
- Keep the work force always happy, but disciplined.
- Minimize the overtime work by production workers
- Minimise employing external (contractual) labour
- Never default on payment of statutory dues and taxes, payment to government agencies and obligation towards stake holders (financial institutions, share

holders, lenders who have advanced loans, payment for materials and equipment received at site.)

2.5 Typical Priorities

- Compliance with all statutory rules and regulations.
- Ensuring that the manufacturing plant is completely safe for personnel working inside the premises as well as those outside
- To minimize the generation of waste (and maximise reuse and recycle as far as possible) and proper treatment of all the effluents to prevent environmental pollution
- **To use the modern techniques (and analytical approach) for achieving the goals of safe, pollution free, efficient working of the process and equipments by ensuring their reliable performance by quality maintenance**
- To build up reputation in the market by regular supply of products of specified quality and quantity as required by the clients in time (as per delivery schedule) and at reasonable cost by proper packaging and transport.
- To manufacture the products at minimum cost by maximizing efficient working. The consumptions of all inputs like raw materials, energy, water etc. shall be minimum.
- Operate the plant at optimum/rated capacity. Only in exceptional situations the plant may operate below rated capacity
- To coordinate and control the activity of other departments and manage technical, human, commercial, logistical and financial matters.
- Ensure plant operation and maintenance by qualified and experienced persons
- To plan the product mix for the present and the future. This needs planning of all products and their individual production rates, production cycles; installation of process units, machineries and auxiliaries required; provision of safety equipments, plant layouts for smooth working, storage arrangements for raw materials, other necessary inputs and finished products, handling systems for all materials, energy management etc.
- To allocate the generated revenues very judiciously for effluent treatment, for necessary maintenance of safety devices, process units and machinery; procuring supplies of raw materials and other inputs; paying taxes and duties to the government, welfare of all working personnel, carrying out innovation for improvement and research for new technologies; and return of capital with interest as per terms of agreement with all creditors and stake holders.
- To set aside reasonable amounts for replacement of old (corroded) process units and worn out machinery; and as a contingency fund for any unforeseen situation. This is very important for a chemical industry because the rate of corrosion is much more than in case of general fabrication shops or engineering industries. **The distribution of profits to the promoters/owners (and the equity share holders) is generally done after above allocations**.

  – To support and encourage **innovation, research and development** for suitable design modification and improved construction of the process units and machinery so that

  (i) they can be easily monitored, operated, inspected and maintained by plant personnel,
  (ii) there is provision of proper safety devices, warning alarms and trip systems
  (iii) ease of detection and rectification of faults
  (iv) ease of repairs
   (v) replacement of faulty components
  (vi) to ensure immediate availability of stand by units/ machinery by proper inventory of the correct spares.

  – Creating facilities in the premises for the integrated maintenance for ensuring safety, control of environmental pollution, energy efficiency, reliability and availability of critical plant units and machinery.
  – Encouraging modifications of equipment design.

- Decision making for progress of the organisation in various situations
- **To address requests from various departments for guidance and coordination.**

2.6   Integrated Management of Chemical Industries

The aims and policies of the organisation should be made clear to all the concerned departments. Management should consider the priorities of various departments *(since all of them contribute to the successful running of the industrial unit)* and the situations faced by them while working in their respective areas.

They should be guided to work towards these aims within the framework of the policies by resolving the differences in their priorities (which could be occasionally at variance with each other as explained before).

The main departments are:

- Marketing
- Production
- Maintenance, Utilities and Instrumentation
- Purchase
- Safety and Environmental pollution control
- Logistics,
- Human resources
- Stores and Inventory control
- Research and Development
- Commercial, Finance, Taxation and General Administration

2.7   Regular Interactions

These shall be arranged among all concerned departments:

- To discuss the targets set for them and the current situations.
- To look carefully in to the divergent news and priorities of different departments in the organisation (production, raw materials purchase, marketing, maintenance, safety, stores and inventory control).
- To call for detailed presentation from each of them regarding their individual priorities and difficulties in regular review meetings; and ask to submit suggestions for improvements.
- Management shall carefully study everyone's view of working in the organisation and their suggestions. Some useful suggestions can be generated after consultation with various departments for improving the design or construction of processing units while planning for future procurements, capacity expansion, diversification to more products etc.
- Suggestions for the modification in designs and constructions should aim to make the operation of the process plants/machinery safer, to improve product quality and the product output.

2.8   Guidelines and Instructions

Senior Management may then prepare guidelines and instructions in order to guide the various departments towards meeting the aims of the organisation while remaining within the frame work of organisation policies and statutory rules.

The guidelines and instructions for the various departments should be such that all of them understand each other's difficulties/situation and cooperate with each other.

These instructions can be revised after review of the performance of the organisation at regular intervals.

Efficient Integrated Maintenance and Energy Management thus requires close understanding of difficulties of each department, their mutual cooperation and clear guidelines/instructions from senior management to all concerned for carrying out the departmental work within the frame work of organisational aims and policies.

2.9   Revision of Policies

The Board of Directors may *revise* certain policies or aims in consultation with senior staff from various departments if these are found to be *impractical* after careful analysis of the performance of the organisation at regular intervals.

However there are certain external factors on which the organisation can have no control such as outages of external power supply grid, interruptions in water supply, changes in government taxes, problems with railways and roadways transport sectors, sudden heavy rains or storms; besides international scenarios which can affect the market for finished products and supplies of raw materials fuels, special catalysts etc. These can affect the working of production, marketing, maintenance, stores and logistics departments of the organisation.

Managements will find it very useful to anticipate such uncertain situations (which may or may not actually arise) during running of the plant; and be prepared for them by keeping available or planning various options to choose from.

Instructions or guidelines shall be prepared (in advance if possible) by carefully looking into the likely effect of such situations on working of the various departments. These instructions/guidelines shall be worked out in consultation with their senior representatives and the feasibility of implementation shall be assessed in light of the performance of the organisation at quarterly or bi-annual intervals.

Clear, unambiguous communications are essential for safe, pollution free, energy efficient working of the chemical production units in order to meet the specifications of products and delivery schedules as desired by the clients.

2.10   Departments in the Organisation

2.10.1   Marketing

Aim   To maximise market share of the organisation by increase in sales volume

Priorities
- Retain existing clients
- Maximise sales
- Get new customers
- Meet specification and delivery schedule given by client
- delivery of products shall be in ready to use packs which are easy to handle (can save time and efforts required by the clients while using.
- The packs should contain accurately weighed/measured quantities. There shall be no short supply.
- products may be delivered in container lots if required by client

Situations to be discussed
Own plant is to be stopped for some urgent repairs, there is a shortage of certain important input, annual shutdown is due very soon and Client wants large amount of product in a tight schedule but available stock is less or not as per specifications given by client.

Guidelines/Instructions
- *The* delivery schedules must be met to maintain good relationship with client
- build up sufficient stock of product as per required specifications before annual shut down period to fulfil demands of client. *In general the marketing personnel should consult the production and maintenance departments before accepting more orders to avoid insufficient supply to client.*
- purify the available off-spec material to the grade as required;
- increase rate of production to meet demand;
- get the product manufactured on contract from outside parties

2.10.2 Production Units

Aim
- to manufacture the product as per specification and in quantity required by client;
- to manufacture as per product mix planned at minimum cost *(the product mix depends on demand for products made and profit margin).*
- *Clients who pay in time and place regular orders are to be given priority.*
- to achieve safe operation with zero pollution level.
- to achieve maximum operational efficiency; zero breakdown and to become independent of imported raw materials, catalyst, and stabilisers,

Priorities
- there shall be no accidents
- to minimise generation of wastes -ideally achieve zero pollution level
- product should be manufactured as per required grade at minimum cost

Situations to be discussed
- some key production units and machinery need urgent cleaning or repairing such as filters, reactor, condensers, catalytic converter, gas absorber, condenser, compressors, and refrigeration units, process feed pumps, heat exchangers. They are in need of augmentation of capacity *even for operation at rated capacity;*
- certain operating conditions cannot be allowed to increase beyond limit (in case higher output is to be obtained) i.e. cannot overload any equipment
- increase in production rate can add load on maintenance, and stores and procurement departments for which they are not equipped at present.
- **Obligations**: Inform correct specification of inputs (with allowable tolerance limits) to purchase department and also inform about the implications/harmful effect on plant units if the specification is not met. *Extra operations may be required or more breakdowns may occur if raw materials do not meet specifications.*

Production department must check on daily basis:

- Condition of process units, reactors, filter, etc. and their maximum and minimum production rate for keep them running without stopping.

- Available stock of products which meet quality specification given by the client.
- Consumption rate of raw materials required for different products should be checked against available stock of raw materials.
- Fresh requisitions should be made well in time so that the planned product mix will be manufactured and maximum clients can be supplied with the products required by them when they want.

Guidelines/Instructions
- to operate the plant at lower production rates if possible, till repairs are carried out,
- to take up urgent repairs till fresh stocks of required inputs arrive;
- to get the product manufactured on contract from other parties either fully or at some intermediate stage,
- to request clients to adjust the delivery schedule;
- to purify the available inputs further before using in process.
- *Revise operating procedures like sequence of charging raw materials, temperature settings, agitator speed, concentration of reactants, flow rates if possible. However these changes must improve product quality/rate of production, energy efficiency without compromising safety and pollution control.*
- stop the production process (partially/completely) for a short time and carry out cleaning/repairs on urgent basis then restart as early as possible. Now try to make up the loss of production but without compromising safety and pollution control

2.10.3   Safety Department

Aim To achieve zero accidents and ensure safety of manpower, equipments, and environment.

Priorities All dangerous situations must be avoided, and key equipments and safety device must be maintained properly. To make every one aware about safety aspects related to equipments handling and their do and don'ts.

Situations to be discussed
- if required spares or facilities are not available or under repairs;
- key equipment are not working safely (malfunctioning of cooling system, support columns, vent condensers, flow controllers.)
- operating conditions may exceed beyond guidelines for safe working if production is continued at higher rate;
- leakage from any pipeline, reactor and valve;
- adverse weather is making working very difficult;

- personal safety appliances and safety dresses, gas mask, safety clothes, shoes, helmet, and gloves must be provided to all.
- emergency escape facility is not proper;

Everyone should feel responsible for safety since accidents may occur due to lack of attention, lack of work skills and knowledge, equipment failure, overload on units and equipments, high temperature and high pressure in equipments, congested space between adjacent units.

Guidelines/Instructions   Safety must be given top priority always.

HAZID analysis must be done before erection
HAZOP analysis should also be done before commissioning, and after plant start up.
Ensure:

- follow up of corrective actions.
- All safety systems as advised by statutory authorities and insurance companies shall be implemented. (e.g. auto visual alarms for high level of pressure and temperature – safety should have first priority among all activity)

There must be proper coordination of production, maintenance and safety departments with updated manuals for Standard Operating and Maintenance Procedures.

Establish practice for issuing work permits before taking up maintenance work on process units (inside or outside), electrical facilities, in presence of inflammable or dangerous materials

2.10.4   Environmental Pollution Control

Aim   To achieve zero pollution level

Priorities
- meet all statutory norms
- air pollution control and zero liquid discharge from the plant by proper treatment
- safe disposal of hazardous waste (which cannot be done at site) through authorised agency
- reuse/sale of waste generated as per law

Situations to be discussed
Effluent Treatment Plant (ETP) and Air Pollution Control (APC) cannot handle effluent load beyond a certain overload limit. Therefore, production units should not be operated in a way that may cause excessive overloading of APC/ETP.

Guidelines/Instructions
- augment ETP and APC capacities if possible

- store additional effluent in special tanks and treat it after attending ETP units.
- do not discharge untreated effluent under any circumstances.
- dispose off hazardous wastes safely through statutorily authorised agency only. This could be a Common facility for storage, treatment, incineration.
- maximise recycle and reuse of effluent and waste as much as possible.

2.10.5 Maintenance Department

2.10.5.1 Mechanical Maintenance

Aims
- to achieve zero breakdowns and to increase equipment life;
- to restore performance of process unit or machinery (if not working safely or inefficiently) at the earliest with minimum cost.
- to keep standby units always in a state of readiness for immediate use when the running unit performance is unsatisfactory or it needs repairs.
- **This needs looking in to** design and fabrication, quality checks at all stages, erection and commissioning carried out for existing units.
- **addressing** important issues such as whether operation is within safe limits, regular condition monitoring, timely repairs through proper maintenance, inventory control of Spares.

Priorities
- compliance with statutory rules and regulation.
- Do not allow breakdown of any key equipment which may endanger safety, pollution control, product quality and manufacturing of product itself.
- to increase equipment life by scheduled maintenance, using proper spares & lubricant, and reducing vibrations

Situations to be discussed
- some safety device needs replacement for which process unit must be stopped
- urgent repairs (by welding for arresting leak) are required for reactor shell, gas ducts, liquid pipeline; electrical cables, control circuit
- bearings of blower/compressor is making abnormal noise/vibrations
- storage tank is leaking (hence must be emptied before repairs can be done)
- production and marketing department are not allowing stoppage of key equipment for taking up maintenance work.
- required spares are not available in the specified material of construction (thermal stability, and **chemical resistance**).
- maintenance work is yet to be completed before start-up date;
- skilled personnel are not available;
- very difficult to work in present harsh weather conditions;
- arrangements are still incomplete for some major maintenance work planned (for carrying out during the annual maintenance).

- It will adversely affect the production as well as marketing. Hence necessary arrangements–spare parts, scaffolding and temporary supports, chemicals for cleaning and flushing, lubricants and skilled person must be available before the annual maintenance is taken up.

Guidelines/Instructions
- Provide standby units for key machinery. Running machine can be taken up for repairs after commissioning the stand by machine.
- Maintain sufficient stock of important spares always,
- carry out at least some repairs while plant is still running (without compromising on safety and pollution control activities)
- The plant may be stopped for urgent repairs if the delivery schedule and product quality required by clients can be met.
- repairs/replacement of machinery/part of process unit/piping, etc. shall be undertaken immediately if it is dangerous and can cause injury to personnel.
- certain repairs may be postponed if extension is available to operate the concerned unit for some more time.
- repair kits shall be available round the check (24×7) to plant personnel to carry out at least temporary repairs.
- these kits may consist of quick setting cement, clamps for piping, plugs of different sizes to stop small leakage, etc.

Maintenance engineer should closely coordinate with purchase department for
- provide a list of regular consumable items such as lubricants, fuses, gaskets, nut bolts, flanges, and their specification for keeping ready with sufficient stocks.
- use spares from Original Equipment Manufacturer (OEM) as a matter of company policy. However reconditioned spares may be used/in house repairs done in case of very urgent need (specially when spares from OEM are delayed)
- develop substitute components or operate the plant under less drastic condition if possible.

2.10.5.2   Electrical Maintenance

Aim
To ensure safe working always without any break downs of electrical system

Priorities
- All electrical installations must work safely
- No accidents due to any fault in any electrical equipment,
- no short circuits, fires in power supply cables, bus bars
- electrical power consumption must remain within limits of maximum demand
- compliance with all statutory matters
- no interruption in power supply to any key equipment

Situations to be discussed
- Sudden demand for extra power from production department (to meet demand for product)
- Heavy rains, storms, severe summer resulting in breakdown
- Frequent interruptions/low voltage or low frequency in power supply from external grid
- Corrosion of electrical equipments, power cables, lighting arrangements

Guidelines/Instructions
- Production department must inform increase in requirement of power well in advance
- Operate machines and reactors in a staggered manner
- Use DG set to provide power in case of emergency and run some units
- Keep watch on Maximum Demand indication
- Control all emissions of corrosive gases, dust and spillage of chemicals so that the cables, wires, motors are not affected

2.10.6 Instrumentation

Aim
To ensure safe efficient working of the process units and machinery

Priorities
Provide correct indication, recording and control of process conditions, levels in process vessels, detection of pollutants in exit streams

Examine and attend frequent damage to sensors, drift of set point, response of control valves is slow, incorrect indication of process parameters (temp, pressure, level, flow rates.)

Guidelines/Instructions Check correct location and installation of sensors (including protective sheath,) sampling line choking, long length, proximity of sensor to strong electric or magnetic fields or high temperature units,

Vibrations from nearby heavy machinery can disturb settings. Hence check calibrations and set points, provide locking (fixing) arrangement on control actuators

Reset and calibrate set points of flow, pressure, high/low level alarms.

Independently check flow by rise/fall of level in day tanks per hour against flow rate indicated by instrument

Reduce production rates, operate at lower flow/temp/pressure in case instruments are not indicating correctly (attend /replace immediately) if there is a chance of mishap.

2.10.7   *Provision of Utilities*

Aim To become independent of power supply from external grid during stabilised plant operation.

To make arrangements to receive external grid power as per estimated requirement and keep arrangements ready for supply of emergency power. Make sufficient provision for water and fuel supply for initial plant start. Recycle maximum amount of treated water.

Priorities Reduce consumption, maximise recycle and recover water, power; sale of steam (if some steam is still available after in-house use); select a fuel which delivers maximum heat per unit cost (refer to Chapter 10.5.8 for more details).

Situations to be discussed
- Boiler capacity is inadequate;
- cooling tower needs cleaning;
- enough make up water is not available;
- Refrigeration plant not able to take cooling load in-spite of NH_3/Freon addition. This situation may cause insufficient cooling of equipment and reduction in production rate.
- Safety and production will be affected. Escape of uncondensed, toxic or inflammable vapours must not be allowed to take place. It may result in fire or some other type of accidents.

Guidelines/Instructions
- Recycle maximum water, use effluent after treatment
- Arrange contract manufacturing
- maximise heat recovery/postpone some unimportant operations
- Operate at different conditions for condensing vapours, cooling tower settings,
- Use higher flow rate of chilled water if lower temperatures cannot be achieved.
- Add slabs of ice to chilled water in case of emergency if there are chances of escape of toxic or inflammable vapours.

2.10.8   *Purchase*

Aim To purchase all inputs (of raw materials, catalysts, stabilisers, machinery spares and other input) as per specifications at lowest price

- Purchase department should purchase less energy consuming equipment with better shelf life and low maintenance cost.
- To purchase all the raw materials as per specifications given by production engineer. The quantities shall be procured to match the requirement of product mix.

- The materials procured should have at least the minimum amount of required component. *However this may not be acceptable in actual situation if there are some impurities which could be very harmful to the process units, catalysts or machinery in the plant.*
- Procure in a packing size or container which is easy to use directly.
- Inventory to be maintained. There shall be never any shortage of raw materials due to which production rate will have to be reduced. Also excessive stock of raw materials should be avoided as it blocks funds, can be dangerous:
- **Reorder point** – to be properly estimated for all important raw materials/inputs

- Minimum quantity of raw materials: The maximum expected consumption per day/per unit product for the planned product mix shall be known or estimated. It should be multiplied by the maximum lead time (time required for actual arrival at site from the day of placing order). Add a safety margin (about 20–30%) if supplier is far away and consider this as the lowest safe stock level. Fresh supplies are to be ordered whenever this lowest level is reached.

Purchase department may even try to procure inputs which are even better than specified quality. But the procurement must be approved by all concerned -and senior management.

Priorities
To procure from cheap, reliable sources (local suppliers may be preferred for reducing cost of transport and reducing inventory.) To purchase efficient equipment and spare parts acceptable to all concerned.

Situations to be discussed
Cheap material is available but does not meet given specification exactly or contains more impurities. These can affect product quality, production rate, equipment life, energy efficiency.

- Procuring spare parts from OEM may delay the maintenance work if not available in time.
- Raw material quality is as per specs but delivery may be delayed
- Raw material quantity available is not sufficient to meet demand for the production
- Raw material quality and quantity are as per requirement but the vendor is far away. There will be more cost of transport
- Purify available raw material/products by installing more process units
- request client to accept product in small lots.
- Supplier insists to buy large amounts (ship loads)
- Large quantities of raw materials may be purchased only if there is future demand--marketing department must try accordingly.

Guidelines/Instructions
Suggestion/*Remark: (i)Production and maintenance engineers should inform tolerance limits of impurities **or** next acceptable grade of raw materials (ii) spares* store keeping should inform purchase department about current inventory level of spare

parts, expected requirement and re-order point for replenishment (iii) senior management should carefully look in to the effect on safety, environmental pollution, product quality, production rates, machinery life when the next lower grade of raw materials or reconditioned spares are used.

2.10.9 Logistics

Aim Arranging containers, tankers, suitable vehicles for the transportation of raw materials (if the supplier is unable to provide) and finished products as required for procurement and for delivery of products according to schedules.

Priorities
- arranging vehicles which are suitable for transporting the chemicals in a standard size container;
- meeting the planned delivery schedule to supply at minimum cost to the client
- find safe roadway (away from inhabited areas as far as possible) for efficient way to transport the materials without any mishaps on the way.

Situations to be discussed If transport of inputs to the production site is slow the production rate may get disturbed. Delivery to clients may subsequently get delayed if proper vehicle/transport arrangement is not available.

Guidelines/Instructions
- to avoid such problems, arrange outside transport and ask the marketing department to request client to accept delivery of smaller container in a longer time period.
- always build up buffer stock (depending on shelf life of inputs products, and spares) at production unit or at point of delivery so that normal working is not affected by transport bottlenecks.

2.10.10 Human Resources (HR)

Aim To maintain discipline, keep all personnel motivated, use all manpower available most efficiently, to train newly recruited personnel and ensure clear communication among all personnel. This is necessary for the successful running of the organisation with proper integration of working of all departments.

Priorities
- Employ trained and qualified people only
- Assist senior engineers to develop safe comfortable shop floor conditions for efficient working in order to ensure accident free operations

- Assist management in consultation with production department to optimise manpower employment by proper span of control
- Solve genuine grievances and problems related to human relations
- Organise refresher courses for old employees and training programme for new employees; arrange programmes for skill improvement at different levels
- Arrange external workforce for (i) repetitive work like unloading of raw materials (ii) loading of finished products during normal running of the production units or (iii) for temporary activities during annual overhauls of the plant (iv) during emergency situations
- Develop systems for clear unambiguous communications among all concerned persons.
- *Example: Requisitions for getting supplies of raw materials (for production) should have clear specifications, quantities required, dates on which required etc. Any difficulty in fulfilling this request should be made clear to the person who has originated it, as well as marketing and senior management. The production programme can then be revised.*
- *Clear instructions should be given to production workers and maintenance crew if a process unit is to be stopped for repairs on a particular day and time to enable them taking up appropriate steps for the same. Tools and spare parts can be kept ready on the spot in advance when clear communication is available.*

Situations to be discussed
- If unskilled untrained persons are hired they may cause accidents
- Worker's unions are complaining that increasing the production rate or diversification to more products will increase their work load.
- Sufficient numbers of qualified/trained personnel may not be available
- adequate unskilled/semi-skilled manpower is not available for large volume of loading, cleaning, washing (repetitive work)
- skilled labour not available

Guidelines/Instructions
Use available personnel on overtime wages or on off-days since they are familiar with the process units and machinery (but no statutory rules shall be violated regarding employees and the available personnel shall not be over loaded with work).
A word of caution-Mishaps can occur if persons are too tired and hence not able to pay sufficient attention to the process units or machines.

- employ external contractual labour for simple, repetitive jobs so that available skilled manpower can be used for more important jobs.
- ensure safety at work place with good lighting, less noisy conditions
- ensure that all dust, toxic and irritating vapour are sucked out from work place by powerful exhaust fan and discharged through scrubbing system
- put up sign boards for safe; Standard Precautions and Operating conditions
- upgrade their skills, arrange refresher courses, encourage and reward good suggestions

- employ only qualified, experienced and physically fit persons for operating key equipments and maintenance of complicated machines
- insist on good housekeeping and standard practices as work culture. This improves product quality, equipment life and energy efficiency,
- motivate the work force by developing suggestion schemes, quality circles, and rewarding good work done

2.10.11   *Innovation, Research & Development (R&D)*

Aim

To improve quality of products, to develop safe methods for operation and maintenance, and develop new products.

To increase efficiency of the process plant and energy recovery units.

To try new methods, process, another raw material which may be cheaper/available locally

Priorities To improve quality of product, improve production methods for better energy efficiency and for reducing maintenance problems, use of cheaper/different and raw materials which are available easily; also to try a new product i.e. diversification

Innovation Innovation shall be for modification of operating and maintenance procedure and hence will generally need less funds than purchase of new equipments

Situations to be discussed

- Trial of new method for manufacture, new equipments or cheap raw material can be a matter of contention between production, maintenance and purchase departments.
- Trial run of new material cannot be carried out during stoppage of plant for maintenance.
- However purchase and production department may or may not agree for postponement of such trials and may ask for postponement of maintenance instead.
- Trials of new modified method should be done without disturbing the existing operations as far as possible. The production, maintenance, marketing, and human resource department may object to such trials if there are chances that it can affect product quality, safety, effluent generation, life of machinery or production rate itself.

Guidelines/Instructions *Management should convince all concerned that the R & D is being done in order to meet the aims of the organisation* (more safety, less pollution, less cost and high efficiency) *while working within the policy frame work.*

Senior management shall discuss with production and maintenance departments and permit these trials if they do not disturb existing operations, safety or pollution control (excess effluent is not generated) *or, a calculated risk can be taken in consideration of the likely benefits.*

Safety can be achieved by using better materials of construction, by increased equipment life and less drastic operating condition.

Highly efficient process achieves more conversion of raw materials to products and hence there can be less load on ETP. The more efficient system will need less energy for manufacturing operations.

Example
Ring shaped Caesium promoted catalyst in a sulphuric acid plant results in less energy consumption by the air blower and more efficient conversion of SO2 to SO3 as compared to conventional pellet shaped catalyst (*due to reduced pressure drop, and higher equilibrium conversion.*) *It can save power consumption by blower and alkali consumption by air pollution control unit.*

2.10.12   Stores and Inventory Control of Machinery Spares

Aim   Inventory of all required spares parts, consumable (fuel) and materials to be minimised.

Priorities
 (i) Minimum inventory of inflammable/toxic/dangerous materials and
(ii) Availability of all important inputs as per requirement

Situations to be discussed
- Stock of spares is limited and there is delay in arrival of spares from vendors.
- This can result in loss of production and poor maintenance
- Spares are not available (when required/when plant is shut down for annual maintenance). Delivery may be delayed from OEM. However reconditioned spares or spares from some other supplier is not to be used as per organisation policy.
- Equipment available in a packed box, undergoing repairs, or lying idle in stores for want of some component cannot be considered as a standby unit.

Guidelines/Instructions
- Keep continuous track of requirement of important spares. Ensure more than one reliable vendor for supply
- Standby equipment must be in a state of readiness always i.e. as an installed standby which can be immediately started when required
- Reconditioned spares may be allowed for equipments which are not very critical.

- Production may be curtailed till spares from OEM arrive at site for key equipments so that unsafe conditions do not develop.
- Use old spares procured long time back if possible by servicing them in-house
- Try another source for spare parts if there is a long delay from OEM.

2.10.13   Commercial Matters, Finance, Taxation, and General Administration

These are essential activities for smooth functioning of the organisation and their views, priorities and difficulties must be looked in to during the Review of Performance of the organisation.

Aim To maximise the profitability and return on investment.

Priorities to allot funds for:

- Safety equipments, Effluent Treatment Plant, Air Pollution Control
- Disposal of hazardous waste
- statutory permissions
- important payments such as government dues, taxes and installments to financial institutions
- payment for power, fuel and water supplies
- Essential maintenance and spares.
- procuring essential inputs for keeping the production units running
- salaries to employees and external contractual agencies maintaining delivery schedules and product quality.
- Funds allocation – as per organisation aim/policies sanctioned by management
- Obtaining certain equipments on lease
- Research and development
- Depreciation funds
- Modernise for more efficient operation and less breakdown

Situations to be discussed *Problematic situation*: Funds are not available for purchasing of inputs like raw materials, power, water because statutory dues, repayment of loans taken from financial institutions, banks, workers salaries are to be paid soon.

Guidelines/Instructions
- Commit sale of future production at present prices and get advance payment--to be approved by senior management
- Arrange short term bridge loans–these can have high rate of interest, but will be useful to prevent shutting down the plant for want of urgent necessary important inputs in case funds are not available

- Venture capital from external party -try to obtain on a profit sharing basis if one is confident of innovation or better technology developed in-house but funds are not available for implementation

Equipment leasing additional equipments may be obtained from external party who may not be interested in investing money, but can give equipments for improving the working of the plant. They may opt to share the benefits from the equipment till their investment is recovered (with interest). The ownership remains with the external party till this is done. This arrangement will be found very attractive for energy recovery units or new machines (as they will have less maintenance problems). This can be discussed by all concerned and a decision taken accordingly.

2.10.14   Project Planning, Future Expansion

Aims
- Export entire manufacturing plant on a turn-key basis or provide Consultancy services (only know how or know-how with key equipment),
- Acquire idle plants and revive them;
- To expand own production capacity and diversify to new products
- Revival of old plants may need less capital and time to start production provided the condition (and expenses for repairs/replacement) of all old units has been assessed diligently.
- Build pilot plants based on own R & D, try in own manufacturing units and develop reliable technology before export

Chapter 3
Integrated Maintenance: Aims, Responsibilities and Activities

The production and maintenance engineers have to work in close cooperation to run the plant safely and in an efficient manner. Their active participation can provide good suggestions to the management for procurement of necessary inputs (when assessing new technology).

Valuable assistance from Maintenance engineers
Maintenance engineers can be asked to assist in preparation of technical parts of contracts for Purchase order while purchasing process units, machinery and even an entire manufacturing plant. They can be given responsibility to look in to the safety, energy efficiency, reliability and maintainability of the equipments when offers are received for process plants and to assist for preparation of Purchase Orders.

3.1 General Guidelines

Both Production and Maintenance shall study the engineering information in the offers on the process for manufacture (with details of reactions and operations), Design basis, Process Flow sheet, broad specifications of process units, equipment and raw materials and utilities, plant layout; site infrastructure required and pollution control facilities for effluents internally. They shall also look in to:

- the safety and operating efficiency (with respect to yield of raw materials, power, and other utilities) aspects of the process, requirement of manpower etc.
- Whether it is a **proven** technology appropriate for manufacture of the planned products for the scale of operations planned
- There should be no environmental pollution (where the generation of effluents should be minimum and adequate treatment facilities should be offered)
- Can it operate efficiently in range of 70–110% of rated capacity?
- Should have robust equipments with minimum maintenance requirement

© Springer Nature Switzerland AG 2019
K. R. Golwalkar, *Integrated Maintenance and Energy Management in the Chemical Industries*, https://doi.org/10.1007/978-3-030-32526-8_3

3.2 Internal Discussion among Production and Maintenance Engineers

Discuss the following matters internally before the Purchase Order is drafted. Maintenance engineers shall give their inputs and thus assist in preparation of PO

- Study the present and future product mix planned; and the offers made by various vendors, suppliers for supply of the manufacturing plant or know-how.
- Generally the following engineering information is given in the offers. This should be studied by the production and maintenance engineers.
- Whether the offer is for (1) a plant on turnkey basis with entire responsibility on vendor till performance guarantee test is completed (2) only for knowhow (3) know how with some key equipments.
- Design basis considered by the vendor.
- Process Flow Diagram (PFD) and description of Process for manufacture (with typical reactions and operations)
- Main process units and machinery
- Generation of wastes/effluents and treatment facilities which will be supplied
- Plant capacity (MT/day) and (MT/year).
- Consumption of raw materials and utilities required per unit output.
- Plant layout proposed
- Manpower requirement
- Time required for implementation (for setting up the unit, commissioning it and stabilising the production run till the plant is able to deliver required performance)
- Scope of supply offered by the vendor
- Arrangements to be made by the purchaser

3.3 Internal Study

3.3.1 Plant Capacity

- Design basis for rated capacity as MT/day and MT/year.

Example: Rated capacity: 100 MT/D × 365 days = 36,500 MT/year

Actual working days	*=365 − 25 (annual shutdown and various stoppage)*
	=340 days
Actual production	*340 × 100 = 34,000*

Hence plant capacity should be $= \dfrac{36500}{340} = 107.35 MT / day$

Add 15% margin to take care of derating due to various operational problems (loss of catalyst activity, heat exchanger scaling etc.).

∴ Plant should be designed for 107.35 × 1.15 = 123.45 MT/day say 123.50 MT/day

This should be clarified before releasing purchase order.

- Can the plant be operated at different production rates as per market demand?
- Can it be stopped/restarted frequently?
- Operation: whether it will be Automatic, Manual, Semi-automatic.
- Whether highly skilled manpower will be required?

3.3.2   Operational Safety

- Study the PFD and process for manufacture.
- Make a list of the main process units and machinery which will be operating at high pressure, high temperature, corrosive conditions and will be required to handle dangerous materials.
- **HAZID** studies must be made in-house for identifying unsafe conditions. All hazards should be identified by the production and maintenance engineers by carrying out a HAZID study.

  - The details of design basis, details of equipments subjected to high pressure, temperatures, fabrication drawings, materials of construction, corrosion and safety allowances required should be discussed among senior plant engineers
  - **To check whether adequate safety features have been incorporated in the plant design?** It should be internally assessed whether the process units and plant machinery/equipments have been designed by the supplier for all hazards identified by them (as per their HAZID study). **All process units should have sufficient built in safety features**
  - **Analyse carefully the information** (given in the offer) **from maintenance point of view**: Regarding those process units and machinery in the plant (i) (which will operate at severe operating conditions) (ii) which will be used for handling of dangerous materials at high temperatures and pressures in view of their dangerous properties (toxic, inflammable, corrosive nature) (iii) process streams with hard suspended erosive particulate matter (can cause erosion of machinery, protective linings of process units), continuously or intermittently since they can pose many problems for maintenance.
  - This will give a clear idea for any frequent breakdowns; and the jobs (major repairs or even replacements of process units/machines) which will have to be taken up during the Annual Maintenance

Typical safety features necessary
 (i) Reinforcements for making stronger shells, provision of safety vents and valves, rupture discs, instruments for monitoring process conditions, protective lining of refractory bricks/acid resistant bricks.
(ii) Audio visual alarms and safety interconnections should be available for warning if any dangerous condition is likely to develop. Typically these could be for high pressure and temperature in reactors; high/low level in boilers, overhead tanks etc.

These are necessary for safe operation and longer life of the equipments. Safe operation can also improve employee moral and they become more attentive. As a result the product quality will be better and equipments will also last longer.

Internally also examine carefully whether it will be possible to carry out the process with less drastic conditions of pressure, temperature, concentration of reactants, lesser flow rates.

3.3.3 Design of Process Units and Machinery

(whether the following have been considered by the vendor)

• Properties of materials to be handled
• Selection of suitable materials of construction. Check in the offer whether any special Materials of Construction will be used for key equipments.
• *Look in to their limitations, their cost and their expected life. These matters should be discussed internally.*
• Process design and Equipment Design: should also be done after considering (i) local climatic condition, chances of sudden heavy rains, snow, storms, severe summer, earthquake, shortage of water (ii) Operating conditions (normal, minimum and maximum pressure, temperature, flow rates, likely composition of process streams to be handled etc.).
• The equipments should be able to work satisfactorily in the given site conditions (quality of raw materials as available, power, water, climate)
• Limitations on their working specially should be addressed in the plant design
• Load factor on individual machine (will the machines be operated a just a little below their peak capacity or always very near to their maximum capacity?) This should be assessed **internally** after looking into the capacity of each machine and the material requirement for the process units.
• Facilities for treatment of raw material should be included in scope of supply (in case the raw materials are not available exactly as per specifications). This is to ensure product quality, production rate, equipment life, safety or environment.

It will be clear that the process units and machinery should be properly designed with due considerations for the operating conditions and materials to be handled.

This will reduce erosion and corrosion of the units and thus have longer equipment life.

Provision of adequate safety features and devices increases morale and confidence of the operating personnel and they become more attentive (not afraid to go near the process units, observe and operate the feeding system, check levels and temperatures, operate outlet valves etc. This reduces chances of operation beyond limits prescribed; and can reduce breakdowns.

3.3.4 Material Handling Equipment

- Details of Material handling equipments (e.g. bucket elevator, belt conveyor) which will be provided by the vendor
- **Will these be suitable for the plant and can they work satisfactorily at given location for present and future product mix?**
- *Bar chart should be prepared in-house by the purchasing organisation for their own activities (in consultation with maintenance, production, stores, finance departments).*
- Provision of alternative arrangement for power in case of emergency
- The equipments should be constructed/assembled for ease of observation of critical parts; lubrication of bearings, replacement of coupling halves, removal of internal parts which are subjected to wear and tear; erosion and corrosion The offers made by all vendors for technology/process and equipments should be looked in to carefully (technically evaluated) by production and maintenance departments in light of their own analysis and working of specification as above.

  - Submit a report to senior management. *Request for assistance to be obtained from own consultants or external independent agencies for getting a safe process design and technology.*
  - This report should clearly mention their apprehensions. The management should ask the vendor to explain safety features incorporated and to make appropriate changes in design, and then send a revised offer.

3.3.5 Essential Matters to Be Written in Purchase Order PO for Project

More conditions can be added to the PO as per need.

- This depends on whether project is purchased on (1) Turnkey basis with entire responsibility on vendor till performance guarantee test is completed (2) Only knowhow is to be bought (3) know how will be purchased with some key equipments.

- Rated capacity as MT/day and MT/year. Please see 3.3.1 above.
- Annual expected running period days/year
- Plant stoppage required for annual maintenance
- **Product quality**: Product quality must meet specifications given by clients.
- *(These specifications shall be prepared after considering whether the products can it be sold directly/further purification will be required)*
- **Battery limits:** Scope of supply of the vendor should be clearly given.
- **Effluent Treatment Plant/Air Pollution Control** must be included in scope of supply. **However, al**l other necessary pollution control facilities (disposal of any dangerous waste material not specifically sent to ETP/APC should also be included in scope of supply.
- **Exclusions from supply** shall be clearly understood while preparing the PO.
- *Obligations by purchaser should be very clear (for providing facilities for stores, power, water, security etc.) This will define the items to be arranged by the purchaser for completion of the project. In case of any ambiguity the matter shall be clearly discussed with the vendors.*
- *Specifications of raw materials, utilities which are to be used.*
- Performance Guarantee for production and maximum limits of consumptions of all inputs per unit of product of specified quality must be clearly specified and confirmed from the vendor.

Vendor should give details of:
- All process vessels and equipments operating at high temperature, high pressure, and those which will handle corrosive/dangerous materials.
- Pumps required to handle process streams with suspended/erosive particles
- Whether these units will work continuously or intermittently.
- Whether machines and drive motors will be available as pre-assembled, carefully aligned units or will be supplied separately
- Load factor on individual equipment/machine (will the machines be operated little below their peak capacity or at maximum capacity always?).
- Safety features like alarms, safety devices which will be provided for all such vessels, critical equipments
- Safety interconnection arrangements between process units
- Details of operating conditions for all process units and machineries with maximum pressure and temperature expected, pH, flow rates, MOC offered, properties of materials (which will be handled by the equipment or process unit),
- Material and Energy balance
- Details of heat recovery units, provision of boilers, economisers, steam turbines, electrical installations, cable routes, Variable frequency drives, lighting (flame proof lighting in areas with inflammable vapours)

Detailed Fabrication Drawings should be provided by the supplier/fabricator for the Process units and machinery on placing order. These should be approved by Statutory Authorities

Radiography (100%) for all welded joints will be required when dangerous materials are to be handled by the process unit.

- Sectional elevation drawings and Plans at different elevations should show working space around all units for cleaning and maintenance
- **Piping and Instrumentation drawings** should also give similar details for cleaning and maintenance
- All material handling equipments which will be used (such as hoists/bucket elevator/pneumatic conveyor/screw conveyor/belt conveyor) and to confirm they **will be suitable for the plant.**
- **Guarantees offered for smooth working of all these units and machinery at site**
- **Erection and commissioning should be done by vendor**, Necessary manuals must be supplied by the vendor.
- **Bar chart should be provided by vendor for all items in their scope of supply (maintenance department should check this)**.
- Supply of drawing for civil foundation or load data.
- Supply of foundation bolts for all equipments
- Statutory approvals for pressure parts to be arranged by vendor before the equipment is dispatched from his manufacturing shop.

  - Test certificates for pumps, MOC, motor, cables, catalyst to be given by vendor.
  - Vendor should give details of material of construction of all such units and protective linings provided.
  - Technical experts of fabricator/vendor should be present to supervise the erection and commissioning activities even when only knowhow is given (since these are very crucial for ensuring equipment life)
  - Vendor should make available standby units as installed units (or as spares if there is no space in the plant to install the standby unit). Vendor should be asked to modify the layout if possible.
  - Vendor should agree with the conditions mentioned in the Quality Assurance Plan prepared by purchaser also. (**see Sect. 3.3.9 QAP below**)
  - Vendor should recommend Spares and supply them during procurement.

3.3.6 Plant Layout for Safe Operation and Ease of Maintenance

Production and maintenance engineer should discuss this with vendor and final layout shall be prepared only after an agreement is reached.

- Storages of inflammable, toxic, corrosive items should be away from inhabited areas in the premises (which have more personnel in those areas). Storages of fuel oils, LPG, Propane etc. should also be away from such areas.

- **Safe and secure location** The plant layout should be such that no equipment is exposed to harsh weather conditions (heavy rain, high ambient temperatures, snow,), strong radiant heat near furnaces etc.
- Process units shall not be exposed to corrosive vapours, dust from nearby grinding units, spillage of corrosive chemicals from overhead pipes, heavy vibrations from machinery.
- Sufficient space must be available around key equipments for observation from all sides; for cleaning, lubricating, and maintenance of important components/ fittings; for replacement of components if required, connections to process piping and ducts, and for correct positioning of sensors of the instruments.
- Good lighting and railings should be provided at the work platforms and ladders for operation of control valves; cleaning manholes and drain valves and for maintenance of the process units
- There should be no obstruction to the movement of material or personnel due to the installation of the process unit or machinery.
- No safety escape route shall be blocked due to process vessels, big ducts, machinery.
- Safety showers, eyewash fountains (with dedicated overhead water supply tanks) should be provided near all locations where dangerous materials are handled.

3.3.7 Performance Guarantee Test

The conditions to be written depend on whether project is offered/purchased on (1) Turnkey basis with entire responsibility on vendor till performance guarantee test is completed (2) Only knowhow is offered (3) know how + some key equipments are offered; with detailed engineering to be done by the purchaser.

- The performance guarantee runs should be for a minimum of 3–5 days (72–120 hours) of continuous run.
- It can be for 3–5 batches of marketable grade products in case of a plant run in a batch wise manner.
- In case this is not possible due to any reason like shortage of raw material, plant commissioning delay, power outage, labour strike, accident, the performance guarantee run can be postponed to an earlier opportunity whenever it is possible. About 10% of the payment shall be retained as retention money. This matter needs to be discussed with the vendor.
- All safety devices and interconnection should be demonstrated to be working satisfactorily. **This condition should also form a part of PG test.**
- Ease of cleaning, repair and maintenance may also be included in the PG test.
- Rated production capacity on per day basis with a maximum of ±5% tolerance limit.
- The consumption of raw material and utilities per MT of product of saleable quality (or quality which has been clearly defined in the purchase order) should

not exceed the guaranteed figures. Penalties may be specified in case of excess consumptions.

- Likewise the quality/complete physical and chemical analysis of the raw material should be clearly defined.
- The methods of measurement for the production and consumption shall be agreed upon before starting the performance guarantee test.
- Rated production capacity on per year basis. This is important to arrive at the profitability of the project because there will be a stoppage of a few days every year for annual maintenance as well as due to unforeseen breakdown in between.

3.3.8   Annual Maintenance Contract

- Purchase Order may include separate terms and conditions for (i) regular visits to plant for inspection, guidance and repairs every month or three monthly intervals (ii) visits to plant for inspection, guidance and repairs during annual shutdown/overhaul. Monthly/quarterly visit by OEM engineer or annual maintenance contract should be entered into with supplier
- Observations regarding deterioration of performance shall be informed
- Contract may be given for major repairs [handling, cleaning, repairing, servicing and supply of spares]: the Purchase Order should include a condition that certain specified level of performance will be guaranteed after repairs.
- Improvement/enhancement of performance required in the equipment if plant expansion/modernisation/diversification is planned.

3.3.9   Quality Assurance Plan QAP

- QAP must address the basic idea to obtain equipments/machinery items for the plant which have appropriate safety features (as required by law), which will comply with statutory rules and regulations, ensure safe working, meet required performance and will have a long useful life without breakdowns.
- These should be discussed among senior plant engineers, stores in charge, finance, department, senior management and the conditions shall be finalised.
- Details of the requirement of Quality Assurance Plan should be made clear to the vendor.
- The machinery shall be constructed for ease of observation of critical parts; lubrication of bearings, removal of internal parts which are subjected to wear and corrosion; and ease of repairs.
- It should be possible to trace all spares and inputs bought by vendor (valves, steel plates, heat exchanger tubes, view glasses, level gauges, instruments.). All such items shall be stored properly at Vendor's Works so that they are not exposed to

dust, rains, sun corrosive conditions etc. Items of higher standard may be procured if the items available currently do not meet specified ones.

- Purchaser shall have the right to depute qualified and experienced persons during the entire process of placing order to actual delivery, erection and commissioning at site. This will include stage wise inspection during fabrication and final inspection before the equipment leaves Vendor's works.
- **General Assembly drawing** for all equipments and process vessels. These should show battery limits, Materials of Construction, thickness of walls, bottom and top cover plates, nozzles (for all incoming and outgoing process streams, probes for instruments), valves, safety devices, thermal insulations,
- **Design Codes** according to which the detailed design of the machinery has been done
- **Safety and corrosion allowance** (for process vessels), dirt factors (for heat exchangers), additional reinforcement features (external ribs, pad plates) which will be added.
- **Fabrication drawings and procedures** for fabrication (including welding) concerning the vessels for handling fuels, dangerous materials and those which will operate under pressure must have approval of statutory authorities.
- *This is applicable even for those vessels which may get pressurised due to any reason (heating from a nearby source, due to failure of cooling systems...)*
- **Fabrication codes which will be followed**.

Some of the well known fabrication codes are

 (i) **ASME Sec I: Boiler and Pressure Vessel Code,**
 (ii) **ASME Section VIII: Rules for Construction of Pressure Vessels,**
(iii) **IS 2825--1969 for Unfired Pressure Vessels with Class I (for lethal materials) and Class II (in chemical industries with radiography for all longitudinal and circumferential joints)**

Equivalent international codes should be considered for fabrication in the countries where the plant units are to be installed.

Vendor to specify which codes will be followed during fabrication. These must be approved by statutory inspectors/competent authority in the country.

- Guarantee for ability of the equipments/items to meet (required) performance

Requirement of Original Documents

- Test certificates for materials of construction and all bought out components, welding electrodes used
- Procedure for fabrication which will be followed by the vendor
- Certificates for welder's qualification and experience
- Tests to be done at vendor's works which will be witnessed by purchaser.
- Films and necessary records for Radiography of welded joints, hydraulic pressure tests as per statutory conditions should be available.

- Vendor should give details for programme for stage and final inspection to be carried out in presence of technical representative from purchaser
- Records for Post Weld Heat Treatment,
- There should be traceability of all spares and inputs bought by vendor (valves, steel plates, heat exchanger tubes, view glasses, level gauges, instruments.) bills, inspection reports
- Safe methods of storage at Vendor's Works to be confirmed: (protection from rain, dust, corrosive vapours)

3.3.9.1   Cooperation by Purchaser

- A reasonable deviation proposed by the fabricator may be considered by purchaser (designer, plant engineer). It can be either accepted or an alternative may be suggested without affecting safety or violating statutory conditions.
- Technical teams of the purchaser should prepare a detailed list of the activities in their scope and a bar chart for the same. These shall be addressed promptly so that the erection and commissioning is not delayed.

3.4   Responsibility of Maintenance Engineers

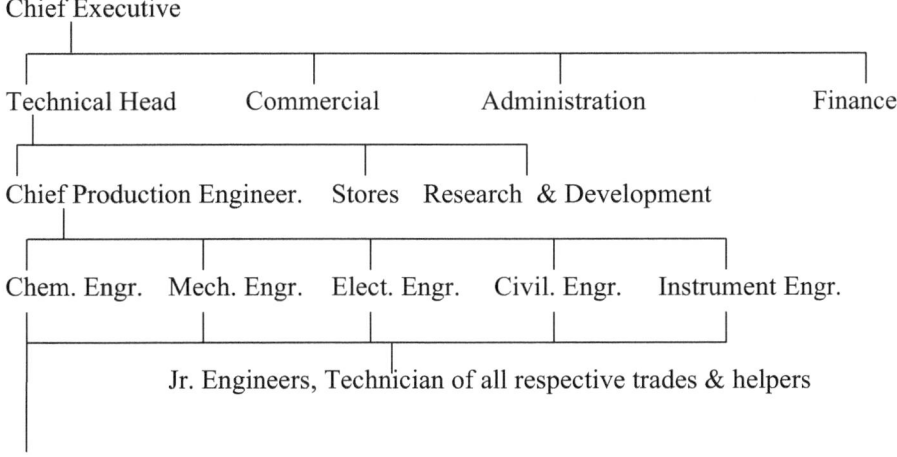

Chief Executive

Technical Head   Commercial   Administration   Finance

Chief Production Engineer.   Stores   Research & Development

Chem. Engr.   Mech. Engr.   Elect. Engr.   Civil. Engr.   Instrument Engr.

Jr. Engineers, Technician of all respective trades & helpers

Plant operating teams to report to chemical engineers

- All senior engineers shall be assisted by junior engineers, technicians of respective trades (chemical plant operators, laboratory chemists, pipe fitters and machine fitters, welders, electrician, riggers, instrument mechanics, refractory and civil masons).
- Marketing, Stores and procurement, Logistics, Human Resource Department, Finance shall have close coordination with production and maintenance teams.
- R & D section shall also coordinate with production and maintenance team.
- External agencies/contractors may be engaged for certain specific jobs as per instruction from senior management (must work within framework of aims and policies of the organisation). They can be engaged for voluminous jobs like unloading raw materials, packing and loading finished products, cleaning of process vessels, tanks, or when sufficient resources are not available in-house.
- They can provide assistance for major maintenance activities like removal of old units, dismantling old structures, shifting and installing new units etc.

3.4.1 Useful Suggestions from Maintenance Engineer

It is necessary to properly maintain all critical machinery to ensure reliable performance of the plant. The Design Engineers should consider incorporating the following suggestions in the plant design for this purpose:

- Metering pumps: Provide strainers in suction line to remove particulate matter which can erode the pump internals. Provide installed standby unit
- *Operate standby pump regularly (on alternate days if possible) to keep in an immediately available condition.*
- Air blowers: Provide air filters in suction line; check regularly.
- pH controllers/indicators: Provide strainers in fluid sampling lines to protect the sensors. Sample coolers may be installed if required for proper functioning of the instruments. (these may not be required if temperature compensator systems are available which ensure instrument working)
- Dial thermometers in liquid lines: Provide stainless steel 316 or Teflon coated thermowells in liquid lines instead of directly inserting the stem of the dial thermometer in liquid lines. Provide screens/grids to prevent entry of big pieces of solids into the machines used for crushing, pulverising etc.
- Provide strainers at suction side of pumps, feed controllers.
- Boiler feed water lines: Maintain sufficient level in Boiler Feed Water tank by float controlled make up water inlet valve. Operate the feed water pumps alternately (running + standby) so that both are available always. Maximise recycle of condensate from process units, steam turbine condenser.
- Boiler performance: Confirm proper working of Non Return valve, conductivity meters in BFW lines and firing of fuel.
- Install cyclones/Dust Collectors Demisters in gas ducts to prevent deposition of dust on impellers/rotating parts of machines installed downstream.

- Ensure uniform distribution of liquids/gases all over the cross section of packed beds by suitable design of inlet nozzles. This will prevent channelling of fluid flow in the units.
- Use centrifugal compressors (if possible) instead of reciprocating types
- Reactor Design modification- Provide sufficient empty space at top for a Fluidised Bed Reactor. This will reduce carryover of particulate matter with the exit gas stream and hence there will be lesser deposition of particles on rotating parts of downstream machines like, Induced Draught Fans.
- Oil firing unit: Maintain sufficient fuel always in supply (day) tank. Install strainers in suction line of oil pumps. Check and adjust setting of oil heater to ensure ease of pumping in cold weather.
- Water treatment plant: Maintain sufficient water in raw water tank for 2 days supply. Regenerate the resin columns well in advance to maintain supply of treated water. Maintain all water pumps properly in working conditions.
- Silos for storing and feeding materials to process units. Maintain sufficient stock by operating filling system (well in advance). Provide a mirror at top to show the level of solids inside if it is cumbersome to climb up frequently.
- Provide and maintain high/low level alarms – for overhead cooling water tanks.
- Use lubricants as per recommendations of OEM only.
- Provide battery back up (for lubricating oil pumps of all critical machinery) in addition to power supply from DG sets.
- Install standby unit for critical equipments (e.g. air blower, oil pump, cooling water pump, Electrical overhead hoist) in a ready to use/change over condition so that the production plants can continue to work without interruption.
- Purchase spares from OEM only.
- Keep spare screens for strainer/strainer units for critical units always in stock.
- Replace filter cloth in Filter Press every 90 days (or earlier). Keep a spare set ready
- Replace rupture disc in safety system every 90–120 days whether damaged or not.
- Keep spare filter leaves always available for pressure leaf filter.
- Replace active carbon in Active Carbon Filter units every 120 days or earlier

3.4.1.1   Provide the Following

- ceramic/stainless steel ferrules in heat exchanger tubes.
- Fiber Bed Mist Eliminator instead of mesh type pads in absorption tower for gas
- protective heat resistant sheath on cables used in high temperature areas.
- dry air injection (by 15–20 mm line) into glands of butterfly valves, shaft seals to prevent corrosive fumes coming out and damaging nearby units.
- dry air injection into nozzles for sight glass to prevent deposition of soot/un-burnt material from inside. It will keep the sight glass clean for better observation.

- strainers in feed lines to prevent damage to glass lined reactors, agitators blades and shaft etc. due to hard particles.
- cooling system for hydraulic oil for operating dampers, actuators.

3.5   Increasing Life of Process Units and Machinery

Process units and machinery in chemical industries are exposed to corrosive conditions which can deteriorate their performance, create early need for repairs or even reduce their useful life. Considerable efforts are to be made and resources have to be spent to maintain their performance to a satisfactory level when the plant is running.

Major repairs to components or even change of the entire unit or machinery may become necessary within a few years of their commissioning due to such continuous exposure to corrosive conditions. Managements have to therefore set aside funds from the revenues generated in order to replace these depreciating assets in time.

It is necessary to take steps to prolong the useful life of process units, their internals and plant machinery. This can be done by correct design, choice of suitable material of construction (which will withstand the operating conditions), incorporating safety features, following proper fabrication codes, carrying out erection and commissioning under expert guidance, careful operation and timely maintenance.

These steps are useful to sustain the performance of equipments and to reduce material and energy consumption. They can result in longer life (*and may need lesser efforts to be done and expenses for major maintenance*)

3.5.1   Some Suggestions for Better Life of Process Units and Equipments

Design of feeding system for raw materials shall be such that excess feeding should not be possible. This is to ensure that runaway reactions will not take place and subsequent high temperatures or pressures will not occur. This can be done by using metering pumps (as they deliver only at a particular flow rate which has been fixed beforehand) instead of centrifugal pumps which can feed excess material.

In addition, a day tank with a limited capacity should be provided for feeding the reactor (instead of feeding the material from the main storage tanks). This tank can feed only a limited quantity. It may be filled twice or thrice every shift as per need.

Heating system An interlock shall prevent start of heating system unless sufficient cooling water is flowing through the concerned downstream process unit (e.g. cooling water flow through the condenser should be started before vapours are generated by heating the process fluid) It will prevent development of dangerous situation also due to build up of excessive uncondensed vapours in the system.

A flame monitoring device should immediately switch off oil supply pump to burner if the flame is extinguished due to any reason. It should also get stitched off in case the combustion air supply is stopped.

Boiler water level control system should trip the firing system if the water level goes very low in the boiler. *As an additional precaution, even the very low level mark should be a little above (40–50 mms) the uppermost boiler tubes*

On-line Conductivity meter must be provided in Boiler Feed Water for continuous check on the quality of water since only demineralised water should be fed to boilers for their longer life and smooth working. Dissolved salts present in water may form deposits on the heat transfer surfaces, cause high temperature of the tubes and eventually cause them to leak. Higher conductivity may indicate acidity of feed water also in some cases. Hence quality and conductivity of the boiler feed water should be checked continuously.

Chlorides in boiler feed water can be corrosive to the boiler internals and tubes. Hence feed water with chlorides should never be used.

Screening of incoming streams
Provide magnetic separators or screens in raw material feeding arrangements; strainers in pipes to protect internal lining of vessels or parts like agitators from erosion due to sharp particles or metallic objects in feed streams. It is useful to protect glass lined process vessels.

Example: strong magnets can pick up tramp iron pieces from input points of rock phosphate feeder to acidification unit. It protects the special brick lining.

Protection for shaft seals
Electric motors or gear boxes for agitators can get corroded due to escape of hot acidic vapours coming out of chemical vapours from the agitated vessel. Hence the shafts must have proper glands (conventional or water cooled), mechanical seals or provision of air injection under pressure to prevent escape of corrosive vapour from shaft seals of agitated vessels.

Induced draft fans are used to suck out gases from the air pollution control scrubbers (which are provided **after process units**) and then exit through a chimney. These fans are generally provided with rubber lining/glass fiber lining as protection from any acidic mist particles which may escape even after scrubbing the gas stream.

The tip speed of such rubber lined/glass fiber lined impellers shall never exceed the limit specified by OEM as the lining can come off at higher speeds. Hence the rotational speed of the fan shall be controlled either by variable frequency drive motors or by properly choosing the diameters of pulleys of fan and motor if VFD is not provided.

Lubrication Important machinery must have sight glass for indicating lubricant levels (minimum and maximum). The name plate must have correct grade of lubricant mentioned.

Instrumentation

Provide the right sensors at the right location in process vessels and pipelines to get a truly representative sample of process stream in real time with no time lag/minimum time lag. Install sample coolers/conditioners in gas sampling lines.

Strainers must be provided in pipes to protect the sensors of instruments.

The instruments in plant should be calibrated against standard instruments; and their accuracy shall be checked regularly. Any drift in reading/zero value must be immediately corrected. This is necessary for operating the process units within prescribed (safe) limits.

Strengthen

Equipment should have stronger outer shells even if they cost little higher.

Agitator shafts should have larger diameters to minimise shearing off (when highly viscous liquids or thick slurries are to be stirred in agitated vessels).

External ribs can be provided for strengthening the shell. External limpet coils are also found useful instead of jacketed construction as they reduce the L/D ratio.

- Calibrate Flow Controllers for feeding raw materials, for flow of heating medium to reactors,
- Provide Variable Frequency Drives for motors
- Reset press release valves at as low a value as possible;
- Change operating conditions if possible [reduce temperature/pressure/flow rates/concentration of process fluids]
- Provide vibration dampening pads for reciprocating machines, compressors etc.
- Dampening pots in discharge line of reciprocating pumps
- Recheck alignment of motors and driven units in case of any abnormal noise.
- Check Dynamic balancing of rotating machines every 15 days.

Stress relief by provision of freedom for expansion

Roller support is necessary for long horizontal equipment which are likely to expand due to operation at higher temperature (after a cold start).

Provide expansion joints/freedom for expansion for ducts. Expansion bellows or 'S'shaped bends may be provided in gas ducts (which are subject to considerable temperature changes) to reduce stress during operation.

Substitute by better MOC

- Carbon steel shaft of pumps can be replaced by Stainless steel-316 or Alloy-20.
- 40% Al_2O_3 bricks can be replaced by 65–70% Al_2O_3 bricks.
- Cast Iron gills protect the economiser tubes from acidic gases (sulphuric acid plant)
- Rubber lined impeller in place of bare carbon steel impeller for Induced Draught fan

Substitute components with those having better life
- External jacket on a reactor replaced by limpet coil which adds to mechanical strength
- Replace Shell & Tube heat exchanger by Disc & Do-nut for better heat transfer and better heat distribution inside for reduced stresses.
- Pellet type V_2O_5 catalyst can be replaced by Ring type Catalyst which has more activity in presence of dust in gases (in a sulphuric acid plant)
- Cast Iron trombone cooler for sulphuric acid can be replaced by Plate Heat Exchanger constructed from Hastelloy C-276.
- CS_2 Plant: (i) Graphite electrodes in place of water cooled electrodes (which can get damaged if water flow is interrupted due to any reason) (ii) Charcoal Calciner drain: Water cooled sliding damper instead of steel sliding gate.

Some additional practical suggestions for improving design/construction
- Refractory lining: Provision of an additional layer of insulating bricks or ceramic paper inside the shell for minimising heat loss from furnaces.
- Interlocking type design for refractory bricks: to prevent loosening or falling away of refractory bricks from lining.
- Provision of anchors and retainer plates: to prevent falling away of refractory bricks from lining
- **For sulphuric acid-oleum plant** (i) Use of 50 mm tubes instead of 25 mm diameter tubes to prevent choking due to acidic sludge; ease of cleaning of the Cold heat exchanger (ii) introduce the return gas from IPAT tube side instead of shell side: for case of cleaning the tubes by brush. (iii) Pad type demister in Inter Pass Absorption Tower can be substituted by Candle type demister for better life of downstream heat exchangers. (iv) Alloy steel spindle and refractory seat for WHRB bypass duct: for longer life; for better process control. (v) Provide Cast Iron gills on heat transfer tubes of economiser (which is provided downstream after converter of sulphuric acid plant) for protection from corrosive gases. (vi) Alloy 20 wetted parts for circulation pumps of acid/oleum: for longer life. (vii) Operating cooling system on siphon instead of pressurised water supply for cooling circulating 65% oleum: To prevent ingress of water into hot areas in case of leak (since it may cause accidents due to excess pressure from steam generated). (viii) Using falling film of water for condensation of SO_3 instead of pressurised water flow: To prevent ingress of water into hot areas in case of leak (since it may cause accidents due to excess pressure from steam generated).
- External cooling water spray on hot units: As additional safety.
- Sight glass and light glass nozzles to observe reactors internals/scrubber internals: for better process control/adjusting flow of scrubbing liquor.
- Wet ESP: Upward flow of gases with rubber lined internals for particulate matter removal: Wet ESP design is better than dry ESP. Can be cleaned online to remove deposits

- **Pretreatment of raw materials**: for better life of process units and machinery internals **e.g. r**emoval of tramp iron by magnetic separator, screening of big lumps rotary screen, filteration to remove sharp hard particles, calcinations/drying of wet material, removal of dust: to minimise masking of active surface of catalyst.

3.6 Spares for Process Units and Internals/Accessories

Efficient maintenance activities are necessary to restore the performance of the Production Plant to the desired satisfactory level and should be carried out in a safe manner at minimum cost. The Production and Maintenance engineers must observe the current performance of the plant carefully and note the plant output, effluents generated, consumption of raw materials and energy, and the quality of products. Any shortcoming shall be recorded and plans for remedial action shall be initiated.

Planning

This shall be done by a detailed study of design of the Production Process once again. It should be followed by reference to the actual working methods, process units and machinery installed (with ready availability of standby units), and recommendations for spares by the Original Equipment Manufacturer and vendors.

Suggestions from experienced operating and maintenance personnel shall be invited and considered for improvement

Detailed working programmes shall be worked out for necessary cleaning, minor and major repairs of process units/machinery; and replacement of components based on these observations and suggestions.

3.6.1 Procurement of Spares

Analysis of consumption of spares in the last 3–4 years shall be carried out to arrive at the future requirements. The request for procurement of spares shall be jointly prepared by production and maintenance engineers. Some additional requirements of certain items can be asked for if any unit is performing much below par (and thus may need more spares) and if it may take longer time for actually getting the fresh supplies at site. Management shall ask the concerned persons to send all such requests for procurement well in advance

Practical guidelines

Consumption records of all spares shall be updated. A careful study of the last 3–4 years will reveal which items are getting consumed in larger amounts.

Sufficient stocks of spares should be available before taking up:

- any major shutdown/plant overhaul
- expansion of production capacity

- modification of process plant
- planning any diversification to more products in near future
- Stock of a process spare (e.g. catalyst, active carbon, filter cloth) shall preferably be at least 15–20% of the quantity loaded in the respective process unit in case of items having a service life of about 5 years; and 35–40% in case if items having a service life of 3 years.
- All spares to be procured must meet the specifications as given by process engineers and maintenance engineers; and OEM.
- They must be tested (either samples or assembly as the case may be) in actual operating conditions and test certificates obtained). The sampling plan must be mutually discussed and agreed between purchaser and vendor. The samples may be tested in an independent laboratory whose results will be acceptable to all.

3.6.2 History Cards of Individual Process Unit

These should have records of serial number, name of vendor, date of installation and commissioning, original specifications of duty conditions, maximum permissible limits of pressure, temperature, flow rate etc. and any other relevant data.

Material of Construction (for units handling corrosive process materials), repairs carried out since commissioning, actual running hours (days), rate of input and output (for units like filter, settler, converter) total throughput since last shut down or major repairs, any specific limits for operating conditions recommended by OEM

3.6.3 Safety Margin for Higher Stock Level

Factors for maintaining additional stock if:

- It takes a long time to procure
- The items are not very costly
- the items will not get deteriorated (will not become unfit for use)
- will not occupy too much of space

The additional stocks may not be built up if:

- the materials are inflammable, toxic, or dangerous materials
- occupy more storage space
- the items are very costly
- become obsolete if not used within a certain time
- elaborate safety precautions will be required to store.
- Insurance charges will be high

These requests should be studied by the Stores department and checked against available items before reporting to senior Purchase officers. The procurement activities can be started on getting the necessary sanction for funds.

These can be done on the basis of following broad classification: *The list of items given below is indicative for a typical chemical plant.* It should be made according to the individual industry with due consideration to consumption of various items for the current rates of production and for the future; as well as for any plant shut down planned, ease of availability, product mix planned, cost of spares, demand for product in the market, consequences if a particular item is out of stock etc. Stores department shall be consulted and present stocks shall be checked.

3.6.4   Considerations for Procurement and Storage of Various Items

- Specifications for quality, chemical and physical properties and packing container must be clearly written in Purchase Orders and mutually agreed.
- All raw materials, chemicals and spares must be tested in own facilities or test certificates shall be obtained from vendor/approved laboratory before accepting them.
- Samples shall be drawn in presence of supplier and sent to approved test house/laboratory with common consent.
- Some samples/a part of sample can be preserved for future reference.
- Sampling plans must be discussed and agreed with suppliers for bulk items like refractory bricks, solid raw materials in bags, filter cloth, resins etc.
- Representative samples of liquids can be drawn from tankers and tested
- Storages in open yard: Solids which are not generally affected by sunlight, rain etc.
- But these shall be kept on sloping concrete floor to drain out rain water. (sulphur, rock phosphate, bauxite)
- Storage in covered shed: solids packed in bags, carbouys, drums shall be kept on raised platforms or wooden pallets to prevent contamination. (salt, anti foaming agents, stabilisers.)
- Liquids can be stored in storage tanks which have dyke walls around them, and warning facilities for high level, and collection of any overflown material (during filling). Overflow nozzles of the tanks can be connected to each other to prevent spillage. Lightening arresters shall also be provided for storage tanks of (mineral acids)
- Gases shall be stored in cylinders, containers with full compliance with statutory regulations. Safety Valves and vents shall be provided.
- In general the chemicals, lubricants etc. shall be kept in original packings/containers in a covered shed having good lighting, sprinkler system and smoke detectors. Separate bays shall be created for easy identification and handling.
- Resins for Water Treatment plants should be always kept wet.

- Membranes and accessories for Reverse Osmosis units should be kept in original containers.
- Sand shall be kept in closed bags.
- Active carbon, alum, $FeSO_4$, polyelectrolyte shall be kept in original packings in a cool dry covered shed protected from rain and surface water.
- Instrumentation spares shall not be taken out unnecessarily. They are to taken out from boxes only for calibration and immediate installation.
- Both ends of Valves shall be closed by blinds to prevent ingress of dust or moisture.
- Electrical motors, starters, light fittings shall be kept wrapped in clean water proof sheets and kept in a warm room (heated by low power heaters or light bulbs)
- Catalyst and bed supports shall be kept in original containers in a warm room if there is a chance of loss of activity due to moist atmosphere.
- Tower packing and internals for Absorption systems: Store in covered shed in original containers well protected from rain, dust
- Fuel oils, Propane, LPG, LDO etc. shall be stored away from the working area and all precautions as per OISD (Government of India) or equivalent regulations in the country must be complied with. Proper insurance cover to be taken and arrangements made for fire fighting, safety for persons in plant and surrounding population.
- Pumps, compressors etc.: – keep in original container with all openings properly closed. They should not be kept hanging. Support on wooden pallets.
- Pipe lines: both ends shall be kept closed (wrapped by impervious sheets) to prevent ingress of dust or moisture. A thin layer of light oil may be applied to prevent rusting. But it must be removed before use
- Refractory materials: should be protected from rains (specially insulating bricks)
- Always take care that bags, carbuoys, wiring, demister candles, instrument parts, catalysts etc. are protected from sharp objects, electrical contacts, high temperatures, spillages from leaking pipelines etc.
- For items not covered above: Please follow instructions from vendor.

3.6.5   Other Necessary Spares for Process Inputs, Equipments, and Auxiliary Items

Safety equipment
Personnel Protective **Safety equipment** dress, safety goggles, gas masks, safety shoes.

Protection of plant units Pressure release valves, rupture discs, alarms, warning lamps, guards for level indicators, flame proof light fittings.

- Safety Showers, Personal Protective equipments.
- Fire Fighting

Hose pipes of rubber, reinforced rubber, PTFE lined, braided with stainless steel mesh in various diameters and standard lengths.

Raw materials
Quality: As per specifications given by production department; while strictly rejecting materials which have any harmful impurities to catalyst, process machinery etc.

Quantity of raw material to be stored should be commensurate with product mix planned, and the time required to obtain fresh supplies (some additional quantity should also be stored as safety margin). This will prevent disturbance in production programme in case of delay in procurement.

Melters, Dissolvers, Mixers, Blenders
Agitator blades and shafts, bearings, shaft sealing glands (assembly), set of baffles, level indicator assembly, sight glass/view nozzle assembly

Material conveyors (belt, bucket, screw, hoists, cranes.)

- Lubricants for all bearings,
- Spare length of belts with fasteners and pulley, spare buckets and discharge chutes, spare screw, intermediate bearings, etc... as per recommendation by OEM

Water Treatment Plants accessories
Spray nozzles for filter, NaCl bags, sufficient anion and cation resins, HCl in tanks or carbouys, Caustic lye 45% approximately), alum, $FeSO_4$ solid
The resins must be kept moist always in closed containers.

Tubes for tube settler, plates for lamella clarifier, rubber lined valves, Dissolved Oxygen meters, spares for air spargers (pipes, nozzles)

Clean graded sand and activated carbon... equivalent to at least one full charge loaded in the respective vessels (Pressure sand filter and Activated Carbon Filter)

- **Reverse Osmosis unit**
  Membrane assembly for at least 25–30% of total number of assemblies in use. (assuming a service life of 3–4 years) These should be preserved as per OEM instructions in a cool safe place.
- **Ultra filteration candles ...as above**

Air Pollution Control accessories APC
- Demister candle
- Pads for mist separator
- Assembly/spare flap for flow control valve for Venturi scrubber
- pH controlling system spares
- Absorber tower internals (packings like intalox/berl saddles, plain and partition type raschig rings, tellerette and pall rings resistant to dilute acid and alkali)
- Complete assembly of spray nozzles for scrubbing liquor
- Level indicator for scrubbing liquor tanks

Get samples of following chemicals analysed through approved laboratory or in own laboratory. Alternatively, obtain test certificates from supplier of chemicals who should be a reputed party:-.

Chemicals for Effluent Treatment and Air Pollution Control facilities (lime, alkali, activated charcoal, graded sand for filter) Special Chemicals: (examples are hydrogen peroxide, Na_2SO_3 etc.)

Alum, Ferrous sulphate, $CaCO_3$/CaO, NaOH/Ash, sand, CO_2, DCP etc.

Raw materials and other inputs for treatment,
Spares for pre-treatment (filter cloth for filter press, stainless steel screens for filters)

Filter aids (for increasing efficiency and output of filteration, settling agents for impure liquid raw materials,)

Pressure Leaf Filter (steam jacketed type)–spare filter leaves (generally of stainless steel construction), bearings for operating screw for opening, steam traps, safety valve, pressure gauges, **O –Rings** for fixing filter leaves on exit manifold

Filter press Filter Clothes of various specifications (100,300,500-mesh; Polypropylene)

Spare Plates and special gaskets for filter plates, drain valves, pressure release valve at liquid inlet and vent valves, pressure and temperature gauges, spare operating handle.

Hot Gas Filter
CI support column pieces and grids, refractory bricks, graded mass of crushed refractory (made from bricks with not less than 45% alumina) filtration media, quartz pebbles

Process units and Machinery
Check stampings and mark of approval from accredited testing agency for all items being procured. Make arrangements for in-house testing or visit supplier's works.

- Reactors: Pressure release valves, rupture discs, alarms, warning lamps, level indicators and repairs to internals
- gas-liquid contact packings for absorption tower: Various types of random packings are used with packed towers *Most commonly used packings are Pall Rings, Berl saddles, Intalox saddles, Raschig rings, Tellerettes, honey comb type which should be resistant to operating conditions and high temperature of incoming gases.*
  Additives to electrolytes for electrolysis plant
- screens for centrifuges,
- complete set of fittings of sight glasses for reactors, level gauges for process vessels
- internal spares for distillation units
- spare plates for PHE with a complete set of gaskets

Converter
Catalysts and catalyst bed supports, spare grids and partition plates for converter sections

Fresh catalyst equivalent to at least 25–30% of one full charge loaded in (which is based on the assumption that service life of catalyst is 3–4 years) It should be stored as per instructions of vendor/catalyst manufacturer.

Catalyst and catalyst bed support (ceramic balls/graded crushed fire bricks/stainless steel/support grids made from special alloy.) as per recommendation from plant designer and history card records, spare grids and partition plates for converter section

3.6.6   Product Quality Control (Necessary Inputs for Maintaining Product Quality)

- anti foaming agents:- oil based, water based, powder based defoamers
- stabilisers: (for increasing shelf life of products, for preventing polymerization, for light fastness. These are specific for the concerned products being manufactured
- anti-caking agents to enable free flow of solid granules
- special property modifiers (pour point depressant, agents to prevent polymerisation)

Refractory materials
high temperature resistant and insulation bricks of various sizes and shapes (rectangular, circular, manhole arch bricks) as per design/process vessel geometry, graded refractory filter media and quartz pebbles for filters insulating and high alumina castable cements

Acid/Alkali resistant materials
acid and alkali resistant bricks and cements, potassium silicates, acid resistant bricks for bottom layer and arch bricks for gas entry nozzles, manholes of acid towers; for dyke walls around storage tanks

Civil works
Bricks, clean sand, steel rods, and other items, portland cement

3.6.7   Auxiliary Facilities/Items for Process Plant

(i) **Thermal Insulation**: Pads of glass wool, mineral wool, chicken netting, aluminium 18–20 SWG cladding sheets, calcium silicate blocks

Check stampings/marking on gas cylinders, outlet valves and mark of approval from accredited testing agency for the following. Obtain certificate for gas analysis

(ii) **Nitrogen cylinders** to be used for blanketing of inflammable materials, as preservative for food industries (must be 99.98% pure), for flushing out toxic and inflammable vapours from vessels or pipe lines before taking up maintenance work

(iii) **Oxygen cylinders** for special oxidation reactions, gas cutters,

Check stampings and mark of approval from accredited testing agency for Pipeline and Valves with test certificates for material of Construction, pressure ratings:

(iv) **Pipeline and Valves**: Various types of valves. Ball valve, Globe valve/Gate valve, Non-return valves etc. used for different services like water, oil, steam jacketed, acid, alkali handling purpose in different end connections flanged and screwed in various sizes 15/20/25/40/50....200 mm

(v) **Glass piping**–PTFE bellows, special pipe supports

(vi) **Various Pipes**: MS "C" class, sch 80, A 179, A 106, SS 304, 316 of HDPE/UPVC/Rubber/PTFE pipe with external reinforcement which are in standard length along with bends of 45°, 90°, 180° with short and long radius,

Fuels

Various types of fuels that should be available could include Coal, propane, acetylene, Light Diesel Oil/Furnace Oil/naphtha/LPG etc. These are as per operating practice at the plant.

All cylinders must have markings as per statutory rules, test certificates and expiry dates clearly mentioned on them. **Local Statutory regulations and recommendations for safety arrangements for storing such items must be strictly complied with**.

A set of fittings for oil firing arrangements (oil firing burner nozzles, oil pumps, flow controllers, oil heaters).—check physical dimensions and name plates against purchase order.

Obtain test certificates for pumps, flow controller, and oil heaters.

3.6.8   Electrical Spares

Check stampings and mark of approval from accredited testing agency for the following:

At least one item of each type in use such as motors, starters, fuses of various ratings, flame proof light fittings, power and lighting cables, spares for circuit breakers, bus-bars.

Electrodes and couplers

Graphite electrode for electrically heated furnaces

Electrolytic electrodes for electrochemical cells

3.6.9 Refrigeration Units

Keep in stock a quantity of refrigerant equivalent to one full charge in the system being used at the plant and all other essential spares as per OEM advice.

Include annual maintenance contract in Purchase Order

3.6.10 Diesel Generator Sets

Include annual maintenance contract in Purchase Order

Spares shall be always procured and stored as per OEM;

Spares for electrical switch gear as for auto switching ON and for auto load sharing arrangement

3.6.11 Instrumentation Spares

Obtain test and calibration certificates from vendor. Install and commission in presence of vendor's engineer.

(i) as per OEM recommendation for individual units
(ii) Spares required for various process units such as filters, water treatment, boiler level control, air pollution control, gas analyser for exit gases, pH controller
(iii) One or more sets of calibrated and ready to use (with test certificates) pressure gauges, pressure controllers, pH meters, temperature gauges, thermocouples of various types (for corrosive, high temperature working conditions) and ranges
(iv) Spare thermistors, rotameters, conductivity meter, pressure sensors, level sensors, control valves, temperature probes, thermowells, compensating cables, dials for local indication and spares for control room instruments, gas analysers

3.6.12 Cooling System Based on NaOH or LiBr

Stock of concentrated solutions of NaOH or LiBr, circulating pumps, spares for steam ejectors and plates for PHE, valves and rotameters.

3.6.13 Heat Recovery Boiler and Economiser

Ferrules for boiler tubes, set of spare mountings, valves, fittings, safety valves, feed water pumps and motors, instruments for level control, TDS analyser etc. as per advice from OEM.

C I Gills for economiser tubes

Packing bags, carbouys, drums (bare or with special lining materials) for dispatch of finished products

3.6.14 Process Control and Material Testing Laboratory

Set of spare apparatus and analytical grade chemicals for raw material, finished products, process water, boiler water, treated effluent and exit gases; and to keep track on operation of process units.

3.6.15 Disposal of Hazardous Wastes (Please See Appendix (IV) Also)

It is very important to dispose the hazardous wastes generated (if any) in a proper manner. Government authorities allow disposal of such wastes through parties having the necessary storage and disposal facilities which are established and operated as per statutory guidelines. However certain primary treatment is to be given to the waste before transporting to the authorised disposal plant.

The chemicals required for providing such primary treatment (neutralising agents) and Special packing drums, bags, spares for handling machines shall always be available. Services of tanker operators or own transport fleet shall be available for disposal of hazardous liquid wastes. These vehicles must have all fittings and safety devices as prescribed by law.

3.7 Careful Production Planning and Operation

Standard Operating Procedures

It is essential to have a proper Standard Operating Procedure for smooth running of the plant. Maintenance department can assist Production department for designing SOP for the plant.

3.7.1 General

The following shall be considered carefully in consultation with chemical and maintenance engineers.

- Product mix planned for the present and future.
- *Stock of saleable finished product already available* should be taken in account
- Available stock of all important spares and inputs (filter cloth, belt drive, instrument spares, raw materials, fuels, stabilisers, packing containers etc.) necessary for the production rate and product mix
- Condition of Effluent Treatment Plant, Air Pollution Control units.
- Condition of all critical units/those which have to operate at high pressure, high temperature, dangerous materials etc. These should be in safe operable condition
- Confirm whether the plant can be run as per desired production rate or will have to be run at a lower rate due to shortage of some necessary input.
- Confirm whether the plant is to be run at a higher rate to meet urgent demand in market. However it should never exceed safe operating conditions. Maintenance department should insist on this.
- All safety interconnections, pressure release devices and safety vents (to release the gases away from working area through a safety seal with dilute alkali/suitable absorber)

The operating procedure to be actually used may be somewhat different than SOP in view of actual conditions at site. Check all the important equipments before starting the plant:

Confirm (For each individual unit where applicable)
- satisfactory working of ETP, APC; availability of chemicals for ETP, APC.
- arrangements for fire fighting, stock of raw and treated water, stock of fuel,
- Working of Diesel Generator set which will start automatically in case of power outage and supply power to important units as per programme set before hand. *Manual over-ride may be provided if necessary.*
- Statutory permission period for operation of pressure units should not have expired.
- Availability of key personnel.
- Availability of Laboratory facilities for checking raw materials, progress of process and working of ETP, APC etc.
- Availability of weighing, packing machines and facilities for dispatch.
- Estimate total electrical load and possibility of exceeding maximum sanctioned demand. **This must be checked very carefully**.
- Safety valve, rupture disc, safety interconnections.
- Cooling system (refrigeration units for volatile materials) should be checked first before heating of process units is started.
- Satisfactory and smooth operation of valves in gas and liquid lines for process control.

- Correct positioning of thermocouples and their calibration, working of indicators and controller instrument which work on getting signals from thermocouples.
- Visual inspection from all sides – any leak or dent on shell (specially units working at high pressure, high temperature or handling dangerous materials).
- Setting of electrical relay for preventing overload of drive motor.
- Working of ammeters for each motor to keep watch on load.
- Operate feed system for raw materials (take trial through a bypass route for the process unit), or run the mechanism without putting in materials if possible.
- Circulation of process liquid through the system (to observe constant level/return flow of liquid to confirm there is no hold up due to choke in exit pipe, drain nozzles).
- Calibration of individual circulation tanks attached to process units
- Holding capacity/idle volume of lower parts of process units like condensers, reactors. (material from previous batches can contaminate next batches)

Study the sequence of starting the plant units

For cooling/heating, admitting feeding/charging of raw material for the particular unit,

- running of agitator (check direction of rotation and free movement of shaft first), condition of baffles, agitator blades.
- close manholes, inspection holes, vents (may be kept open initially) as per instructions from production engineers.

3.7.2 Maintenance Engineer Can Propose Some Changes in Operating Conditions in Consultation with Production Engineers

It is always recommended to change the operating process parameters in small steps rather than suddenly increasing or decreasing feed rate, concentration, temperature, pressure.

- Sudden changes in such conditions may develop thermal/mechanical stress in the equipment. Welded joints of the equipment can become weak and subsequently may crack or leak and affect the product quality. These leak could be dangerous if corrosive or inflammable materials are being handled.
- Do not dump a large amount of raw material (bags) in reactor – If too many bags of raw material are charged into reactor together or in quick succession they may damage agitator, shaft, bearing, and shear off blades of reactor due to sudden load; and can result in equipment failure. Therefore, it is recommended to add raw material bags one by one in reactor so that load on the agitator will not be suddenly increased
- Keep ready arrangement for cooling (water flow) – It's necessary to have cooling arrangement in state of readiness before any heating operations are commenced

during plant commissioning. The cooling water must flow at sufficient rate to pass through glands, jacketed reactor, seal pots, etc. It provides cooling to the gland packing and prevents escape of corrosive vapours which can damage gear box, drive motor etc. It can also prevent escape of dangerous uncondensed vapours from vents. Make provision to receive cooling water from overhead tank also (Elevated Storage Reservoir)

Manuals giving details of process units, machinery, GA drawings, process streams (pressure, temperatures, concentrations, flow rates, pH) must be given in the SOP. All details of process must be fully explained to and understood by all the concerned personnel.

Simplify the operating method in clear steps to avoid any confusion.

Detailed step by step procedures and precautions to be taken should be explained to operation teams. Dos and Don'ts shall also be made very clear.

Reduce frequent changes in production rates and product mix start-up. Try to run the plant at a rate where efficiency is maximum. Running at very low rate of production wastes energy for operation of bigger machines (blower, compressor etc.) and also heat loss from surfaces of heat exchangers, hot ducts, reactors, furnaces. Similarly overloading the plant i.e. operating at much more than rated capacity can waste energy due to heat loss from hot surfaces. Hence, operate within an optimum range (generally 70–110%).

Reduce frequently stopping and restarting the process unit.

Try to have longer production runs for a product before changing over to another product (if different *products* are to be manufactured in the same process units). This will minimise changes in operating conditions every now and then.

Avoid short batches It is generally necessary to completely stop the reactor for draining out, cleaning and restart after every batch of a product. When it is used to manufacture different products (which may need some other raw materials) may be introduced. Running the plant in small batches will require frequent stoppage

Short batches should therefore be avoided because energy and manpower are required for starting the reactor for a different product every time.

Small amounts of products can also remain inside and get wasted during every cleaning of the reactor. *These small quantities may even contaminate the next batch.* Hence one may use either separate reactors for different products or use a reactor to produce a single product in more batches consecutively (to meet the demand in the market).

Typically this can be done for building up sufficient stock of the particular product as per demand by running the batches consecutively.

Other products can then be produced in next few days, and so on. This is to reduce the loss of materials, manpower, time and energy due to frequent changeover to different products from same reactor.

Avoid frequently starting and stopping the plant Starting of a plant generally consumes fuel and power for bringing up the process units to the normal operating conditions (from a cold start). During this period very little product may be obtained. It may take many hours after *every start-up* to stabilise the plant and steadily get the product of required specification as per rated production capacity.

However more energy may be consumed (when calculated per tonne of product) if the plant is stopped and restarted frequently.

Process units (*which have to run at high temperatures during the process*) should not be frequently started and stopped as this can create thermal and/or mechanical stress in their refractory brick lining which this can reduce their life.

The loss in energy can be reduced by avoiding frequent stoppage and restart for product change over (since it needs readjusting the flow, temperature, control valves).

Running a little below peak capacity
It is recommended to run/operate at only 75–85% of rated capacity *if possible*. This allows working of the mechanical and electrical systems well within safe limits (rotating machines, HP unit, and high temperature units.)

Provide Strainer at suction of pumps with small clearances – The strainer protects the internal parts of the pump (e.g. gear pump, reciprocating piston pump) if the liquid to be pumped has hard suspended particles which can erode to pump internals; or increase clearances.

3.7.3 Operation–Dos and Don'ts (Advice Given by Original Equipment Manufacturer)

Avoid suddenly pressurising/depressurising process vessels Sudden pressurisation of equipment should be avoided because it may adversely affect the life and safety of equipments, and consequently the quality of product *if the internal protective lining develops cracks.*

Similarly, the decrease in pressure shall be carried out only slowly to reach the required condition. It can reduce the mechanical stress on the concerned equipment and hence will be useful for longer life

Avoid suddenly heating or cooling glass lined process vessels
Follow instructions given by OEM to prevent development of cracks due to thermal shocks to the lining. Heating and cooling shall be done at slow rates.

Strictly operate as per instructions given by OEM and always well within the maximum limits prescribed by OEM. Operating in this manner can result in better life of the equipment (which also saves cost of frequent repairs for restoring the performance)

3.7.4   Some Preventive Steps for Longer Equipment Life

Lubrication
- Charge in recommended grade and quantity of lubricant at first start. Later on inspect samples of lubricant at intervals/after running the machine for certain hours as per instruction of OEM.
- Replace old lubricant by fresh correct grade of lubricant suitable for the operating conditions of the machine as per instruction of OEM.
- Inspect the machinery internals if the lubricant samples indicate wear of some component
- Filter the lubricant *if possible* while machine is running.
- The lubricating greasing nozzles shall be easily accessible

Vibration
Excessive vibration of a machine may weaken either own supporting structures or adjacent structure for other units. Hence a high capacity compressor should always be installed on ground floor rather than first floor since it may vibrate when run at high speed. Machines like crushers grinders should be installed away from acid resistant brick lined units, glass equipments etc.

Some corrective steps to reduce vibrations are: (i) provision of strong reinforcement links for the structures. (ii) Check the vibration/dynamic balancing of impeller of blowers before assembly. Correct immediately if any imbalance is noticed. (iii) Eliminate causes of any excessive vibration in the process unit (specially glass lined reactors, glass piping, acid resistant brick lined process tanks) – Please see Appendix on **Vibration for more information)**

3.7.5   Installation of Special Equipments

Glass lined unit:

- These should always be provided with good rigid support and anti-vibration pads. Vibrations from nearby compressor/pump should not reach to such vessels.
- Rubber pads/sheet can provided at clamped fixture for glass piping.
- Poly Tetra Flouro Ethylene PTFE expansion bellows can be used in glass piping for protecting them from vibrations.

3.7.6   Management Support for Innovation, Research and Development

R&D efforts should be encouraged for improving the overall working of the plant. Safety, controlling pollution, product quality, rated output, and ensuring long life of process units and machinery shall be the priority areas. The efforts shall be taken up within frame work of company policies.

Funds shall be allocated for development of better designs, better MOC to improve equipment life. Maintenance engineers can give good suggestions:

(i) innovative methods for operating the plant in a safe and pollution free manner
(ii) reducing certain drastic operational conditions (e.g. high temperatures, high pressures); and maximising yield of products–which reduces wastes and environmental pollution
(iii) substitute the currently used corrosive and erosive materials by better materials. This can improve the life of process units, piping, instrumentation sensors and machinery used in the plant.
(iv) development of resistant materials of construction for processing units

3.7.7 General Guidelines for Maintenance

Timely Maintenance is essential for all equipments to ensure that the required performance is obtained safely, without causing environmental pollution and by efficiently using all inputs. A regular schedule of maintenance can minimise breakdowns and keep the equipment working for a long time satisfactorily. This is necessary to obtain the required products as per desired quality, and also to get the expected yield from the inputs. Proper maintenance also reduces generation of effluents (which may contain dangerous/toxic compounds)

Standardize spare parts for similar machines so that inventory can be reduced by having common spare for more than one machine.

Procure and replace spare parts from Original Equipment Manufacturer (OEM) or as per specifications given by them.

It is necessary to always keep available the correct spare parts so that maintenance jobs are not postponed for want of spares.

3.8 Main Activities of the Maintenance Department

• To confirm that all safety devices and interconnections are in working order.
• To confirm that all units and stand by equipments of Effluent Treatment Plant, Air Pollution Control, fire fighting, boiler feed water system, emergency power supply arrangements are in working order and will be available immediately when required.
• To look in to reports made by production personnel every day and act immediately on any unsafe condition of process units or machinery; as well as repairs required for uninterrupted production operations. This will include material handling systems, process reactors, heating units, refrigeration plants, storage tanks, product dispatch arrangements etc. (but not limited to these only).

- *Maintenance engineers should develop a check list of their own for routine monitoring of all important plant units in consultation with production engineers*
- To advise stores to procure spares as per specifications, in required quantities and on the dates given (for routine and preventive maintenance; as per programme for major repairs; as per planned annual shutdown; for any modification planned).
- To inform stores management about good vendors, local distributors and good workshops nearby who can supply spares or a recondition old spares in case of emergency at short notice.
- Advise the process operational team for correctly checking equipments prior to starting, taking trail runs, running them thereafter and correct procedure for stopping.
- To caution process operational team if any operating condition is exceeding the safe limits of pressure, temperature, vibration, speed etc.
- Be always ready for carrying out emergency repairs immediately, especially for attending leaks of dangerous materials, any damage to material handling system or machineries like compressor, pumps, reactors, ETP, APC.
- To advise production engineers and to assist senior management for (i) planning and (ii) execution during expansion, modernisation, diversification of the plant (iii) when it is desired to take over an old/sick unit by reviving it.
- ***These jobs can be done by minor repairs of old units, upgrading of existing units by some modifications, or obtaining new balancing equipments of higher capacity.***
- **To check existing infrastructure** in the plant for material handling, boiler capacity, cooling tower, refrigeration plant, electrical installations (main transformer, power cables, motors and starters), storages for fuels and products, dispatch arrangements, and any machine which is operating at its maximum/almost at peak capacity and **whether it can be loaded further**.
- To advise management if any further loading can result in dangerous situation or may cause breakdowns.
- To work out detailed terms and conditions for allotting work to external agencies or Contractors (including performance guarantees for the new machines) jobs and to supervise their work along with production engineers by proper coordination.
- To further improve existing systems of material handling, monitoring and operating process, utilities and ETP, APC for safe and smooth working. Suggestions can be invited from experienced operators and suitable guidance taken from Human Resource Department (HRD) also.
- To train technicians and junior engineers in safe maintenance practices; and conduct Refresher Courses for senior personnel (with assistance from HRD) by inviting experts from Technical Universities.
- To minimise the idle period for production plant by better selection of equipments, preventive and predictive maintenance, meticulous planning and execution of activities required for annual shutdown in close cooperation with production, stores and HRD under overall guidance from senior management.

- To comply with all statutory rules and regulations as well as instructions regarding pressure vessels, boilers, storages of dangerous materials and fuels, and assist all concerned departments for implementation of these directions.

3.9 Site Fabrication Facilities

Advise senior management if it is necessary to have facilities for fabrication of new piping systems, support structure, patch welding, replacement of major parts of reactor shell (or cooling jacked) etc.

Consider the in – house fabrication from cost – benefit points of view, such as time and cost required for the job and whether own technician, welders, fabricators possess the necessary skill. This is to be weighed against sending the process unit/machinery to external party for fabrication as this will need more time, cost of transport and some process know how or details of plant design may leak out.

Some Machines which can be purchased for own maintenance shop:

- Plate bending Machines (Capacity – Width of plates, Thickness of plates)
- Gas Cutting sets
- Welding Transformers (AC and DC)
- Drilling, Milling, Shaping machine
- Lathes, Shapers, Profile Cutters

3.10 Budgeting for Maintenance Jobs

3.10.1 Observations

Both Production and Maintenance Engineers shall make the following observations when the plant is running and also plan for future.

- Actual Production rate as compared to rated capacity
- Actual operating conditions (to be compared with design conditions/instructions given by OEM). These could be pressures, temperatures, flow rates, compositions of all process streams at inlet and exit, conversion efficiency, absorption efficiency, filteration efficiency etc. for each reactors/absorber/heat exchanger i.e. all important process unit and equipments and are to be compared with design/expected performance
- Consumption of raw materials, power, steam, cooling water etc. per unit product (output) and compare with standard norms.
- Current drawn by each drive motor, and identify overloaded/under-loaded motors
- External visual inspection/physical condition (any leaks, fugitive emissions, abnormal noise, vibrations, obvious damage)

- Any malfunctioning of safety devices
- potentially dangerous condition which needs correction;
- generation of effluents and whether they are being treated properly
- Planning for change in product mix
- Planning for expansion of production capacity

This will enable identification of equipments which may be unsafe or inefficient. Process and Maintenance engineers shall accordingly prepare of a list of equipments which are to be attended.

Unsafe and/or inefficient operation should be improved by attending to such plant units and machinery. It will reduce the bottle necks in the plant when proper maintenance is carried out for such units.

It will be seen that some of them may need only cleaning/minor repairs and these jobs can be done with less efforts; and may need less funds as well.

List out separately
 (i) those which need major repairs
(ii) those which are to be replaced by a new one

3.10.2   Estimate Cost of Maintenance

- cost of isolation, creating passage for movement by dismantling ducts, process units in the way (this needs stopping the plant and hence can cause loss of production and difficulty to marketing department)
- restoring dismantled units, pipes, ducts
- cost of cleaning the large sludge/waste stuck up inside process units, storage tanks
- treatment of effluent generated during cleaning
- estimate requirement of items/materials (spares, fittings, consumables)
- cost of new unit itself (ex-works cost as quoted by vendor, packing and insurance, taxes and levies, cost of stage and final inspection, transport and unloading at site)
- number of joints/leaks to be attended in old unit
- connecting piping and ducts to the **new unit**, and provision of expansion joints (if any) (*Cost of welding will depend on pipe diameter, bends and position/height from ground level*)
- erection work required for replacement of old units by new units
- making new civil foundations (if existing ones cannot be used for new or additional equipments to be provided),
- tools and tackles, scaffoldings, extra supports (temporary) hiring of heavy duty cranes;
- contractual manpower to be engaged from external agency,
- jobs to be done at external workshops if they cannot be done in-house
- removal of existing old thermal insulation, anchors and providing new anchors

- obtaining permission (**to restart the unit**) from Statutory authorities after repairs are carried out as required.
- provision of external thermal insulation after getting clearance from Statutory authorities

3.10.3   Contingency

It is possible that some unforeseen repairs may have to be carried out after inspection of units Some good suggestions may also come at a late stage while the units are opened and are being cleaned and inspected. These suggestions may be difficult to implement when the plant is running. Hence provide about 15% additional funds in the budget.

3.11   Some Typical Accidents and Their Prevention

Some preventive steps are obvious and hence not written
- Fires caused by welding/gas cutting.
- Heavy items falling down on persons standing directly below. This area should be cordoned off
- Persons caught in moving machinery [belts, blades of big fans, agitators). *Loose clothes may get caught in belts, chains, drive coupling. Do not allow such dress.*
- Persons getting exposed to toxic gases, inflammable vapours, corrosive liquids without first checking these and entering inside closed vessels. Check by gas detectors and provide fresh air continuously by breathing apparatus.
- Persons falling down from height, getting electrical shock, exposed to hot gases from furnaces, external hot surfaces where thermal insulation has come off. Provide all safety devices as recommended by safety officer.
- Supporting structures/chain pulley blocks giving way—*generally due to excess load. Use chain pulley blocks having capacity three times the weight to be lifted. Test and confirm.*
- Machineries like pumps or mixers are started while persons are still working on them/downstream units or while some maintenance job is incomplete. **Put lock on starter and remove fuses from mains**.
- Pumps are started while some pipe is still being replaced or is open, or gasket in flanged joint is being changed or bolts are not yet tightened fully. **Put lock on starter and remove fuses from mains**.
- Piece of internal component/fitting, shaft getting damaged while opening an equipment (for inspection, cleaning; replacement of internal part) and hitting the working persons. **Provide safety dress and safety helmets**.

- Fires due to using steel tools (not using non-sparking alloy tools), cutting by chisels, hammering or by friction due to pulling on hard surfaces. **Always use non-sparking tools**.
- Earth connection has got disconnected due to some reason (accidental hit by movement of something being carried in the premises) causing the build up of static electricity. **Check continuity of earth connections regularly**.
- Persons wearing shoes which are not having rubber soles. (in areas having inflammable vapours) These can cause sparking and hence ignite the vapours.
- Working inside a process vessel/working at a height without breathing app/24volt lamp/rope/safety belt, hamlet, safety catching net at lower level.
- Vehicles started in area where inflammable vapours are present. *Flame arrester on exhaust pipe is no guarantee that it will not cause fire.*
- Isolating Valves before Safety Valves has remained closed (after the safety valve has been repaired or replaced)
- **Limit Snitches not provided at all or are not working properly**. These should be provided at all equipments where the to and fro movement of actuator for control systems, lifting of weight by hoist etc. is not to be allowed beyond certain limit; It shall not be possible to ordinarily by pass them unless specifically permitted by plant engineer for a certain limited period/purpose only. These must always work properly.
- **Provide** Y- Type drain valve for use at bottom drain nozzle of a process vessel.

3.12   Practical Matters/Issues to Be Looked into

These are some issues which can cause misunderstanding among operating and maintenance staff. These relate to complaints by Maintenance Engineers for careless operation by Process Operators; or complaints by Production Engineers regarding delay in taking up maintenance work/improper repairs carried out to the units etc.

- Process conditions (*pH, flow rates, pressure, temperature, filtering out of hard particles.*) are not controlled carefully by the operators which can cause corrosion or erosion of the units
- The operating procedure or charging (feeding) sequence is not followed correctly.
- The equipment or machineries are run in overloaded condition for **longer** time than permissible limits (*overloading for only a short run time may be allowed by the original equipment manufacturer in some exceptional cases*).
- Equipments are installed in open (exposed to snow/rain/sun) or in dusty conditions
- Spares not procured from OEM/insistence by finance department or higher management to use cheaper spares or from another vendor.
- Improper grade (of cheaper) of lubricant procured.

- Not paying enough attention (heeding) warning signal like vibrations, abnormal noise, frequent tripping of drive motor; hot spots on shell [indication of damage of internal protective lining]; frequent failures of bearings; belt drive slippage causing slower speed of driven unit (resulting in longer reaction time); exit gas analysis showing loss of reagents [or polluting the atmosphere]
- Maintenance personnel hurriedly leaving site without waiting for trial run before restarting the repaired machinery or plant unit and subsequent stabilization of operations.
- Maintenance crew is asked to hurriedly complete the job and restart the machinery which can again result in breakdowns.
- Maintenance crew is asked to carry out repairs while the plant/concerned machinery is running or without waiting for availability of proper tools or spares on the spot.
- *Insistence by senior executives to follow a particular method of operation, raw material, or change in operating conditions of temperature, pressure, pH, concentration of solids in solution being processed.*
- *Such instructions may be issued to try out some change desired by the seniors; but there is a chance that something may go wrong.*

Chapter 4
Monitoring the Process Plant

4.1 Condition Monitoring of Process Units

Process equipments and their accessories are designed and constructed to carry out processing of raw materials for meeting the production targets in a safe, efficient manner without causing environmental pollution. Hence it is necessary to monitor their condition very carefully and regularly; and to look at the symptoms which can indicate:

 (i) development of dangerous conditions
 (ii) chance of a sudden breakdown
 (iii) deterioration of performance – e.g. loss of production, product quality
 (iv) chances of pollution
 (v) possibility of fire

4.1.1 Some Typical Symptoms

Following observations during condition monitoring of individual units can indicate internal damage or its likely occurrence in near future.

- hot spots on shell (deterioration/damage of internal protective refractory lining)
- abnormal noise or vibrations are noticed (failure of bearings of the machine).
- Draw of excessive current/abnormal smell is emitted from electric motor, body is getting very hot (indicates burning out of windings of motors)
- oozing out of fluids from shell (a major leak from process units may occur)
- abnormal noise or vibrations of agitated units (damage to internal fittings like agitator blades or shaft)
- presence of particulate matter in filterate. (damage to filtering cloth in filter press)

© Springer Nature Switzerland AG 2019
K. R. Golwalkar, *Integrated Maintenance and Energy Management in the Chemical Industries*, https://doi.org/10.1007/978-3-030-32526-8_4

- presence of sludge particles observed in wash liquor when ID fan internals are washed (excess deposits on internals can damage impeller of Induced Draft fan)
- dust in gases entering the boiler (scaling on boiler/economiser tubes).
- acidic mist particles in exit gases (partial damage of demister pads).

Example Typical Condition monitoring of an entire process plant

- Overall production rates and product quality.
- Operating Conditions such as flow rates, temperature, pressure, purity, concentration of various dissolved chemicals and suspended solids in liquid streams
- Consumptions of raw materials, utilities (steam, power, water)
- Consumption of catalyst, filter aid, water treatment chemicals.
- Quantity of effluent streams generated and analysis of each stream.
- By product steam/power generated.

A weekly/fortnightly summary of all such items shall be prepared and compared with standard norms of production and consumption.

Any deviation from the standard norms shall be investigated immediately.

The observations made during condition monitoring of individual process units and machinery shall also be taken in to consideration and studied carefully before arriving at some conclusion regarding the likely reasons for deterioration of performance.

Such analysis is useful for determining the efficiency of the entire unit and its profitability, and planning for maintenance jobs to be taken up soon or during annual maintenance.

Example Compare the following operating data of a Sulphuric Acid Plant (based on raw sulphur) with standard norms of performance (daily output) when the plant was new. It will enable finding out which unit is not working efficiently or may need urgent attention.

- Production Rate per shift and per day.
- $SO_2\%$, SO_3 and acid mist particles in the exit gases from chimney
- Electrical power consumption per MT of acid produced
- Consumption of raw Sulphur per MT of acid produced
- steam generation per MT of acid produced
- Feed water consumption by the Boiler as a cross check
- Surface temperature of Sulphur burner, Converter, Gas heat exchanger shells etc.
- Current drawn by all motors shall also be monitored

4.1.1.1 General Visual Inspection: (of All Process Units and Machinery)

- Civil foundation and base plates, any loosening of foundation bolts
- I-beams/angle irons and other structural supports

- Covering shed on the concerned unit to protect from rain, direct sunlight
- Protective canopy on open vent lines (to prevent ingress of rainwater in the equipment).
- Weak portions of work platforms, railing, ladders
- Condition of external insulation and cladding. There should be no chance of ingress of rain water/snow from gaps in the cladding.
- Poor lighting at the equipment (makes working at the unit/observation of instrument dials very difficult)
- Difficulty in operation of valves due to inaccessible locations.
- Any tilting of the unit.
- Any spillage of corrosive liquids from pipe lines/other process vessels located above.
- corrosive fumes or dust coming from nearby chemical reactors or vessels which are open at top.
- Condition of external shell-corrosion or leak of any oozing liquid/leak of gases from cracks in welded joints.
- Loose flanged joints for inlet/outlet nozzles for process fluids.
- Corroded gussets at inlet/outlet nozzles for process fluids.
- Corroded gussets for cleaning manhole nozzles.
- Sight glass/view port which have become opaque.
- Smooth working of all valves in connected process streams
- vent valve for release of pressure
- Pressure release lines from vents/safety valves shall be away from the equipment and working areas.
- Provision of safety showers and eye-wash fountains in nearby areas to the equipment.
- Provision of collection trough below the equipment to take care of any leak/seepage.
- Provision of sloping platform towards storm water drain if a collection trough cannot be provided.
- Provision of work platform, (railing, ladders) for operation of valves, attending to thermocouples, pressure taps.
- Proper earth connection for the shell and pipelines which are handling inflammable materials
- Secure covered electrical connections for motors, closed junction boxes
- No loose hanging electrical wires to be allowed in working area.
- Provide only flame proof lighting where inflammable gases/vapours could be present
- Availability of an installed standby unit which can be immediately commissioned in case of need (this allows taking up of maintenance work on the other unit which has developed fault).
- Corrosion resistant coating/high temperature resistant paint should be provided on support structure, bottom plates (before erection of unit) and side walls..
- Properly connected and laid out connections for thermocouples.
- Waste removal and disposal from the units

- Air pollution and water pollution control units (settlers, clarifier, filter press, cyclone, venturi, packed/spray tower, Bag filters, ID fan etc.)

4.1.2   Examples from Some Chemical Industries

These examples also include (but are not limited to) *some additional observations to be made before starting the plant and when the plant is running.*

Following are some additional observations to be made before start of the plant:

4.1.2.1   Sulphuric Acid Plant

Melter
- Condition of screen at charging point and charging hopper
- Setting of safety valve on steam supply line at inlet.
- Condition of agitator (for agitated melters only) and gear box.
- Condition of insulation/brick lining.

Working of steam traps for pumps, sulphur transfer lines up to spray gun inlet
  Working of centrifugal pumps—check flow through recycle line
  Sulphur Filtration system (*please see Pressure leaf filter*)
  Metering pump – Check (i) calibration by measuring delivery rate through recycle line (ii) working of safety valve (iii) dampening pot in discharge line (iv) strainer at inlet line (v) working of suction and discharge valves

Furnace
- Condition of covering shed to protect from rains.
- Condition of sight/view ports. (some designers provide transparent mica sheet at view port and a small 15–20 mm air injection connection to keep the view nozzle clean always)
- Operation of burner inlet valve and steam trap on feed line for burner.

Waste Heat Recovery system
- Quantity of treated water available in the feed water tank.
- Working of boiler feed water pumps,
- Boiler: water level indicator and controller, blow down valve, collection tank for blow down water, safety valves, vent valve, main steam stop valve, Non-Return valves (where advised), steam pressure gauge etc.
- Working of valves in boiler exit and boiler bypass line.

Hot Gas Filter
- External thermal insulation and cladding
- Connection of thermocouples at inlet and outlet.
- Working of drain valve at bottom (to check for presence of acidic condensate)
- Working of valves for pressure taps in gas inlet and exit lines

Converter
- Provision and proper connection of thermocouples at inlet and outlet of every stage
- Operation of drain valve at bottom plate of the unit (to check for presence of acidic condensate) and valves for pressure taps

Heat exchangers (shell and tube type)
- Connection of thermocouples at inlet and outlet of hot and cold sides.
- Operation of drain valve at bottom of shell.
- External insulation and cladding.
- Operation of valves for pressure taps on shell and tube sides.

Inter Pass Absorption Tower-IPAT
- Levelling of distribution tray for incoming absorber acid/any damage to spray nozzles of incoming absorber acid
- Working of circulation pump for absorber acid (liquid).
- Flow of return liquid (acid) from the tower to the circulation tank
- Height of U-seal at liquid exit from the tower (a larger height of U-Seal can cause hold up of acid in the lower part of the IPAT)

Drying tower (for air) and Final Absorption Tower (for gases)
- Levelling of distribution tray for incoming liquid/any damage to spray nozzles of incoming liquid, flow of liquid as above.
- proper fixing of demister mesh pads in air/gas exit
- Height of U-seal at liquid exit from the tower (a larger height of U-Seal can cause hold up of acid in the lower part of the DT and FAT)

Acid cooler-PHE
- Connection for incoming and outgoing hot(acid) and cold(water) fluids.
- Circulation of cold fluid through the unit and pressure drop.
- Positioning of temperature sensors or dial thermometers at inlet and exit of hot and cold sides.
- In certain plant designs the PHE is used for recovering the heat from hot acid in the form of hot water. This needs an additional PHE for preheating cold water from circulating hot water.

Air pollution control units
Venturi scrubber – working of gas flow control valve at throat, flow of return liquid from the venturi when circulation pump is run.

Packed tower – working of liquid inlet flow control valve, flow of return liquid when circulation pump is run.

Working of alkali addition system for pH control of scrubbing liquor

Observations for condition monitoring when the plant is running.

Melter
- Setting of safety valve on steam supply line at inlet.
- Working of agitator (for agitated melters only) and current drawn by agitator motor.

- Working of steam thermodynamic traps (inlet pipe to the trap should be hot/exit pipe from the trap should be at lower temperature. Quick test—a piece of sulphur will melt on inlet side but will not melt on exit side)
- Rate of melting when plant is running
- Presence of froth (in case of high moisture content in raw material)

Furnace
- Condition of covering shed to protect from rains.
- Observe the flame inside through view port (appearance of flame),
- temperature of exit gases
- composition of exit gases
- Operation of spray burner and steam traps on feed line.
- Hot spots on shell (will indicate damage to refractory lining)

For maintenance of furnace
- This can be done during annual plant shut down.
- Cool the unit to ambient temperature before entering inside.
- Open cleaning manholes. Make provisions for fresh air ventilation before going inside.
- Check the inside brick lining.
- Remove the clinkers and ash deposit on the refractory lining.
- Examine if there are any cracks or damage to refractory.

Precaution
- Use breathing apparatus, face shield, and good helmet because some loose brick can fall on the head of person inside.

Waste Heat Recovery Boiler
- Operate both boiler feed water pumps alternately
- Working of water level controller.
- Conductivity of feed water.
- Working and position of valves in gas exit and gas bypass line.
- Gas inlet and exit temperature.
- Gas press at inlet and exit
- Steam pressure
- Steam generation to be determined from feed water consumption
- **Calculate heat balance to assess condition of heat transfer surface (rusting, corrosion, deposit of dust on surface).**

Hot Gas Filter
- External thermal insulation and cladding
- Gas inlet and exit temperature
- Gas pressure at inlet and exit.
- Presence of particulate matter in exit gases.
- Presence of any acidic condensate when drain valve at bottom is opened.

- *Satisfactory working is indicated by steady building up of pressure drop, while channelling/damage to filter media or support is indicated by lowering of pressure drop with time*

Convertor
- Connection of thermocouples at inlet and outlet of every stage
- Gas analysis at inlet and exit.
- Temperature at inlet and exit of each pass.
- Pressure drop in each pass.

Heat exchangers
- Connection of thermocouples at inlet and outlet of hot and cold sides.
- Operation of drain valve at bottom tube sheets (check acidic condensates).
- External insulation and cladding.
- Pressure drop on shell side and tube side
- Gas inlet and exit temperature of hot gases.
- Gas inlet and exit temperature of cold gases.
- **Calculate heat balance to assess condition of heat transfer surface (rusting, corrosion, or deposit of dust)**.

Inter Pass Absorption Tower-IPAT
- Flow of incoming acid on distribution tray (through sight glass)
- Working of acid circulation pump.
- Acid mist in exit gases.
- Presence of acidic condensate from drain point on exit gas line, if any.
- Gas pressure at inlet and exit.
- Gas temperature at inlet and exit
- Any seepage from bottom side.
- Any leak of acid/gas from side walls
- *Condition of FBME (Fibre Bed Mist Eliminator) can be inferred after above observations.*

Drying tower (for air)
- Air pressure at inlet and outlet.
- Moisture content in exit gases (*must be as per norms*)
- *Improper drying is indicated if exit gases contain more moisture. It could be due to less tower packing/channelling/low flow of absorbing liquor*.
- Acid mist in exit air—indicates damage/displacement of mesh pad;

Final Absorption Tower
- Flow of incoming acid on distribution tray(through sight glass)
- Working of acid circulation pump.
- Acid mist in exit gases.
- Presence of acidic condensate from drain point on exit gas line.
- Gas pressure at inlet and exit.
- Gas temperature at inlet
- Any seepage of acid from any side.

Acid cooler-PHE
- Any leak from connections for incoming and outgoing hot and cold fluids.
- Any excessive pressure drop on cold fluid/hot fluid side through the unit (may indicate choking of flow channels, damage to strainer screens at inlet)
- Flow rate of cooling fluid/current drawn by motor for cooling liquid pump
- Temperature of hot fluid at inlet and outlet.
- Temperature of cooling water at inlet and outlet.
- **Calculate heat balance to assess condition of heat transfer surface**.

For Maintenance of PHE
- Isolate from both hot and cold sides. Provide blinds in the connecting piping. Do not depend if the valves are closed since they can leak.
- Drain out residual acid
- Flush out residual acid
- Neutralise by dilute 5% alkali
- Again flush by water
- Now carefully disconnect all piping and remove one by one plate without damaging gasket.
- Clean all passages/grooves/strainer at inlet.
- It is advisable to get these job done through technicians from OEM.

4.1.2.2   Single Super Phosphate Plant

Observations to be made before start of the plant

Pulverising unit – Screen at the inlet of unit and feed control damper, drive mechanism, free movement of grinder wheels (if possible to rotate).

Rotary screen – condition of screen and its free movement,

Provision of hood and dust removal arrangement

Arrangement for collection and transfer of powdered material to feed silos

Feeding of ground material to reactor at controlled rate

Acid dilution system and feeding to reactor check for any leak of acid, working of rotameter and valves

Mixer-reactor – condition of protective brick lining and mixing screw/paddles

Cylone separator – Inlet and outlet (with removable piece) gas ducts, rotary air lock valve for removal of powdered material.

Bag filter – Condition of bags, support for fixing the bags, working of ID-fans, working of timer and air compressor, availability of spare bags

Venturi scrubber – working of gas flow control valve at throat, flow of return liquid from the venturi when circulation pump is run.

Packed tower – working of liquid inlet flow control valve, position and clean condition of spray nozzles, flow of return liquid when circulation pump is run

Recovered acidic liquor recycling system (from scrubber)

Cooling tower and water circulation pumps

Gas exit chimney—gas sampling points, drain valves

ID fan.. condition of rubber lining, drain point on casing, free movement of impeller, wash water arrangement, pressure taps in gas ducts before and after the blower, driving system (belts, bearings, motor), speed control

Air pollution control units–flow of scrubbing liquor through each unit/current drawn by motor of scrubbing pumps.

Systems for transfer of product to shed, EOT cranes, filling, stitching weighing of bags.

Single Super Phosphate Plant: *observations to be made when plant is running*

Pulverising unit – feeding rate and product output rate; screen analysis of product

Rotary screen – screen analysis of product

Working of dust removal arrangement

Arrangement for collection and transfer of powdered material to feed silos

Controlled feeding of ground material to reactor

Controlled rate of acid dilution and feeding to reactor

Mixer-reactor—smooth working of mixing screw/paddles

Cyclone separator – pressure drop in unit, working of rotary air lock valve for removal of powdered material.

Bag filter – working of ID-fans, working of timer and air compressor, working of rotary air lock valve for removal of powdered material.

Venturi scrubber – pressure drop in unit; working of gas flow control valve at throat, flow of return liquid from the venturi in to the circulation tank and its analysis

Packed tower – pressure drop in unit; working of liquid inlet flow control valve, condition of spray nozzles, flow of return liquid and gas analysis at exit

Recovered acidic liquor recycling system (from scrubber) collection tank and transfer lines

Cooling tower and water circulation pumps

Gas exit chimney—gas sampling points, gas analysis at inlet and exit, drain valves

ID fan working.. drain point on casing, free movement of impeller, wash water arrangement, gas pressure before and after the blower, load on driving motor, speed control

Systems for transfer of product to shed, EOT cranes, filling, stitching, weighing of bags

4.1.2.3 Alum Plant

Observations to be made before start of the plant

- Screen at inlet and feed controller for bauxite crusher/grinder, and dust collector cyclone
- Bag filters: (RAV) Rotary Air lock Valve, timer and air compressor for purging the dust, condition of bags, fixing of individual bag on nozzles.
- ID Fan for dust control.
- Steam supply from fired boiler or from WHRB

- Steam coils in evaporator to be tested 1.5 times the maximum working pressure.
- Condition of main reactor with acid resistant brick lining.
- ID fan for removal of fumes and scrubbing system
- Clarifier and its accessories (rake mechanism, sludge removal)
- Systems for transfer of product to shed, filling, stitching, weighing of bags

Filter Press
- Connection for incoming dirty fluid and outgoing clean fluid.
- Provision of pressure gauges at inlet and outlet.
- Operation of drain valves.
- Operation of feeding pump.

Alum plant *observations to be made when the plant is running*
- Screen Analysis of powdered material and working of dust collector cyclone
- Removal of dust by bag filters: working of RAV, timer and air compressor for purging the dust, pressure drop in system
- ID Fan for dust control.
- Steam supply from fired boiler or from WHRB
- Working of main reactor with acid resistant brick lining.
- ID fan for removal of fumes and scrubbing system
- Clarifier and its accessories (rake mechanism, sludge removal)
- Working of evaporator for concentrating the dilute solution
- Moulding arrangement for alum slabs
- Systems for transfer of product to shed, filling, stitching, weighing of bags

Filter Press
- Any leak from connection for incoming dirty fluid and outgoing clean fluid.
- Pressure drop on dirty fluid side.
- Flow rate of clean fluid at exit.
- Recycle of filterate to reactor
- Operation of dirty liquid feeding pump.
- ***High pressure drop and low flow of filterate indicates choking of the filter cloth sheets (even when valves are fully opened)***
- Collection of sludge and disposal (drying bed and land fill)

4.1.2.4 Nitric Acid Plant

Observations to be made before start of the plant

- Main storage for liquid ammonia and safety valves provided.
- Ammonia evaporator (to be pressure tested before start of plant)
- Steam pressure regulator for evaporator.
- Air compressor with accessories (air filter and drive mechanism)
- Air pre-heater (to be pressure tested before start of plant)

- Control valves for air and ammonia to feed the gas mixture as per desired composition.
- Main ammonia oxidiser (convertor) checks: shell thickness by ultrasonic thickness tester, and position of thermocouples at inlet and outlet.
- Catalyst recovery system.
- Cooling water system
- Nitrogen di-oxide absorber column.
- DM water plant.
- Power recovery unit from exit gases.

Nitric acid plant *observations to be made when the plant is running*

- Ammonia evaporator
- Steam pressure regulator for evaporator.
- Air compressor with accessories (air filter and drive mechanism)
- Air pre-heater (temperature of pre-heated air to converter)
- Instrument to analyse the gas composition and Control valves for air and ammonia to feed the gas mixture as per desired composition.
- Main ammonia oxidiser (convertor)temperatures and flow rates
- Catalyst recovery system.
- Cooling water system
- Nitrogen di-oxide absorber column.
- DM water plant.
- Transfer of product acid to storage tanks
- Power recovery unit from exit gases.
- Exit gas analysis

4.1.2.5 Carbon Di-Sulphide (CS$_2$) Plant

Observations to be made Before Start of the Plant

- Charcoal conveyor, storage silo, and calcinations system.
- Electric furnace: Inspect the refractory lining from inside, fixing of electrodes
- Sulphur melting, filtering and feeding system to the furnace.
- Condensation system (primary and secondary condensers).
- Cooling tower and refrigeration system.
- Storage tanks for CS$_2$
- Distillation system.
- Sulphur recovery system from waste gases.
- **Carbon di-sulphide (CS$_2$) plant**: *observations to be made when the plant is running*
- Charcoal conveyor and calcinations system.
- Electric furnace: power and sulphur feed, gas pressure
- Sulphur feeding system to the furnace.
- Condensation system (primary and secondary condensers).

- Cooling tower and refrigeration system.
- Analysis if purified CS_2.
- Treatment of waste gases.
- Rate of Sulphur recovery

4.1.2.6   A Typical Water Treatment Plant

Assemblies (with internals) of the following units are to be checked before start.

Observations to be made when plant is running are:
- Raw water filter – smooth of operation of valves and running of feed water pumps, any leak of water from pipes or flange joints, flow and quality of filtered water at outlet of filter, pressure of water at inlet and outlet.
- Anion, Cation resin columns: Pressure of water at inlet and outlet, flow of treated water, analysis of treated water, conductivity of water, any escape of resins with exit water.
- Polishing tower: flow and analysis of treated water; conductivity of water

4.1.2.7   Reverse Osmosis Unit and Ultra Filtration/Micro Filtration Units

Observations to be made when plant is running are:
- Pressure of water at inlet and exit (pressure drop through the unit)
- Discharge pressure of feed pump
- Analysis and flow rate of feed water, exit water and rejected water.
- The membranes, filter media etc. are to be cleaned as per advice from OEM.
- *Consult OEM of membranes for replacement if the plant output is falling and pressure drop is increasing.*
- Check the turbidity in filtered water every hour
- Clean filter media when the pressure drop is high/exit flow rate has reduced.

4.1.2.8   Process Boilers

- Flow rate of liquid (to be distilled) at inlet and exit.
- Temperature of the liquid at inlet and exit.
- Flow and temperature of heating medium (if it is a process gas) at inlet and exit of the process boiler. *Position of bypass valve in the gas duct to be noted.*
- Pressure of steam (if it is used as a heating medium) at inlet of boiler
- Setting of safety valve on steam supply line
- Flow of condensate at exit.
- Pressure and temperature of chemical vapours leaving the unit
- Any mist carried over in vapour lines.

- Production of distillate fractions at different temperatures when the process boiler is being used for fractional distillation.
- Calculate heat balance to assess the efficiency of heat transfer *(condition of heat transfer surfaces can be judged accordingly).*

To be properly checked before starting a plant and strictly monitored during running.

4.1.2.9   Air Pollution Control System

- analysis of gases at inlet and exit
- Pressure drop in individual units and in the system
- Consumption of power
- pH control system
- working of individual units, scrubbing pumps,
- Dry/Wet ESP working of the unit, particulate matter in the inlet and exit gases, power control system, power consumed

4.1.2.10   Effluent Treatment Plant

- Analysis of incoming and treated effluent (BOD, COD, TDS, TSS, pH.)
- working of lamella clarifier/tube settlers, air compressor, aerators, filter press
- analysis of recycle streams
- power consumed by the system

4.2   Likely Reasons for Unsatisfactory Output/Performance

Plant engineer and maintenance engineer should try to find out the likely reasons if any unsatisfactory output or performance is noticed during operation of the plant.

- The condition monitoring of the plant before start and during running should be studied for this purpose. The study should be specifically done to assess the potential for: development of any dangerous situation, fire, environmental pollution, reduction of production rate and deterioration of product quality.
- **History cards of the units** *(please refer to Sect. 4.6. Record of maintenance work also)* should be referred to for study of the design and drawings, materials of construction, operating and maintenance records, and Dos & Don'ts from manufacturer.
- *Check the actual operating conditions of various units and the deviation from recommended conditions/maximum permissible limits.*
- *Examine all operating records to see the extent to which these conditions had exceeded the recommended limits, e.g. (i) high pressure at inlet/high tempera-*

ture/high speed/high acid concentration etc. (ii) on how many occasions (iii) the length of time the equipment was run in such conditions.

- **Short listing of the likely reasons** (which have already resulted or can result in deterioration of performance) can be done after above study.
- **Root cause of unsatisfactory performance** When the above records are studied further in the light of current situation (product mix, quality of raw material being used, ambient condition, and any postponement of preventive maintenance or small repairs) one can arrive at the root causes or the most likely reason *which may lead to a breakdown.*

4.2.1 Some Typical Causes for Unsatisfactory Performance

Poor quality of raw material:
- High moisture content
- Excess dust
- Big lump size
- Presence of harmful impurities
- Free acidity

These can lead to corrosion, erosion of protective lining of process vessels. They can also lead to less rate of melting in melters, more load on ETP/APC; more power consumption by grinding unit; loss of conversion efficiency due masking of active surface of catalyst by deposit of dust, choking of filter leaves, more energy consumption by calciners or dryers.

Frequent power outage
These can result in need for frequent re-start of process plant which requires resetting of controls, readjusting of feeding rates to reactors, heating and cooling systems etc.

There are also chances of overflowing from some process tanks, damage to agitators due to sudden jerk, draining of cooling jackets …These will need considerable extra preventive equipments and precautions (and their regular maintenance as well)

Problems due to power outages:

- Disturbance in plant operation
- Corrosion of process unit
- Environmental pollution
- Production loss.
- Deterioration of product quality
- Poor energy efficiency due to frequent start up required.

Some more reasons (for unsatisfactory performance)
- Unusually harsh weather (heavy rains, storms, snow, extreme summer)
- Congested plant layout (not enough space for working or maintenance)

- Unsafe, dusty working area, poor ventilation, presence of corrosive or irritating vapours, poor lighting, confusing instructions for operation, obstructions in escape routes,
- Malfunctioning of instruments
- Poor quality of available water (will need more water treatment facilities)

Marketing department
- Pressuring the production team to overload the plant to meet some urgent demand.
- Pressuring the maintenance team to postpone maintenance work

Stores department
- Not able to procure critical spares from OEM (can provide only reconditioned spares).

4.3 Typical Conditions Exceeding the Design Values or Recommendations from OEM

- High temperature during operation of Furnace, catalyst passes of converter, absorbing acid temperature from towers.
- High pressure at inlet of SO_2 condenser, SO_3 condenser, chlorine condenser, ammonia condenser.
- High concentration at inlet of following: (i) Ammonia oxidiser (convertor) for nitric acid plant.(ii) high SO_2% at inlet of converter
- High flow rate at inlet of following: (i) rotary dryer, (ii) SO_3 condenser (iii) Sulphur recovery unit from H_2S.
- High pressure at inlet (i) viscose pressure at spinning machine (ii) Filter press (iii) Reverse osmosis (iv) Excess steam pressure at jacket on reactor (v) Excess steam pressure in steam coil for process boiler (vi) at inlet of fibre bed mist eliminator (FBME).
- High speed which is exceeding limits at (i) agitated vessels, (ii) rotary dryer; (iii) rotary screen (iv) air blower
- Excess feeding of oil to firing units: (i) damage to refractory lining (ii) explosion (iii) carry over of un-burnt oil, (iv) overheating of crude petroleum at inlet of downstream unit.
- Excess feeding of power in electric furnace may damage the refractory lining.

Thus, when the limits are exceeded there can be damage to the equipments or internals.

4.4 Technical Planning for Maintenance Work

Preliminary study

- Study the original design, drawings and duty conditions specified
- Limitations on maximum outputs (permissible overload conditions)
- The report of condition monitoring.
- Study history cards, Do's & Don'ts
- Study the most likely reason/root causes for unsatisfactory performance/ breakdowns.

4.4.1 Planning for Maintenance Work

This is to be jointly done by production engineers, maintenance engineers, marketing, stores department in consultation with all concerned and submitted for approval by senior management.

- Product mix planned for present and for future
- Required improvement in performance
- The likely benefits of the improved performance
- Cost required for carrying out proposed changes (spares, manpower, chemicals, consumables, assistance from external agencies, consultants, readers can add to this list as per their own site conditions. Please refer to 4.8 also.)
- Time required to execute
- Safety issues of the job to be taken up and additional resources and preparations required
- Formation of separate teams of senior responsible executives if major jobs are to be done

Work list is to be prepared in advance; and is to be divided into routine maintenance, major repairs, replacement of parts, replacement of an entire equipment/ process unit.

Actions to be considered (and taken if possible) for restoring performance:
- (i) Change operating conditions of flow rates, pressures, temperatures.
- (ii) change setting of controllers for heating/cooling/flow rates/pressures etc.
- (iii) install strainers, filters, steam pressure regulator, control valves.
- clean heat transfer surfaces, attend baffles in agitated vessels
- Consider preventive cleaning, maintenance of process tanks, strainers, filters, etc. (if it is permitted to run the plant through a bypass route provided for the concerned equipment)
- Make minor changes to equipments if possible; and major changes only when very necessary.

Some typical examples are:

Blower
- Operate at higher speed with same motor
- Provide new impeller and shaft assembly (with new bearings)
- Purchase a new blower itself of higher capacity with bigger motor if necessary.

Water Treatment Plant
- Increase the frequency of regeneration by additional manpower and chemicals.
- Add more resins
- Provide one more RO/UF unit.
- Install a complete new plant of higher capacity.

Furnace
- Increase the length, add more refractory
- Increase air flow and raw material feed rate
- Provide additional chequer work wall/baffle wall.
- Add pre-heated air/inject more fresh air
- Improve atomisation of burner

Jacketed reactor
- Increase steam pressure at inlet (for a heated unit – only if safe to do so)
- Increase cooling water flow.
- Use brine in place of cooling water.
- Provide external Heat Exchanger.
- Increase agitator speed/change design
- Install new bigger reactor.

Acid cooler
- Increase cooling water flow.
- Add more pipes to the trombone cooler.
- Provide PHE with additional plates, if required

Catalytic converter
- Increase SO_2 percentage in feed gas
- Add more catalyst
- Provide another convertor.
- Provide new bigger convertor.

Heat exchanger
Provide/arrange:-
atmospheric cooling duct.
external cooling fins.
Air injection into the hot side
Re-tubing of the unit (if many tubes are leaking)
New unit (preferably a disc and donut type). Please refer to Chapter 10 Sec. 10.6.

This has lower pressure drop, better gas distribution on shell side and path of connecting gas ducts need not be altered to suit the nozzles. Orientation of nozzles can be made to suit the ducts.

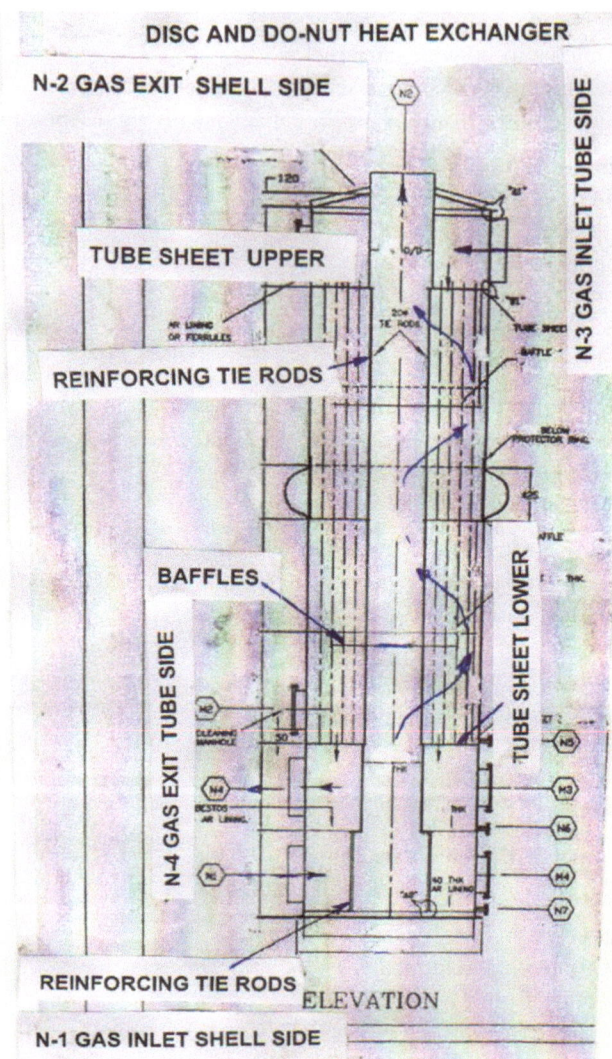

DISC AND DO-NUT HEAT EXCHANGER

N-2 GAS EXIT SHELL SIDE

TUBE SHEET UPPER

REINFORCING TIE RODS

BAFFLES

REINFORCING TIE RODS
ELEVATION

N-1 GAS INLET SHELL SIDE

N-3 GAS INLET TUBE SIDE

N-4 GAS EXIT TUBE SIDE

TUBE SHEET LOWER

Absorption tower
- Better internal glazed packing (Intalox saddles-obtain data from manufacturer for surface area per unit volume, and voidage)
- Raise the packing height (height of tower shell itself may have to be increased)

Fired heaters
- Install bigger oil burner and blower.

Steam heated unit
- Increase steam pressure if the jacket or the coils is designed, fabricated and tested accordingly.

- Install agitator if possible.
- An additional heat exchanger may be considered through which the material to be heated can be circulated. This increases the total heat input to the system.

Effluent treatment plant
- Provide one more equalisation tank.
- Add more aeration facilities with bigger blower.
- Clean the pressure sand filter more frequently
- Add more active carbon.
- Add an ultra filtration unit or RO unit for further treatment of the already treated water.

4.5   Examples of some Typical Cleaning Jobs of Process Units

These can be considered as maintenance jobs for improving the performance

- *Removal and screening of Catalyst from converter:* Refilling with a make up quantity of catalyst as per instructions. This job can get delayed if simultaneously work is done on repairs to convertor shell, catalyst support grid, and thermal insulation.
- *Removal of internal packing and fittings from reaction/absorption tower:* Please refer to maintenance of IPAT, wherein the internal tower packings with sludge deposits are to be taken out, cleaned and filled back.
- *Cleaning of heat exchangers:* Cleaning of tubes by brush and of shell by descalant solution followed by pressure testing.
- *Refractory lined units:* Removing clinkers from internal walls and damaged bricks and repairing the same. Repair by providing new bricks, as recommended by refractory manufacturer. Provide a layer of castable refractory as per need.
- *Removal of sludge from storage tanks, underground melters:* Elaborate safety precautions are necessary to dig out the material and its safe disposal. Work permits shall be issued only after ensuring that it is safe to work inside.

- *Total Volume of work, likely difficulties while carrying out the work and cost of such work must be weighed against improvement in performance of the units.*

4.6   Record Keeping for Maintenance of Plant Units and Machinery

Keeping proper records of all important plant units and machinery:-.

These should be maintained as per guidelines suggested below. The formats can be modified by the Plant Management, Senior Process and Maintenance engineers, Store Keepers, Purchase Officers to suit their own organisation....

Details of Product(s) being manufactured
Trade Name
  Chemical Name
  Specifications of the product (as required/informed by clients/as required for marketable grades)
  Chemical Formula –
  Materials to be handled in the plant (*during manufacturing process; and during cleaning and flushing of process units,*)
  Important Properties of all such materials/solvents/cleaning agents to be considered are:
  Melting and boiling point, density, viscosity, toxicity, corrosive and inflammable nature, physical form at incoming and exit points...

General information
- Process Flow Diagram for the plant showing important units and machinery
- Name of Unit/Machinery (for handling/processing different materials in the plant)
- Location in Plant ---------- Required for ....(transferring, heating, mixing, drying etc.)
- Tag No./Serial No.
- Vendor/Consultant/In- house Fabricated item/Oversees Supplier
- Duty Conditions for the Unit

Rated capacity	Normal	
(In terms of input/ output	Maximum	(i) As per original specification
	Minimum	(ii) As per Purchase Order or any modifications incorporated Later on

- Reference Drawing Number (revisions made and the dates on which the revisions were made and approved for implementation)
- Materials of Construction of parts which will be contact with the materials being handled
- Materials of Construction of Moving parts [which will be subjected to erosion]
- Operating conditions- maximum limits as per **design**/specified by manufacturer of the machine;
- Actual operating conditions
- Safety devices in-built –[internal pressure release valve, rupture disc, high temperature alarm, high or low level alarm....]
- External safety devices [Pressure relief valve, electrical tripping circuit, vibration monitor, strainers, protective screens ...]
- Date of Installation...
- Date of Commissioning
- Standby Units available – Installed at site, Available in stores

4.6.1   Typical Operating Conditions

- Flow rate, composition, temperature, pressure of all process streams (*during handling*) at input and exit points of process equipments and machinery under consideration as recommended by OEM or as per design of plant; and actual values during operation.
- Keep a record of minimum, normal and maximum permissible conditions..
- Normal operating speed, direction and of rotation, Intermittent or continuous operation actual running time per day/per week.

4.6.2   Maintenance History

Date of stoppage	Reason for Stoppages	Minor repairs done	Major repair/ replacement	Cost of repairs	Time taken for the job

Consumptions by the Equipment/Machine [while running at Rated Capacity]

	As per design	Actual values	Reasons for deviation	Remedial action proposed and cost
Power				
Cooling water				
Chilled water/ brine				
Fuel				
Lubricants				
Compressed air				
Steam				

4.7   Considerations for Further Improvement

These shall be based on analysis of operating data collected during plant run time, analysis of breakdowns, and actual performance after repairs. *Original Duty conditions, specifications and Designs shall be again looked in to.*

MOC of parts which have got eroded/corroded shall be suitably changed. The parts may be strengthened and adequate corrosion allowance added. Mild steel parts may be substituted by suitable grades of stainless steel such as SS 304, 310, 316,316 L,321 etc./Alloy-20/Hastelloy-C 276, Teflon(Poly Tetra Flouro Ethylene), fibre glass, HDPE, Polypropylene etc. can also be considered as per operating conditions..

Information on suitability of different MOC in different operating conditions is available on the Internet through various web sites and search engines. However, plant engineers shall prepare test pieces and they shall be tested in actual conditions. Vendors shall be asked to fabricate the equipment by using these MOC as far as possible. Any practical difficulty for using these MOC for manufacturing (cutting, welding, bending during fabrication) shall be discussed with equipment designers and fabricators; along with the cost and time required.

4.8   Cost of Maintenance

While determining the actual cost of maintenance, all expenses incurred directly (*which can be attributed to a particular equipment or process unit*) and indirectly should be taken in to account.

4.8.1   Direct Costs

- Cost of spares, new equipments or expenses for opening, cleaning refitting of old units removal of external cladding and glass wool/mineral wool
- Insulation.
- Isolation from other process units (by putting in blinds in pipes and ducts)
- New nuts, bolts, washers gaskets [of special material for acid resistant to fittings]
- Rust remover chemicals for putting on old (existing) fasteners so that they can be opened with less effort.
- Chemicals required for cleaning (e.g. shell side of heat exchanges; jackets on process reactors; removal of deposits form demister pads/candles) and corrosion inhibitors to prevent damage to steel parts while chemical cleaning is in progress.

- Arrangements to circulate cleaning chemical; and disposal of spent liquor after the cleaning operation is over.
- The residual material may be toxic, inflammable, viscous and may have to be neutralised properly before flushing out.
- Manpower employed [Own technicians, semi skilled and unskilled workers].
- Contractual Manpower [if obtained on a temporary basis from external agencies]
- Consumption of Oxygen, Acetylene, other fuels for cutting and heating operations.
- Consumption of welding electrodes and power for welding.
- Disposal of residual liquids/solids inside the process unit [reactors, heat exchangers, dryers, filters, evaporator] which is still remaining inside even after draining out/evacuating.
- Cost of special inspection and tests[Pressure Test, Spark Test, NDT, Radiography] if they are not available in-house), charges to be paid to external agency for testing which may require chemicals, special pumps and blowers/compressors, instruments for detecting faults and certification after the job is completed.
- Cost of documents, bank charges, fees to be paid to consultants; external experts.
- Cost of any laboratory analysis work done through external agencies when confirming (improvement) quality of products after plant is restarted.
- Performance Guarantee runs carried out as per contractual conditions when major repairs or replacements have been carried out.
- Obtaining quotation from at least three parties finalising the best one.

4.8.2 Indirect Costs of Maintenance

- Removal of thermal insulation and cladding.
- Dismantling the equipment (if repairs cannot be done at installed position) and connecting process fluid piping, steam and condensate piping, or firing/heating arrangements, electrical connections to the drive motors.
- Hiring heavy duty cranes, high capacity chain pulley block.
- Making arrangements for scaffolding and work platforms (if required)
- Taking out instruments and their probes [Pressure gauges, thermocouples, Flow meters, pH meters) and storing them separately till the main activity is over.]
- Safe disposal of scrap/waste which could be contaminated with chemicals being handled.
- Running the main plant below rated capacity if it is not possible to take up maintenance work immediately or, some spare parts are not yet available (even though ordered already)
- If the procedure for stopping and isolation of the plant, depressurisation or cooling, flushing etc. takes more time during which the production rate of the plant will be (very) less.

- Expenses and time required, as well as other inputs like fuel/power required [for heating and curing the protective lining] for mechanical trials.
- The plant is generally not run at full capacity immediately after the major repairs, or after replacing some major unit. The loss of production (and consequently) higher consumptions of power, fuel, water, raw materials per unit of output) should be considered as indirect cost of production.
- Wages paid to additional manpower(employed temporarily for the repairs)
- Salaries being paid to employees on roll who may remain underemployed or idle while some major unit is being replaced.
- Damage to nearby piping and ducts; cables and foundations of other units etc.
- Providing new thermal insulations and cladding on hot surfaces again.
- Re-installation of probes/sensors and thermocouples and attending to any damaged connecting compensating cables.
- Re-calibration/testing of instruments and again connecting them to their respective sensors. Field instrument also need to be checked and installed.
- Changing gaskets and fasteners in connecting piping and ducts.
- Cleaning of area around the unit and dispose off any item which cannot be repaired/salvaged and sold off or used elsewhere in the plant. Amount spent on such items should also be considered as indirect cost of maintenance.
- Any additional cost incurred to speed up the job or procuring the required spares even at a premium [higher cost than normal] by using express transport such as using company vehicle instead of public transport or procurement from abroad by air.

Chapter 5
Planning for Maintenance Work

5.1 Coordination of Maintenance Activities

Following shall be considered before taking up maintenance work on a process plant:

- The unit which is to be taken up for maintenance a list of all activities to be carried out and spare parts required.
- The procedure for making available this unit for maintenance (isolation, flushing, venting detailed safety precautions, evacuation procedures and adequate provision of safety equipment)
- Various types of manpower required for the job; mechanical, electrical, refractory masons, instrumentation and process technician.
- Whether external additional help is also required.
- Expected cost
- Time required for carrying out the job.
- Time for which the plant will be idle (this will include the time for cooling, flushing and actually starting the work plus restart of the unit).

5.1.1 Typical Symptoms Which Indicate Chance of Breakdown

Many times the above different teams of technicians will not be able to work simultaneously (if a welding job is being done on the equipment shell, the refractory mason cannot work inside, and when the catalyst is being filled inside the instrument technician cannot work on thermocouple, or the welder cannot work inside.)

- Reduced output from the machinery/plant unit[Less flow rate, less pressure, temperature (in case of that exchangers) (in case of pump)
- Abnormal noise
- Vibration

© Springer Nature Switzerland AG 2019
K. R. Golwalkar, *Integrated Maintenance and Energy Management in the Chemical Industries*, https://doi.org/10.1007/978-3-030-32526-8_5

- Release of smoke/vapour from shaft seals.
- Jamming of agitator shafts
- Poor quality of product (contaminated with particles of metals, rubber lining PTFE, gland pieces; un dissolved or excessive amount of un-reacted raw materials
- Abnormal temperatures of all/exiting process streams.
- Excessive pressure build up in the unit.
- Frequent release/blowing off vapours from pressure relief valve.
- Frequent activation of warning alarms/hooters.

5.2   Check List for Planning of Major Maintenance

- Which plant unit/machinery is to be taken up for maintenance work?
- Whether the job can be done while the unit/machine is running?
- Whether it will have to be completely stopped?
- Whether complete isolation from all connected at incoming and outgoing process streams will be required? (This may need providing blinds in pipe ducts, removal of electrical fuses or contacts)
- Will it be necessary to disconnect and remove from position?
- Can the repairs/replacement [replacement of bearing/impeller/damaged gauge glass of level indicator etc] be done at present location itself.
- How long it will take to complete job and when will the (repaired) unit be available for use again?
- Will it need special safety precaution before the repairs are taken up? [Through flushing of all resided acid/evacuation of inflammable or toxic vapours]
- What will be the effect on normal production activities? is a standby unit available? If not, whether the maintenance (repair) work will interrupt production units or another arrangement can be available to maintain production? [by drawing the process materials from a storage tank/storage bunker/overhead storages while the repairs are being carried out]
- What is the expected improvement in performance of the unit? Can the original performance be restored? If not, to what extent the improvement will be possible?
- Is there a possibility that some more repairs/replacement of more parts may become necessary during the job? This may become apparent after opening the unit. *Maximum repair and modification work shall be done in-house only since know- how of plant equipments or sensitive information may get look to outside persons.*
- Own technicians should be encouraged to develop new designs, modify equipments instead of getting such jobs done by external agencies. Good suggestions by own technicians and engineers shall be rewarded as soon as possible. It may be difficult/time consuming to frequently visit external workshops where major repairs are being carried out.

- Are all necessary spares, gaskets, internal parts available already or these have been ordered [possibly to a reliable supplier] and will be delivered in time for the maintenance work.
- curing of any protective lining portion repaired or; and
- Expected cost of all inputs and spares.
- Expected cost of isolation, dismantling, flushing; actual efforts (including fees to be paid to external agencies for any special work like machining, welding, fees to special Manpower and Consultants called to carry out and supervise the job]
- How much scrap/waste of material/effluents will be produced?
- Expected time required for the job- This should include time for isolation, flushing, actual repairs, testing the repaired/replaced items, curing of any protective lining portion repaired or; and mechanical trial runs to check the unit. After this certain standard restarting procedure may have to followed – which could be different from the procedure followed for initial commissioning.

Some remedial actions
- Can the faulty part be repaired or it has to be replaced?.
- Can the part be repaired while the machine/process plant is running?

Is it necessary to stop the machine completely or can it be kept running at reduced capacity while the repairs are being done?

5.3 Advancing the Shutdown

However, the plant may have to be shutdown a few days earlier in certain circumstances;

- The raw materials as per required specifications are not available or are available at very high costs only from a distant supplier which is not acceptable. There will be additional high cost of transportation to plant site.
- Trials of new equipments/another raw material can be taken only after the plant is stopped, cleaned, overhauled and **restarted. Results of trials in present condition of the plant will not be reliable**.
- New equipments have arrived, all preparations are in place for erection and commissioning and trials are to be taken before further payment is to be made to the vendor.
- Certain decisions are depending on the successful trials only (e.g. whether to change the converter for increasing production or observe the outcome after changing catalyst only? Since only changing the catalyst might give the desired results)
- Production cost has increased and the present product is not profitable. Diversification to more profitable products is required. Hence stop the plant for making necessary changes.

- Plant has become unsafe or more polluting or is consuming more energy, water raw materials.
- Some key inputs are not available/become scarce at present eq. power, fuel, water, key raw materials and these will be available only after some days/weeks.
- Transport bottleneck – floods, damaged bridges (hence fresh inputs are not available).
- Labour unions are not ready to work unless more steps/repairs are done for plant safety, lighting and dust control is improved.

5.4 Postponing the Plant Shutdown

- Due to pressures from Marketing on Production units since an important profitable demand for product is to be fulfilled
- Trials of some changes in operating procedures, new raw materials (procured from another source) new pumps, Hoists, conveyor is in progress.
- Sufficient preparation for major repair is not yet ready[supports/Building shed/foundation]
- Some important spare part is not yet available at site (it is on the way).
- Government has directed to supply some item urgently.
- Not possible to open up major units (to change tubes in heat exchangers; to change catalyst to replace packing in absorption tower by better ones…) due to rains or high humidity.
- Water scarcity and/or labour problem expected after some days.
- major absenteeism expected in next few weeks due to vacation season.

5.5 Carrying out Maintenance Jobs In-House

5.5.1 By Own Technicians/Engineers

- Condition Monitoring of equipments by own technicians.
- Routine jobs like lubrication, attending to leakage from ducts, spillages from drums,
- Replacement of internals like impellers, stirrers; baffles, bearings,.
- Minor repairs to Conveyor belts, screw conveyors, spray nozzles, valves in piping.
- Where elaborate safety precautions are required/equipments have residual toxic or inflammable materials in considerable amounts which need careful emptying out and flushing thereafter to remove all traces of such dangerous materials. Such jobs cannot be given to outside parties.

- It is preferable to get the job done through own personnel if there is a chance that technicians from outside parties may observe the working of adjacent process units which may be carrying out operations of a special nature/information which could be a trade secret based on a special know how.
- Concerned unit is handling (special) materials and needs special precaution or methods to handle any spillages. Technicians from outside parties may not be trained in this.
- Available tools are sufficient to carry out the job. [opening and dismantling of the machines to be attended; followed by repairs, replacement involving cutting/welding etc.]
- Own manpower is sufficiently trained to carry out such jobs.
- Repairs to protective lining/instruments/Control valves which can be done in-house.
- It will be difficult or very time consuming to use manpower from external agencies if they are not familiar with the equipments to be repaired.
- When proper coordination with other departments in the organisation is required. Own technicians are generally familiar with these matters.
- A small permanent work force is normally employed by the company. They are experienced and trained in carryout routine job.
- When the entire work can be done if all necessary spares and trained manpower is available in house.

5.5.2   Through Contractors/Externals Parties

- For attending major leaks/spillages or jobs where additional manpower is required which is not available in house. Battery limits and Scope of work must be clearly defined in Work order (in the Contract document).This is to be done only after discussion with senior experienced engineers, operating staff, purchase (procurement) engineers and finance department.
- *(Estimate time, materials, spares/bought out components required, work content—dismantling, flushing, washing, cleaning, working at a height, working continuously).*
- Major repairs/replacements are to be done [turbine internals, big agitators, complete re-tubing of heat exchanges or boilers, of large pieces of gas duets,....]. **Knowhow/tools/equipments are not available in-house**
- Job is to done very quickly[units installed at a height, heavy parts hanging from tops when there are chances of major interruptions is production process]
- Available tools and tackles are not adequate and additional tools like heavy chain pulley blocks, more welding sets, cutting tools, lifting devices etc. are required.
- If the maintenance jobs are of a routine/repetitive nature and elaborate safety precautions are not required.
- The units/machinery is installed at places away from process units where sensitive operations are being carried out.

- Own manpower is already engaged in other jobs and cannot be engaged in the repair work.
- Generally for voluminous work not requiring very skilled labour during annual shutdown/replacement of heavy equipment.
- *Properly defined contract needs to be made before commencing the work.*
- *Any extra work needs to be defined for the scope and payment terms, performance guarantee*
- **The jobs may need some supervision by Plant Engineers—specially for important machines or where considerable safety precautions are required to be taken.**

5.5.3 Site Fabrication Jobs (May Be Done Through External Agencies)

- Units having big size—storage tanks, absorption towers, converters..
- Items which are very difficult to transport from vendor's works to plant site.
- Items which do not need elaborate machining, casting (in a foundry)
- Gas ducts, structural supports, pipelines,work platforms fitting to process units
- Base frames, gantry for lifting pumps/heavy parts
- Side rails in sheds
- I-beam supports below storage tanks; below big process units.

5.5.4 Award of Contract for Maintenance Work

Decision is taken during meeting of Senior Plant Engineers, experienced technicians, Finance Department, Logistics and Stores department than it will be better to get the maintenance work/repairs or modifications through external agencies.

It is necessary to follow an internally designed system and Standardised procedure for the organisation before awarding a contract for repair, fabrication, or any other work to outside parties

- List out objectives of the job to be undertaken and decide priorities among the various objectives, Since it may not be possible to complete all necessary activities for meeting the objectives.
- Estimate the cost of maintenance by working out (i) details of all required activities and the resources required [spares, funds, chemicals] and the cost-benefit analysis. (ii) Estimate time required for each activity; their inter-relations (dependence on each other) and whether some of them can be carried out simultaneously (i.e. parallel activities) (iii) the time for which production activities may have to be stopped completely or partially slowed down.

5.5.5   *Quality Assurance Plan for Contract Work*

Quality of work to be assured by contractor.

- Study in detail the levels of difficulties involved – working at height, in dusty atmosphere, inside closed vessels in hazardous conditions p in presence of toxic or inflammable items)
- When adjacent units/those in close proximity cannot be stopped.
- Other additional scope of work which may be required such as shifting of materials from stores (bricks, heavy pipes, cement bags, etc.)
- Work being done by another agency simultaneously on the same equipment/ process unit e.g. welding of shell + repairs to brick lining + checking thermo wells of a Converter/Furnace. There can be disturbance from each other and controversy may arise.
- Work done by one party cannot be tested unless other party has completed their job[e.g. welding of shell and level indicator calibrates)
- Example: *Filling fresh catalyst inside converter cannot start unless shall welding and repairs to internal brick lining are complete.*

5.5.6   *Conditions to Be Included in a Maintenance Contract*

- Regular visits by Original Vendor [Free of Cost/Chargeable basis]
- Annual Maintenance Contract
- Guarantee period during which no charge will be levied by OEM (Original Equipment Manufacturer).
- Contract to arrange extra manpower if required at short notice so that the job is not delayed
- Contract to assist in restart after repairs till the equipment is working satisfactorily.
- Improvement in performance to be watched and compared with guarantees offered by the party who have done the job.
- Also, compare with original specifications and work out estimated cost to restore the performance to required level [if it is different from originally specified]

5.5.7   *Controversies During Maintenance Work by Contractors/ External Parties*

- Equipment/Machinery has not been isolated properly.
- Equipment/Machinery have not been thoroughly cleaned.

- Scope of work is not defined correctly. It is getting increased as the work is being carried out (on opening a unit, unexpected internal damage is noticed).
- Payments are not made in time. No advance is given to mobilise the site [for bringing necessary tools, scaffoldings, welding sets etc.].
- Sufficient manpower is not employed by Contractor. Hence work is being done at a slow pace. This generally happens when the Maintenance Contractor takes up many assignments simultaneously at different locations.
- Key personnel employed/working for Contractor are on leave.
- Adverse weather conditions (heave rain, snow, storm) can damage the units under repair or disturb the work pace. But the organisation [Plant Management] wants to blame the contractor for the delay.
- Occurrence of accidents or fire, release of toxic gases or spillage of chemicals have taken place due to negligence by personnel/Supervisor appointed by Contractors.
- Some adjacent or nearby unit gets damaged [due to spillage of chemicals, fire, wrong earth connections for welding; water spray used for cleaning is not controlled properly.
- Plant Management have not provided security, store room, water, power, unskilled labour, living rooms, canteen, site office conveyance etc.
- Existing procedures for receiving materials, spares in to stores and subsequent issue to work site are too time consuming or creating difficulties in accounting for cost of repairs/work being done by contractor.
- Security Personnel are creating problems for movement of personnel employed by the external agencies (contractor) in plant. On the other hand, these persons are loitering in prohibited areas.
- Normal work in other parts of premises is disturbed due to interference by Contract labour, by borrowing certain tools (which they should have arranged themselves) and scaffolding etc.
- Certain unrelated- but adjacent – units are required to be shut-down for long durations as a safety measure, and this has not been visualised before assigning the contract. This can disturb production activities in main plant. Contractor is not allowing running of these units till his job is complete, tested, and handed over.
- Contractor labour is sometimes idle, due to equipment not released in time; original spare part supply is delayed; power outage, water supply not available. Spent liquor after washing /testing should be properly collected and treated by existing ETP / plant facilities.
- Repair job is complete, but is not being tested or the equipment is not being commissioned [required raw material not available as per specifications, hence the plant is not being started].*This can hold up the last 10–20% of payment since the performance of running unit cannot be checked. Such situations can create controversial arguments which need to be settled amicably between the contractor and Plant Management.*
- Product Quality is not satisfactory when it is being run by the new equipments. [In actual practical conditions *it will be so for an initial run of few days or initial few batches.*
- The equipment is being tested or operated by feeding some other raw material.

5.5.8   Additional Work Given During Execution or End of Contract

- Proper collection, packing and disposal after cleaning of tanks, shifting of brick/ other material from stores, disposal of waste or scrap generated during cleaning, additional welding or replacement of tubes, site fabrication of pipes and ducts…
- To be written in Contract – whether repair are to be done with materials brought by the contractor at their own cost or without materials (all necessary materials will be supplied by plant management). All conditions as laid down in QAP must be fulfilled.
- Test certificates must be produced for all items supplied by the vendor before using for replacement, major repairs or installation in the plant. All such new materials and items brought by the contractor may be tested again (*if felt necessary*) through an approved agency or own laboratory. Terms and conditions for rejecting the items which do not meet specifications shall be clearly mentioned in the Work Order to vendor.
- The plant engineers may accept certain off-spec items (*after repairs or modifications*) for similar use or some other purpose if possible. Such decisions may have to be taken if a major controversy arises in case of total rejection of off-spec items.

Chapter 6
Maintenance of Process Units

6.1 Material Handling System

Feeding of solid materials needs equipments as below:

- Pneumatic conveyors
- EOT cranes
- Hoists
- Belt conveyors
- Bucket elevators
- Screw conveyors
- **Consider properties of the materials:** (bulk density, average lump size, angle of repose, hygroscopic nature, corrosive/explosive or inflammable nature or formation of dust during conveying), moisture content. The system should be able to work while handling such materials at the required rate.
- **Service required**: quantity to be handled per hour/per cycle, distance over which the materials are to be transferred, type of packing (loose powder or in containers) should be designed for safe handling.

  Observations to be made during operation:

6.1.1 Pneumatic Conveyors

Pressure drop in the system, delivery of material at the designated spot, any choke in the duct, power consumption by the blower, arrangement to arrest particles of material escaping at delivery point such as cyclones and bag filters (check satisfactory operation of water sprinklers to prevent fires), earth connection for the system while handling combustible powders

© Springer Nature Switzerland AG 2019   111
K. R. Golwalkar, *Integrated Maintenance and Energy Management in the
Chemical Industries*, https://doi.org/10.1007/978-3-030-32526-8_6

6.1.2 *Electrical Overhead Travelling (EOT) Cranes*

- Material transferred to designated spots and time taken for each trip (to and fro);
- Working and current drawn by motors and their control for vertical, horizontal and sideways movement while lifting bags/containers.
- Functioning of limit switches (to prevent taking up the load to excessive height which may damage the wire rope or the lifting hook)
- Guides for proper winding of wire rope on drum. Condition of wire ropes and lifting hook.
- Functioning of wire rope winding mechanism on the drum so that there is proper winding on the grooves only.
- Condition of link chains and wire ropes used (confirm capacity to carry load with standard weights).
- Automatic brake to prevent the load going down in case of power failure.
- Smooth movement on side rails (any damage to wheels).
- position of stopper at ends

Observations during operation
- Smooth movement while lifting and movement of carrying of load (bags/buckets)
- Total load being lifted—never exceed safe *permitted* limits
- Any abnormal tilt during vertical/sideways or to and fro motion on the side rails.
- Smooth movement of operator's console on the horizontal beam and on the side rails.
- Any abnormal noise from the gear boxes.
- *The operator shall be able to observe the load clearly all the time. Operational Controls pendant switch shall be at a safe distance away and never directly below the load. No one shall stand directly below the load or in the path of the load.*
- Working of alarms/warning devices if excessive load is being lifted.
- Proper lubrication of gears and wire rope.
- Overload tripping of hoisting motor.
- Proper lighting in the entire working area.
- Exhaust system for removal of dust.

Maintenance
- Check proper working of all above systems.
- Weight/Load indicator.
- Overload warning lamps, hooters and tripping mechanisms.
- All wheels (and their bearings) of the moving Operator Console.
- Wiring of pendant switches and all motors.
- Gear Boxes and level of the lubricants in the casings.
- Test the capacity of wire ropes by standard weights regularly and record.
- Physical inspection of entire length of all wire ropes. Apply correct grade of grease.

- Winding mechanism on drum. Clean the drum surface and wire ropes. Apply correct[grade of grease(as recommended by OEM)
- Erosion, Corrosion of Beams, wheels on side rails and protective paint.

6.1.3   Hoists

- Spillage of materials while being lifted-
- Total weight shall not exceed maximum capacity permitted
- Confirm working of limit switch, lifting hook, and wire rope.
- Smooth proper winding of wire rope in the grooves without getting entangled. Check the sliding guide for the rope.
- Weight indicator scale
- Proper vertical movement of the load to be ensured.
- Condition of all strands of wire rope—(shall not be damaged)
- Lubrication of bearings of drum and drive motor
- Cordon off the area directly below the hoist while raising up the load.
- Vendor shall be asked to give test certificates for wire rope lower pulley, lifting hook, hook pin and at least two extra lengths of the wire rope as spare.

6.1.4   Belt Conveyors

These are used to convey powdered materials and solids over considerable distances. This type of conveying system has feeding hopper, belt drive motors, discharge chutes and dividers at delivery end, weighing instruments and set of idlers and roller supports.

Observations during Operation
- Smooth movement (proper rolling of idlers) without any spillage of material on the way, at change of direction when feeding to another belt or at delivery end in to a silo.
- Check working of **manual trip** switches which can stop the movement of belts in case of fire or any major fault noticed.
- Rain and Fire protection system for the belt and material being moved
- Readings of weighing system every 2 hours (should tally with actual material fed)

Maintenance
Check and attend gear box, drive motor, automatic belt tensioning system, any jammed idlers, emergency trip switches and any misalignment of belt (due to gradual shifting away from drive or driven pulley by itself, and any torn pieces.)

6.1.5   Bucket Elevators

Used for lifting up powdered material generally from ground level to silos at a height vertically above.

Observations during operation
- Any overflow from the lower boot.
- Any overflow from the discharge chute at top.
- Any ingress of rainwater in the lower boot (preventive walls shall built around the lower boot to prevent water flooding).
- Jamming of vertical movement of the buckets.
- Any abnormal noise.
- Check earth connection of body.
- Current drawn by the drive motor.
- Any falling off bucket from the system.
- Any loosening of side covers.

Maintenance
- Open the inspection windows and check condition of each bucket, drive chain, any material stuck up inside.
- Lubrication of drive chain and gear box.
- Free tilting movement of each bucket (so it can pick up material from the lower boot and discharge properly at the top into the chute).
- Condition of side cover (any rusted or corroded part to be attended).

6.1.6   Screw Conveyors

Observation
- Smooth movement of screw
- Any overflow/spillage from the conveyor before reaching the discharge end.
- Any jamming of the screw due to excess material inside.
- Any choking of discharge opening
- Flow of cooling water (in case of jacketed screw conveyor)
- Flow of steam in jacket and working of steam trap (in case of heated unit)

Maintenance
- Inspect each flight of the screw.
- Inspect condition of shaft and intermediate support/bearings
- Clearance between flight and casing.
- Clean external jacket if cooling was not satisfactory.
- Attend the steam trap if heating was not satisfactory.
- Speed control mechanism to be checked and attended.

6.2   Crushers, Grinders, Pulverisers

These are used for reducing the size of raw material lumps [bauxite, phosphate rock] neutralising agents [lime stone], before feeding to reactors, ETP units, furnaces etc. These equipments generally consist of feed hoppers, feed controllers, screening grids and magnetic separators (to remove tramp iron pieces).

The crushing elements are generally steel hammers, crushing rollers, grinding wheels Very big lumps or high feed rates can damage the machines and hence protective devices are provided.

- Screening grids to prevent big lumps getting into the machines and feed rate controller.
- Rotary Valve at bottom of hoppers.
- Screens to separate oversize material
- Electrical trip and emergency stop switch for the drive motor (big lump or excess feed causes jamming).
- All such mechanisms shall be carefully observed during operation; thereafter inspected and attended during regular maintenance
- Procure spares only from OEM [Original Equipment Manufacturer].
- Check foundation bolts, support frames etc. which can become loose due to vibrations during operations.
- The working area should have adequate exhaust system for evacuating dust produced. (Cyclone separator, bag filter, an ID Induced Draught fan etc.)

6.2.1   Air Classifier

An air blower is used to blow away (fluidize) and transfer the pulverized material. Bigger particles are separated by cyclone, impingement separators and are returned to the pulveriser. The fines (which are as per required size) are collected. A Bag filter is generally provided to arrest very fine particles (minimise loss of product and prevent air pollution)

Observations during operations
- Pressure drop across various components.
- Current drawn by air blower motor, pulverizer drive motor.
- Connect discharge duct from the bag filter to an empty room. Clean it once a week/15 days].
- Size distribution of product particles.
- Through put of lumps feed to the unit and pulverized material collected.
- Maintenance

Check for any leak from the ducts and attend. Also check the air blower, cyclone, impingement separator, bag filter, rotary air lock valves and connecting ducts.

6.3 Units Operating at High Temperatures (*Furnaces, Reactors, Oxidisers.*)

Following should be carefully observed during operation.

- Shell temperatures at different places. If external cladding/insulation is provided, it may be difficult to do so. In that case, one may observe temperature of external surface of cladding.
- Any leaking spots in the cladding from where ingress of rain water may occur or any escape of hot gases from the furnace/high temperature unit.
- Careful visual observation (through view glasses/observation ports) of the internal refractory living/length and colour of the flame of oil firing units/any abnormal glow of internal surface, deposits of clinkers; visible cracks on internal lining
- Any external cracks on the shell.
- Any isolated hot spots on the shell.
- Setting of fuel supply pumps; and fuel flow rate.
- Working of air compressor used for atomizing the fuel oil.
- Combustion air fan discharge pressure, speed and current drawn.
- Settings/Positions of Primary secondary and tertiary air supply lines.
- Check exit gases from the furnace for carryover of un-burnt fuel (black particles)

6.4 Process Reactors (Fixed Unit with No Moving Parts)

- Always compare actual operating conditions of pressure, temperatures, PH, flow rates of process streams and their compositions with design conditions [normal values; and both maximum and minimum] Keep a record of the deviations observed; the frequency with which the deviations occurred and the duration of time for which the plant was run inspite of those deviations.
- Monitor the production rates, analysis of the incoming and outgoing process streams, the quality of the products, the consumptions of raw materials and utilities.
- Also record the number of time the warning alarms/safety systems had got activated. These records and their analysis will indicate whether the plant was being run as per standard operating precaution.
- Data regarding plant breakdowns, spares used, costs incurred will lead to areas where improvements need to be done.

6.4.1 Reactors with Heating/Cooling Arrangements

- These can have external jackets or limpet coils or internal coils.
- Heating is generally carried out by steam or hot thermic fluids. Similarly cooling is done by circulating cold water, chilled water or brine.

Steam is generally admitted at pressures or 3–5 Kgs/cm^2 and hence the reactor shell, jacket and coils must be strong enough to withstand these pressures. Control Valves must be provided in the incoming steam line to ensure that steam is never admitted at excessive pressure which can damage the reactor shell/external jacket or the internal coils. A safety valve [pressure release valve] adjustable as per required pressure and a rupture disc is also provided. These can protect the heating arrangement from getting pressurised beyond safe limit.

Calibration of the pressure gauges to confirm accurate indication of pressure. Inspection of all relief valves is a must to confirm proper working.

Maintenance – Replace the control valve/attend any damaged internal part/pilot line connection from downstream side of the pressure reducing station. The pressure gauge must be checked and calibrated again if there is a doubt regarding its accuracy.

Pressure Test – Reactor shell, external jacket and heating coils must be treated like a pressure vessel where statutory regulations are applicable.

Repairs – In case of a leak, the unit must be isolated completely from all process streams and steam/heating fluid connections. Blinds shall be inserted in the connecting pipes even after closing the valves (because the valves may leak slightly.) If the process fluid enters the jacket/coil after cooling (due to vacuum produced when the steam condenses) it should be carefully drained out slowly and collected in drums for recycle or disposal through ETP. All remaining traces of the fluid must be flushed out by water/vented by compressed air. The escaping vapours should be passed through suitable scrubber or seal pots. Level of liquid in the process reactor can be checked to get an idea of the likely spot where the leak might have occurred. A high (residual) level indicates a leak at higher spot while a low (residual) level suggests a leak towards bottom side.

The jacket can be opened to observe, confirm and then repair the leak. Gas cutters shall not be used if there are chances of fire or pressure building up due to evaporation of residual fluid in the jacket. After plugging/providing a patch on the damaged/ leaking portion and replacing by new one, a pressure test shall be carried out to confirm suitability for reuse. However, the concerned equipment may be de-rated or discarded if a large portion had to be replaced.

6.5   Absorption Towers

They are used for absorbing components from a gas mixture for production; for pollution control; for simultaneously carrying out reactions, for recovery of chemicals, for purification of gas streams..

These are generally vertical cylindrical vessels whose common materials of constructions are- Mild Steel with lining of special acid resistant bricks, rubber, glass, Stainless Steel-316, Plastic with Fibre Glass reinforcement, FRP, PP, PTFE, rubber, lead.

The absorbing/scrubbing liquor is introduced in the towers at the upper side by means of cast iron liquid distributor, Stainless Steel spray pipes, spray nozzles (which could be at centre or located tangentially at periphery of the tower or both).

6.5.1   Observations During Operation

- Analysis of incoming and exiting absorbing liquid and process gas streams – concentration, temperature, pressure, mist particles and moisture content and other relevant properties.
- Pressure drop on gas side during operation (indicates obstruction to gas flow due to liquid hold up, choking of internal packing in tower)
- Efficiency of absorption for the component to be absorbed (as per design)/actual.
- Examine the likely reasons for inefficient functioning:

  - Channeling of gas in absorber/insufficient gas- liquid contact.
  - Some of the spray nozzles are choked.
  - Absorbing (scrubbing) liquor: Insufficient flow rate, high temperature at inlet/already has high concentration of solute (material to be absorbed)
  - Incoming gas stream not distributed properly over the cross-section [incoming pipe not projecting in sufficiently]

Preventive Methods
- Use only glazed tower packing.
- Use SS-316/Alloy 20 Nozzles/SS-321/Niobium nozzles
- Gas inlet nozzles to be extended up to center of the tower.
- Increase flow of absorbent liquor carefully without causing hold up in absorption tower.
- De-mineralized water shall be used instead of ordinary water (for dilution or solution) to minimize deposits of salt on tower packing.
- Use fibre bed mist eliminator of adequate capacity to remove mist particles of 2.0 micron–4.5 micron size (typically required) for protecting downstream equipments; to minimize emission with exit gases through chimney.

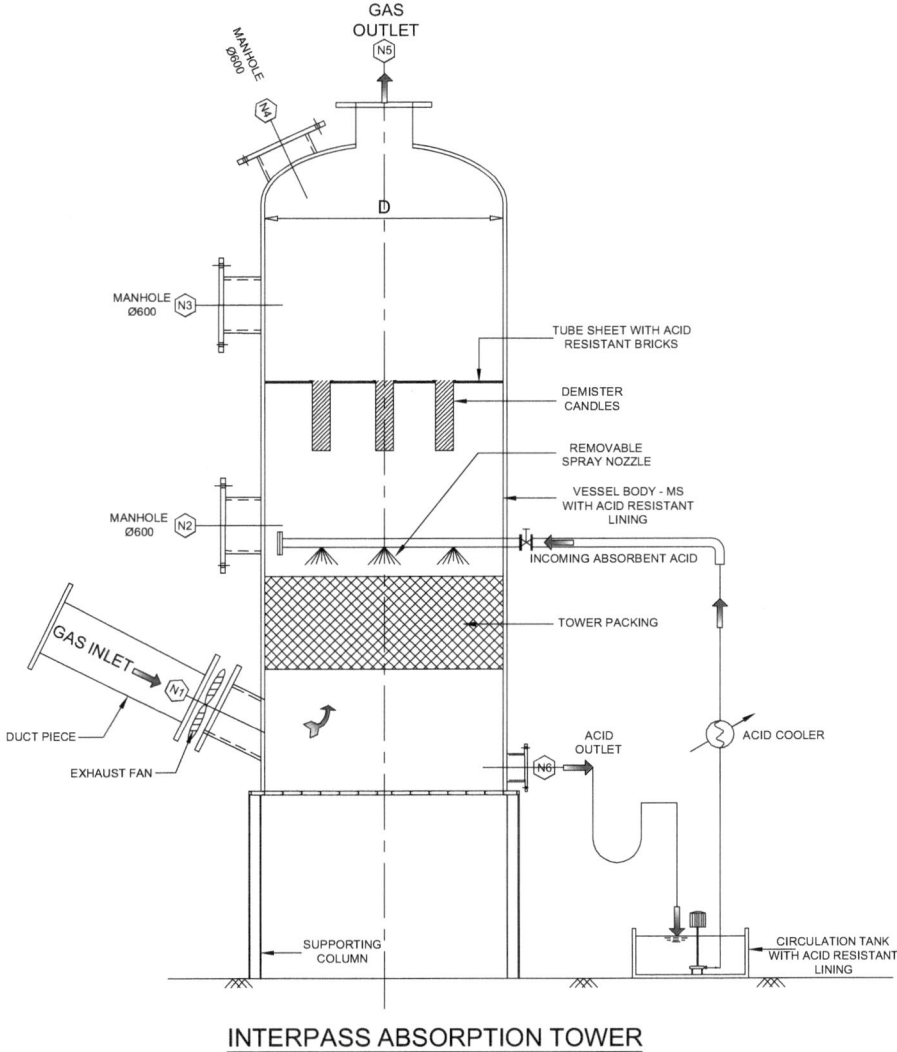

**INTERPASS ABSORPTION TOWER
IN SULPHURIC ACID PLANT**

Example: Inter Pass Absorption Tower in DCDA Process Sulphuric acid plant
Sulphuric acid is a very important industrial chemical and is manufactured in many countries of the world. The modern process for manufacture is known as DCDA process.

Check list during maintenance
- Levelling of distribution tray/distribution pipes for incoming acid
- Condition of weirs on the distribution tray. More acid will flow from the damaged weirs in that side of tower and there will be channeling in the tower (gases flowing upwards from other side will not get sufficient absorbing acid)

- Any choking or puncturing of Fiber Bed Mist Eliminator FBME (if installed inside the tower) or Demister mesh pads if installed in separate vessel.
- Internal brick lining at walls and at bottom side
- Any corrosion of top cover
- Condition of the bolts and gaskets used for fixing FBME on the tube sheet.
- Open top manholes for entry inside the unit. Examine the tower packing inside. Take out internal gas-liquid contact packing having sludge deposit and clean. The waste liquor is to be neutralised in ETP.
- Replace the broken pieces of tower packing by adding new ones. *The new packing (should be as per specifications) and tested for suitability at operating conditions.*
- Check the U-seal in acid exit pipe from the tower. This should be clean. There should be no pieces of damaged tower packing inside as they can obstruct the flow of acid. This can cause hold up of acid in the tower which may obstruct the incoming gas flow.
- A distance piece in the gas inlet duct is to be removed and an exhaust fan is to be fitted on the lower side as shown in the drawing. It will suck fresh air from the top and release from the bottom. This is to ensure that the persons working inside the tower are not affected by acidic fumes.
- Closely observe if any acid is seen to be seeping from bottom plate. Area between supporting I beams should be cleaned thoroughly. This is a serious condition and needs to be rectified. It can take a long time to do so.
- *The author has handled such situation and the details can be available in the book **Production Management of Chemical Industries** published by **Springer International Publishing, Switzerland***
- It is advisable not to wash the IPAT from inside with water as it will form dilute acid inside. It can corrode components of acid circulation system and create lot of acidic mist on starting the plant. This is difficult to absorb fully and can result in environmental pollution. This applies to the other acid towers also.

6.5.2   Fibre Bed Mist Eliminators (FBME)

The FBME shall be taken out and cleaned as per procedure given herewith (obtained from **Evergreen Technologies, Mumbai India**)

It needs a very efficient mist eliminator for removal of acid mist particles from exit gases in the Inter Pass Absorption Tower IPAT. The mist eliminator serves to protect the downstream heat exchanger units and catalyst in next pass of the Converter. Candle Filters or Fibre Bed Mist Eliminators are also used in other chemical plants to arrest acid particles and are thus essential for preventing environmental pollution. The plant engineers must observe the performance of these mist eliminators when the plant is running; and clean them by following correct procedures prescribed by the manufacturer when the plant is stopped for annual overhaul.

Evergreen Technologies Pvt. Ltd., Mumbai (India) are the leading manufacturers of various type of mist eliminators (including candle filters) and have supplied them to many industries. The following procedure is recommended by them.

Monitor the pressure drop across the candle demister every week when the plant is running.

Observe the rise in pressure drop and whether it has become much above its normal value (when it was freshly installed).It needs cleaning when such a stage is reached

This can be due to plugging by dust/rust particles, deposit of sulphur or ferrous sulphate particles (collected acid mist will drain continuously from the candle filter and will not cause an abnormal increase in pressure drop).

A. Procedure to clean if found plugged by -------, deposit of sulphur

- After stopping the plant the manholes on IPAT are to be opened and arrangement is to made for ventilation before entering the tower.
- Now take out the candle demisters carefully and neutralise in a bath of 5% soda ash (Na_2CO_3) solution for 30 minutes or till no bubbling is seen.
- Remove from the bath and very slowly rinse by clean water. Do not use a pressure water jet as it can damage the candles.
- Immerse them in a tank with 0.5% sodium sulphide solution for 30 minutes.
- Horizontal tanks shall be used for best results. A vertical tank may be used if horizontal tanks are not available.
- The solution can be agitated by carefully sparging air in the bath or by slow mechanical means.
- Take out from bath and again slowly rinse by clean water. Do not use a pressure water jet
- Now allow the candle to lie in a horizontal position for at least 4 hours to drain out the liquid.
- After this it can be kept in vertical position for further draining out.
- Allow to become completely dry before installing again on the tube sheet. This will minimise formation of dilute acid (which can corrode the bolts/nuts of the tube sheet) on restart of the plant. Check the gaskets before installation; and change if found damaged.

B. Procedure to clean if found plugged by ferrous sulphate sludge or other particles.

- Remove from the bath and stand upright.
- Very slowly rinse by copious amount of clean water. Do not use a pressure water jet as it can damage the candles.
- Now immerse the candle demisters carefully in horizontal tank in 5% soda ash (Na_2CO_3) solution for 30 minutes or till no bubbling is seen.
- The solution can be agitated by carefully sparging air in the bath or by slow mechanical means.
- Take out, rinse by clean water and repeat the procedure as above.

Check Points
- All gaskets should be serviceable. If not, change the damaged ones.
- All liquid seals should have correct amount of liquid
- All flanges shall be properly tightened.
- Ensure complete drying of candle filter before installation (if required by blowing air through it over a prolonged period).
- Do not operate the plant with a high gas flow rate or temperature above original specifications
- Check corrosion/damage of wire mesh cage for the fibers and repair properly without damaging the fibers.

Evergreen Technologies Pvt. Ltd., Mumbai (India) can be contacted for any further clarification. e-mail: info@evergreenindia.com; website: www.evergreenindia.com

6.5.3   Drying and Final Absorption Towers

- Demister mesh pads in these towers should be closely examined for any damage or displacement from present condition. The damaged pads shall be replaced by new ones. Displaced pads should be fixed securely by stainless steel 316 bolts and thick wires (which are enclosed in Teflon tubes) to the frame.
- These towers should be checked and attended in a similar manner as IPAT.
- Acid circulation tank—completely empty out and examine from inside. Attend the acid resistant lining by carefully removing damaged bricks and provide new bricks. These are to be cured before taking in to use.

Maintenance of Towers/Scrubbers
- Isolate from all inlet and exit process streams. Put blinds in connecting pipes and ducts. Remove fuses from power supply lines for circulation pumps.
- Open manholes and provide exhaust fan as shown in the drawing. Provide fresh air supply for persons who are working inside the tower.
- Start taking out incoming scrubbing liquor pipes lines and spray nozzles. Check their condition and test spray angle by checking outside through a trial run (water pump may be used).

Tower Packings and Internals
Various types of packing's [Intalox saddles, Raschig rings (with plain surface, single and double partitions, with external serrations), tellerette rings, pall rings] are in use. These may get choked due to dust and scales carried with gases; deposits of salt due to reactions or produced due to corrosion of piping or due to erosion of lining materials, insoluble suspended materials. The gas liquid contact surfaces get reduced as a result. Channeling of flow of scrubbing liquor may also occur. This can affect the performance of the absorber adversely. Hence it is necessary to examine

the packings and take them out for cleaning during overhaul. Damaged, badly choked or broken pieces of such packing must be replaced with new ones (from a properly tested lot which will withstand the operating conditions satisfactorily)

6.6 Catalytic Converters (Fixed Beds of Catalyst)

These are process unit having one or more beds of catalysts. The catalyst is supported on stainless steel grids, Cast Iron grids, high temperature and acid resistant brick chequer work etc. ceramic pieces, small ceramic balls of high alumina (can be 60–65% Al_2O_3 or more) are spread uniformly on the grids and the catalyst in the form of pellets, hollow cylindrical rings (with both ends being open) or some special shapes with external fins/petals like daisy. These provide excellent surface areas for the reactions to take place among the components of the gaseous mixture when it flows over the solid catalyst.

6.6.1 Heat Exchange Surfaces

Considerable heat is evolved during chemical reactions when the gas mixture is passed through Catalytic Converters. In certain processes, the gas mixture is to be cooled by removal of this heat, while in some other processes, the gas mixture is to be heated.

Properly designed and constructed heat exchangers are required for these. These can be as external units, or internal coils in the converter. In some cases, cooling air is injected directly above/below/in the Catalyst mass itself as per process design.

- **Internal heating/cooling coils** – These do not need elaborate gas ducts outside the converter as there is not external heat exchanger. This saves area required for the plant.
- If the number of coils is less, it does not increase height of converter substantially. However, lot of efforts and skill will be required to be required to clean (remove) the deposits of ash, dust etc. on the coils as they may have to be pulled out through suitable manholes/opening on the converter shell.
- **External Heat Exchanger** – These unit are provided to heat/cool the reaction gas mixture where larger heat transfer area is required, Hundreds of tube are arranged in these units which can have single pass/more passes has better heat transfer and lesser pressure drop.

However, gas ducts from Converter to these external units and then to the next process units are required.

In case of large diameter duets, there will be need for more area (more space on the ground) for the process plant.

These heat exchangers can be cleaned on the tube side by brush and by chemical cleaning solution (de-scalant) on shell side. Damaged/leaking tubes can be plugged or replaced without entering the converter itself.

The design of the Conversion system depends on the production rate required, the concentration of reactants in the gas mixture, the temperature, pressure required for carrying out the reactions, Plant designers have to consider the pressure drops (consequently the power consumed by the gas blowers/compressors), the quantity of the catalyst required for the production capacity of the plant (reaction rates to be achieved) which depends on the activity of the catalyst its surface area.

Plant operating data is regularly recorded and analysed to determine whether the catalytic converter is functioning satisfactorily. The incoming and outgoing process (gas) streams are analysed by suitable gas analysers to determine the degree of conversion achieved; and the temperature and pressure conditions are also taken into account.

It is generally observed in most of the chemical plants that the conversion efficiency slowly reduces with time. This can be due to deposit of dust particles on the active surface of the catalyst, due to physical disintegration or agglomeration of the catalyst particles or loss of active ingredients. Hence it becomes necessary to restore the converter performance by cleaning or reactivation or replacement of the catalyst.

Samples of existing (installed in the bed) catalyst are taken out during annual shutdowns for overhauling of the plant and analysed. Advice from original manufacturer of catalyst and plant designer shall be taken while replacing old inactive catalyst by new one.

Generally the new catalyst is available at short notice. However, it is advisable to keep a quantity about 15–20% of the total amount (charged in converter) in stock before opening up the converter Fresh amount of catalyst may be ordered/procured in advance if it is not available at short notice. The procured catalyst must be stored as per instructions of the manufacturer *(generally it is supplied in water proof bags or in drums)*.

However these shall be kept on raised platforms in a cool, dry, well ventilated place away from outside dust, smoke, rains etc.

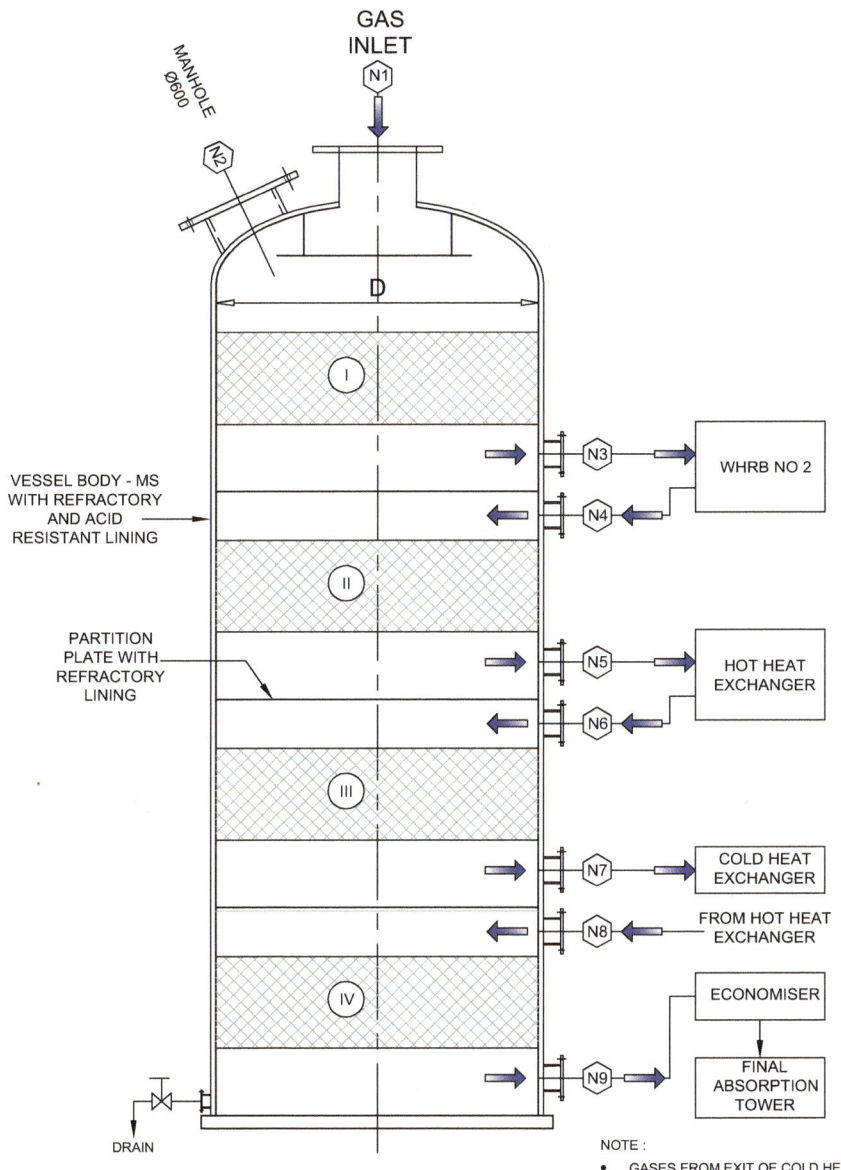

MULTI PASS CONVERTOR

Figure for Multi Pass Converter

6.6.2   Converter Overhaul

On stopping the plant the residual gases from converter shall be removed completely by passing dry/warm air through the unit. This is to be done as per instructions from Catalyst Manufacturer/Plant designer/as per standard manuals provided. The converter shall be isolated from all incoming and outgoing gas (process) streams. Thermocouples and gas analyser probes shall be taken out and then manholes shall be opened for removal & inspection of catalyst beds. It is advisable to suck out the catalyst by means of a exhauster. The gases shall be passed through an empty vessel placed between the converter and inlet (suction) of the exhauster. An induced drought fan of sufficient capacity can also be used as the exhauster. *This method will be found useful for taking out large quantities of the catalyst from the converter without going inside.* Trained men can be sent inside for removal of catalyst by filling in small bags/buckets and taking out.

Catalyst bed supports shall be taken out, cleaned and stored separately. Any attrition loss shall be made up by fresh quantity of tested material only.

All residual acidic gases/chemical vapours must be evacuated completely and breathing apparatus kept available before taking up such jobs. Exhaust fans must be kept running. Fresh air supply shall also be kept continuously available.

Fume & Gas detectors must be used to detect any residual vapours. The working area must be made safe to work. Gas and face shields; safety goggles, emergency portable lighting at 24 volts only *(and not at a higher voltage for safe working)* and mask rope ladder shall also be provided.

Another team member shall be present outside all the time. No one should step on the catalyst beds directly as this can crush the catalyst.

Screening
Take representative samples from different locations of the catalyst bed while emptying it. These samples shall be mixed and a composite sample shall be sent for analysts. A sufficient amount of sample shall be preserved and labelled for future reference.

A rotary screen with an exhaust hood shall be used for screening the catalyst. The screen shall be rotated slowly by a drive motor with variable frequency drive and gear box.

The dust coming out from the screen shall be sucked by an I.D. fan. A Cyclone separator shall be provided to collect the dust. A Venturi Scrubber shall be provided on the suction/discharge side of the I.D. fan depending on the concentration of dust particles. The screened catalyst and the support bed shall be immediately packed in bags/drums and kept protected from dust/rain etc.

Inspection of Thermowells and Pressure Point
These shall be carefully inspected for any damage or leak after all catalyst has been taken out. Recalibration of thermocouples shall be done before starting the plant

again. The thermo wells should project up to centre of the cross section of converter and shall be placed at the interface of support bed and catalyst layer.

In case of high temperature operation, the support grids can expand placed between adjacent grids. The ceramic papers can also be inserted all along the internal periphery. This arrangement can prevent escape to of gases through the gaps while preventing warping/tracking of the grids at high temperatures.
The gases must pass through catalyst beds only.

Support grids
Shall be thoroughly cleaned during plant shutdown and any pieces of bricks, bed supports shall be removed from the grids. Any damaged grid shall be repaired or stainless steel flats shall be fixed if some grid is broken,(but not necessary to replace if the damage is not majors).

Gas Distributor
It is very essential to ensure uniform gas flow over the entire cross section of the catalyst beds. Hence gas distributors of appropriate shape and size shall be provided entry point of each bed. Any damaged to the distributor must be attended during the shutdown.

Protective linings
The internal side of the converter shall and the bottom plate, top roof as well as all portions exposed to high temperature and corrosive conditions shall be provided by protective linings.

The various materials used are refractory bricks; castable refractory layers installed by suitable anchors fabricated from high temperature and corrosion resistant alloys; and layers of acid resistant and insulating bricks.

Some designers recommend entire converter shells to be constructed from stainless steel of suitable grade on the basis of cost- benefit analysis for the plant.

Thermal Insulation
External insulation by glass wool/mineral wool layers is to be thoroughly checked. It shall be covered by aluminium cladding sheets of SWG-16-18 or 20 thickness with proper sealing at all entry and exit points of gas streams, thermocouples, manholes and vents etc.

Special care is to be taken to prevent ingress of rain water or snow from any leaking spots in the cladding because they can disturb the temperature conditions; and thus adversely affect conversion process in the converter.

Condition of Shell, Bottom Plates etc.
A thorough visual inspection as well as measuring the thickness at a number of spots (specially those subjected to high temperatures, corrosion and welded joints).

The areas where there is a possibility of accumulation of acidic condensate shall be thoroughly inspected.

Any such condensate shall be removed/neutralised and sealing done by acid resistant tiles and cement with Potassium Silicate base.

6.6.3   Conversion System with Fluidised Bed

Fluidized bed reactors are used in chemical process industries in which a high heat transfer rate from the solids to the gaseous reactants is required for maintaining isothermal conditions in the reactor (when highly exothermic reactions are taking place)

The catalyst particles are kept in a suspended (fluidised) state in the gases which are passed upwards at a sufficiently high flow rates. The solids and gases get very well mixed as a result and the heat transfer rates are high.

Observations during operation
- Pressure drop in the reactor
- Power consumption by the blower
- Temperature of the reaction mass at a few selected points, and at exit
- Separation of particles of catalyst going out with exit gases.(These must be separated from the gaseous product of reaction and recovered)
- Attrition loss of catalyst

Maintenance
- Collect the catalyst fines which have got generated due to attrition of bigger particles. The loss must be made up in order to maintain the conversion rate, and to ensure sufficient area of contact between gases and catalyst surface.
- **Arrester system** for catalyst particles (Cyclones, impingement separators, collection vessels)
- Manufacturer shall be asked to supply fresh catalyst with test certificates for activity, surface hardness/maximum limit of loss due to attrition
- Gas analysers for inlet and exit
- Thermocouples and thermo wells
- **Gas distributor plates** (perforated, and generally of stainless steel construction with reinforcements/in sections to cover the entire cross section of the converter)
- Inspect the lining of the reactor walls (it is subject to erosion and corrosion); and repair as per need.
- **Blower**: Check the suction screen, speed control mechanism, and clearances between lobes/casing so that it will deliver sufficient flow rate of gases.

6.7   Filter Press FP

Observations to be regularly made during operation

- Flow rate of incoming liquid (to be filtered)—*by monitoring the level in feed tank*;
- pH, temperature, pressure.
- Concentration of suspended solids in incoming liquid.
- Flow rates at various exit points.

- Analysis [concentration of solids in filtered liquid], temperature, flow rate of exit stream.
- Monitor the rise in pressure at inlet- which indicates build up of deposits on the filter media.

Any sudden increase in exit flow or higher concentration of suspended solids indicates damage to one or more filter cloth pieces.

Leak from joints of filter plates indicate damage to gaskets or to plates.

Low flow rate or higher pressure at inlet *very soon after commencing filtration* needs to be investigated.

Some likely reasons could be –

- Low level/emptying out of feed tank for incoming liquid
- Choking of suction or discharge side of feed pump for the filter press.
- The feed pump is not delivering; check the pump itself
- Either suction or delivery valves are not properly opened or choked internally.

Liquid to be filtered has become very viscous due to ambient condition (severe winter, rain water falling directly on the piping); or interrupted flow of heating medium for the piping). This can occur for furnace oil, concentrated solutions with suspended solids, liquid sulphur etc. where electrical heating, steam jacketing etc. are used to keep the liquid at low viscosity.

It is also possible that sticky/oily material gets deposited on the filter cloth and chokes the pores in a very short time. Appropriate filter aids may be added to the incoming liquid. This can form a porous filter cake through which the liquid can continue to flow; but suspended solids are not able to pass through.

Cleaning
- The filter press is to be cleaned at regular intervals so that downstream operations are not interrupted due to less availability of filtered liquor.
- The residual liquid is to be flushed out by process water/Demineralised water in case of special quality required for the filtered liquid (process requirement) or by compressed air depending on the properties of the liquid and cake material. *This minimizes the hold up of liquid in the unit and hence can reduce the loss during cleaning*
- The unit is now to be isolated from input and output sides; and the supply pump stopped [fuses may be removed from power supply line to the motor]
- Now open the filter press by loosening and slowly rotating the main screw which will retract the central shaft. Hand trolleys or a tray of sufficient length and breadth shall be kept below the (filter press) plate assembly before it is opened. The accumulated cake is to be carefully dropped into the tray; and the material sticking to the filter cloth is to be slowly scraped by wooden scrapers. This is to prevent damage to the filter cloth. The cloth is to be removed from individual plates and cleaned by dipping in water. [Avoid high pressure water jets as they can make holes in the cloth pieces] The narrow passages in the plates and the grooves shall be carefully brushed and cleaned.

- Inspect and gently wash the cloth pieces by passing clean water through them. Replace the torn/punctured cloth pieces by new ones.
- Check the gaskets at individual entry and exit point of each filter plate; replace by new ones obtained from Original Manufacturer of the F.P.
- The operating screw for the pusher shaft is to be cleaned and lubricated (greased) as per recommendation by OEM.
- The basin below the F.P. is to be thoroughly cleaned after the job is over (preferably after a few hours of restart since there could be some minor leaks from gaskets. The sludge collection tray may be allowed to remain in the basin).
- The pressure gauge shall be recalibrated if required. The inlet and exit valves, connecting pipelines for incoming (dirty) liquor and outgoing (clean) filtered liquor shall be flushed of all internal deposits.

A record shall be kept as follows:

Dates of cleaning	Flow rate of exit liquid	Analysis of exit liquid	Total amount of dirty liquid processed by the FP	Maintenance jobs done

Total amount of sludge removed since last date of cleaning	Spares changed and their cost	Filter cloth pieces changed	Specifications of original pieces MOC, Gauge Number of openings per unit area.	Specification of new pieces Gauge & MOC	Maximum pressure and temperature permitted [Tearing strength of cloth]

Important precautions

(High pressure) release valve shall be provided in the liquid inlet line to the filter press. An alarm & a rupture disc may be considered. High temperature alarm (optional) may also be provided.

Note

If the frequency of cleaning of the filter press is more than once in 2–3 days either another filter press may be installed or the area for filtration may be increased by providing more plates. It is also possible that too fine gauge filter cloth may have been used. Plant manager shall examine if some relaxation in the gauge can be allowed i.e. cloth with a little bigger opening can be used.

6.8   Pressure Leaf Filter- (Example: For Liquid Sulphur)

Dirty liquid (to be filtered) is passed through a number of metallic filter leaves assembled (mounted) on a common header for collecting the filtered liquid. The enclosing vessel is steam jacketed when operating for liquid sulphur.

The individual leaves, liquid distributor for incoming dirty liquid, external jacket of the vessel and the collecting header are checked before start. Feed pump discharge for dirty liquid is also checked by running through a recycle line. Safety valve on steam supply line is set suitably to keep it hot (to prevent solidification inside). The assembly of filter leaves is pulled out from the enclosing vessel after accumulation of filter cake inside. The assembly of filter leaves is pulled out from the enclosing vessel after accumulation of filter cake inside (Please see the photograph Pressure Leaf Filter).

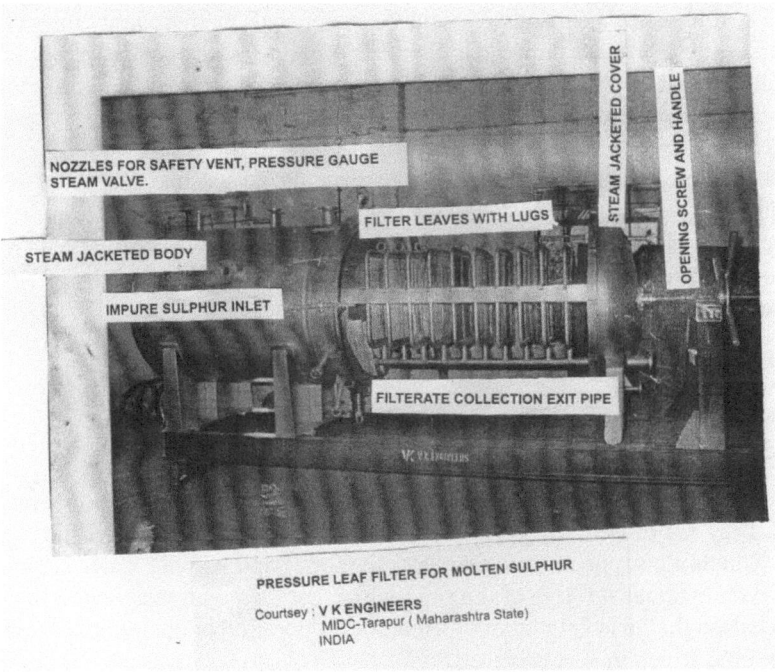

Photograph of a Pressure leaf filter for molten sulphur

6.8.1   Observation During Running

More particulate matter in filterate…indicates damaged filter leaf; displaced leaf from header, leaking gasket where filter leaf is fitted…

Excessive pressure at inlet and low rate of filterate–indicates unit needs to be opened and cleaned. Also check individual filter leaves and liquid (inlet/exit) piping.

6.8.2 Maintenance Work

- Stop pump for feeding impure liquid.
- Remove fuses from power supply to motor.
- Drain out residual liquid back into the feed tank for impure liquid.
- Flush out by steam pressure/compressed air
- Remove accumulated sludge in the vessel and the cake sticking to filter leaves carefully without scratching the leave.
- Lift the leaves carefully by holding the lifting lugs and clean them by suspending in alkali solution [cold/warm] or in liquid sulphur bath as per OEM advice.
- Repair the damaged leaf by replacing the leaking screen entirely.
- Change the side packings at the slots where the leaves are to be fixed.
- Change the Teflon bush for fixing the leaf into the header for carrying away filtered liquid.
- The steam jacketed (main body) and the cover shall be thoroughly inspected, & pressure tested at 1.5 times the operated pressure. Corroded \ leaking portion shall be repaired by welding appropriate patches. Take care to neutralise the free acidity in the dirty liquid by adding suitable filter aids or neutralising chemicals. This can minimise the corrosion.
- Opening mechanism for the cover shall be lubricated and inspected for any damage.
- Check internal liquid distributor for proper flow between all leaves.
- Open and clean the steam traps if required.
- Check setting of safety valve on steam supply line.

6.9 Trombone Coolers

These are rows and columns of pipes arranged in parallel and occupy large floor space. They are evaporative type coolers where in hot fluid is passed through the pipes (which are supported on structural supports) and cooling water is generally sprayed on external surfaces of the rows where part of it gets evaporated [hence an induced drought fan is not required]. Heat for evaporation of the water is taken from the hot fluid which thus gets cooled.

- A water pump sprays water on the pipes.

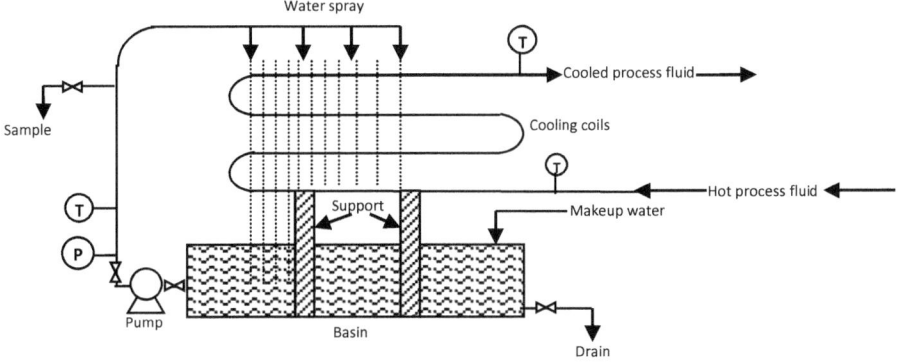

Figure for Trombone cooler

6.9.1   *Observations for Trombone Cooler*

- Working of water pumps (discharge pressure and flow by visual observation)
- water spray on each pipe in the bank
- Temperature of hot fluid at inlet and exit
- pH of cooling water.
- Fresh addition of cooling water is done through a float operated valve in the incoming water line to the basin. [hence basins need to be cleaned periodically]
- External surfaces of pipes can be easily cleaned by scraping or brushing wherever deposits are seen on the cooling surface.
- The liquid to be cooled may deposit some scales inside the pipes which are to be cleaned during suitable shutdown. Heat transfer efficiency of the Trombone Cooler can be maintained by removing deposit of internal scales.
- This can be done by filtering the hot fluid properly before admitting in the pipes or by controlling its composition.

The heat transfer surface can be increased by adding some more columns to the system. More connecting nozzles shall be already provided to the main incoming and outgoing headers (these are to be kept closed by blinds initially).

6.9.2   *Maintenance: (Check and Attend)*

- Water spray nozzles.
- Cooling water pump
- Strainer at suction of the water pump.
- Clean external surface of the cooling pipe. Remove deposits regularly.

- Drain out water from basin if dissolved solids content concentrations reaches 4000 ppm.
- Add fresh treated water (with low TDS content) continuously through a float controlled make up water inlet valve.
- Clean basin of accumulated sludge once in 90 days or earlier as per need.
- Provide rust resistant paint on structural supports.
- Generally protective GI/plastic sheets are provided around the pipes to protect persons around the pipes (in case of a leak from pipes).
- Test the pipes by hydraulic pressure during annual shut down and change leaking pipes, gaskets, etc.
- After satisfactory test drain out the entire water inside the pipes and pass dry air through it for removing last stresses of water (as it can form corrosive acids during operation).
- Enclose the position of supports (which remains dipped in water) in cement concrete with external lining of acid resistant tiles.

6.10   Shell and Tube Heat Exchangers

These are designed and constructed after considering the duty condition (heating and cooling of process streams), corrosion resistance of the tubes, possibility of fouling due to deposits of dust/scales, and providing additional heat transfer surfaces as margin. The number of passes is decided as per fluid velocity required to ensure proper heat transfer coefficient, permissible pressure drops, need for sub-cooling of condensate etc.

However it is generally not possible to increase the heat transfer area once the tubes are provided. Some designers provide extra tubes in the beginning itself to care of fouling later on. These can reduce the fluid velocity inside the tubes. In such case of over designed units some tubes can be kept blocked at both ends initially (to get higher fluid velocity) and can be opened later on. Alternatively, bypass piping can be provided for the heat exchanger.

These units are designed and installed in horizontal, inclined or vertical positions as per need for draining out the fluids.

6.10.1   Allowable Pressure Drop

- **Minimise the cost** of pumping liquids/blowing gases If too low values of pressdrop are specified, there will be need for larger cross sections. This can result in low fluid velocities; causing low Reynolds number (consequently low heat transfer coefficient) which will need larger heat exchanger size.)
- *Hence optimise the pressure drops and ensure satisfactory heat transfer coefficients*

- **Margin for fouling of heat transfer surfaces**: Adequate fouling factors shall be considered (depending on fluid properties, suspended solids, presence of corrosive gases.).

6.10.2   Materials of Construction–Typical

- Shell: Mild Steel. (Indian Standard 2062) or *its international equivalent* is may be considered.
- Tubes: A-179 Boiler quality or equivalent grade for high temperature service. Other tube materials (BS 3059 CDS, A-106) can also be considered and appropriate one for the given application shall be chosen.
- Tube sheets: M. S. (I.S. 2002 Gr. 'A') Boiler quality or equivalent international grade.
- Special MOC is to be used as per compatibility with the fluids at the operating conditions. e.g. Hastelloy C-276 for strong Sulphuric acid/Oleum etc.
- Protective layer of acid resistant or heat resistant bricks may be put on the tube sheets and bottom plates if there are chances of condensation of acid.
- Ceramic/Stainless Steel AISI-304/Stainless Steel AISI-316 Ferrules may be put in tubes and fixed by fire resistant/acid resistant cement.
- Use Stainless steels/Alloy steels as MOC for special applications

Observations
- Temperatures and pressures of hot side and cold side during running of the plant.
- Any build up of pressure drop on tube side/shell side
- Regularly check if there is any acidic condensate inside by opening the drain valves (shell side and tube side).
- Calculate the heat balance to estimate the efficiency of heat transfer.

6.10.3   Maintenance: Check the Following Thoroughly

- Isolate from process plant—put blinds in connected fluid lines
- Drain out safely all remaining process liquid
- Flush out by water/air/Nitrogen (connect exit gases to ETP or scrubber as per need)
- Expansion bellows and the Shell also (when operating temperature is more) multiple pass heat exchanger
- Check shell of multiple pass heat exchanger
- **Tie rods**: replace damaged tie rods for mechanical strength
- **Orientation of inlet and outlet nozzles** (on shell side). Consider possibility of short-circuiting if some baffles are leaking; repair or replace them immediately.

- 100% **Radiography** for welded joints shall be asked for after repairs (when the heat exchanger is to be used for high pressure, high temperature or dangerous fluids, e.g. *petroleum processing).This is to be followed by* **Pressure Tests**
- Clean inside of tubes by brush after opening end cover (plates).
- Check and repair the partition plate at the end covers.

Complete dismantling of the unit (like removal of the tubes) is not necessary for such cleaning.

The unit is thereafter pressure tested to check for any tube leak. The leaking tube can be plugged at both ends in a short time. If more tubes are found to be leaking they can be replaced by new ones.

Any leak from the shell can be suitably welded.

However it is advisable not to operate it at original pressure (if it is to be used for dangerous fluids)

- Chemical cleaning to be done for shell side/difficult to remove scales while adding corrosion inhibitor (rodine) to circulating 3% HCl as descalant. Sulphamic acid can also be used after testing on tube and shell MOC.
- Carry out pressure test at 1.5 times the working pressure. Identify leaking tubes. Plug from both sides OR retube the unit if too many tubes are leaking. Flush and check again. Check condition of baffles and tie rods from cleaning manholes
- Dry out all remaining moisture.
- Restore any insulation and cladding material that was removed earlier
- Connect to process units as before and start the plant.

6.11   Plate Heat Exchangers

These are constructed by assembling a number of parallel plates with narrow grooves (passages) for flow of the liquids. The liquids flow at high velocity when they pass through the grooves; and the heat transfer co-efficient increase as a result. It is necessary to prevent deposits in these grooves and hence strainers must be installed in the incoming liquid lines to the PHE.

Openings are provided at upper and lower corners wherein there are special gaskets between the plates at the openings. When the assembly of the plates is tightened a continuous passage gets created for the incoming and outgoing [hot and cold fluids]

Observations
Pressures of fluids (both hot and cold) and the temperatures at inlet and outlet.

Pressure drops on hot and cold fluid side as compared to the pressure drop when the PHE was initially commissioned and was clean.

High pressure drop indicates obstruction to the flow of liquid which could be due to deposits in the grooves. This is to be considered along with less cooling of hot fluid.

The PHE may have to be backwashed in such situation (under expert supervision from OEM engineer) for removal of the deposits. If this does not improve the performance, then the assembly of the plates of the PHE may have to be opened, cleaned and reassembled. Special care is required for taking out the gaskets and fixing them back again. The adhesive chemical and new gaskets shall be procured from OEM only [Example; Concentrated 98% sulphuric acid is cooled from 70 °C to 60 °C by a PHE constructed from Hastelloy C- 276 plates and Viton gaskets.

6.12 Condensers for Volatile Materials

6.12.1 Observation during Operation of Incoming Vapors/ Gaseous Streams

- Incoming stream:- temperature, pressure, flow rate, concentration of condensable vapours, and *presence of non-condensable gases (to be let out from vents connected to suitable scrubbers*
- Outgoing stream– Flow of Condensed liquid, vapors leaving from vent.
- Smooth, leak proof operation of condensate exit valves and valves for control of Reflux.

Observation during operation for cooling medium:
- Cooling medium- Flow at inlet. The flow at outlet is to be observed through an open funnel if it is not possible to observe or measure the flow at inlet. The temperatures at inlet and outlet must be observed to know about the heat transfer taking place.

    Outgoing condensate stream – *The flow of condensate shall be visible in the sight glass provided in the liquid exit lines.*

6.12.2 General

- Work out heat balance for the unit i.e. heat removed by cooling medium shall be compared with design value for the rated capacity of condensation (when the incoming stream is at the design temperature).
- Alarm for (high) pressure build up in the unit in case condensation is not taking place properly – *this can be due to excess input of vapours; less flow of cooling medium or high temperature of cooling medium at inlet.*
- Check conditions of structural support (whether any corrosion has taken place). *There can be a hold up of the condensed liquid inside the condenser if the outlet valve is closed or exit line is choked. In such situation the load on the support increases considerably.*
- **Shell** – Any leak/escape of fumes; (thickness of shell to be measured by Ultra sonic tester)

6.12.3 Vertical Unit

Visual checks
- Any leakage/broken nozzles for incoming and outgoing vapour/cooling water etc.
- Lug supports
- Vent valves and end plates
- Vertical Supports
- Earth connection
- U–Seals at exit liquid lines
- Safety valves and connecting nozzles, pressure taps and temperature sensor pockets.
- Remove all residual liquids.
- Clean tube side.
- Vent/flush out vapours.
- Check pressure taps and thermowells.
- All valves (for smooth operation)
- External insulation if chilled water is circulated on shell side.
- Chemical cleaning of shell side.
- Brush cleaning of tube side.
- Pressure test of shell and tubes separately.
- Baffle end covers of multipass units.
- Pockets for temperature sensors and pressure taps.

6.12.4 Condenser (Horizontal Units)

Check the following:

- Supports
- Pad plates
- Drain points
- End plates
- Wall thickness, pressure taps and temperature sensor pockets
- Baffles, gaskets/sealing at baffles.
- Any corrosion due to accumulated condensate.
- Pockets for temperature sensors and pressure taps.

6.13 Rotary Dryers

Check the following and attend where necessary:
- Condition of outer shell,
- Speed control for drive mechanism. Drive mechanism can be through spur and girth gear; friction drive; Consult design engineer/OEM for the drive.

- Inside refractory lining, and internal flights. *Take care to repair the refractory lining very carefully and replace the key brick. Cure carefully after this.*
- Feed control and charging chute,
- Rate of feeding wet material and rate of exit of dry material to be checked. A material balance is to be made for confirming that there is no excessive hold up of material taking place in the unit during steady state run.
- Position of burner/inlet point of hot air;
- Temperature of gases at exit point
- Working of cyclone and ID fan at exit (speed control for ID fan).
- Arrangement for removal of dried product (heat recovery from hot product if it is counter current operation)
- Co-current operation is done when the product can get affected at high temperature.
- 90° or lip flights arrangement for tumbling motion of solids (being dried) inside/ hanging chains or weir dams to increase residence time inside.

6.14   Tray Dryers

Check the following and attend where necessary:
- External jacket.
- ID/FD fans, as specified for inside operating pressure.
- Exit duct condition
- Steam pressure regulator/hot air temperature regulator.
- Condition of each tray for loading wet material.
- Loading trolley (which is pushed in and taken out), supports for trays.
- Internal heating jackets/heating stages through which hot water or low pressure steam is circulated for heat sensitive products.
- Ensure proper distribution/circulation of hot air inside and continuous removal from the top of the unit by ID Fan.
- Check the composition and rate of loading of wet feed material.
- Check the composition and rate of exit of dry material.

6.15   Hot Gas Filter

These are generally used in sulphuric acid plant to filter the hot gases to prevent deposition of dust on surface of catalyst in first pass of converter. They can also be considered for filtering dust laden hot gases in other plants. The filtered hot gases can be used for pre-heating of air. This can reduce deposits on heat transfer surfaces of the heat exchanger tubes. The feasibility of incorporating a HGF shall be examined for the individual plant.

Dust particles present in the incoming hot gas are arrested by the graded layers of refractory material made from crushed and screened high alumina bricks.

6.15.1 Observations to Be Made During Operation of the Plant

Pressure drop across the unit at start.

Pressure drop across the unit after every week or earlier.

There should be gradual rise in the pressure drop indicating arrest (deposition) of dust particles. This also results in increased power consumption by the air blower supplying air to the plant.

After a few weeks (could be 12–16 weeks) the pressure drop may become excessive and the unit needs to be cleaned. Either standby gas filter is to be taken into use or, the running unit is to be stopped for screening of the filter media (crushed and graded refractory layers. The material is taken out and replaced by fresh clean material.

6.15.2 Maintenance

During annual shutdown cool the unit and isolate from inlet and exit sides. Take out the entire graded refractory layers and screen them. Inspect the support columns, gas distribution plate and the grids for any corrosion, damage or weak portions. Repair/replace as required. Carefully inspect refractory lining at bottom, side walls and inside top roof. Repair by providing new bricks or castable refractory suitable for operating conditions (high temperatures, acidic gases). Check thermowells and thermocouples; pressure taps at gas inlet and exit.

Thoroughly clean support grid for the refractory layers. There shall be no gap through which some of the material can fall through.

Close these gaps by stainless steel flats or replace the entire damaged section of the grid.

6.16   Multiple Effect Evaporators

- These are used for concentrating dilute solutions while minimising steam consumption. The vapours from an evaporator stage are used as heating medium for the next stage—which is operated at a lower pressure than the previous stage;
- Vapours from the second evaporator stage are used as heating medium for the next (third) stage which is operated at a still lower pressure and so on.

- These can also be operated by using the exit steam from (steam turbines) as heating medium instead of condensing it. This *can increase overall efficiency of the plant since the latent heat of the exit steam from the steam turbine is used for heating in the process plant.*

Some typical applications of MEE are as given below:-.

1. During the manufacture of Viscose Rayon:- Concentration of spent spin bath (for subsequent reuse in spinning machines and for recovery of Sodium Sulphate *as by product*)
2. Sugar Industry:- for concentration of Sugar cane juice.
3. Distilleries & Breweries:- for concentration of dilute alcohol
4. Textile Industry- for recovery of caustic from dilute stream.
5. Paper & Pulp Industry:- for concentration of waste liquors
6. Chemical Industries:-to minimise quantity of effluents/waste water generated
7. Desalination units:- To obtain potable water from salty water

6.16.1   Important Components of a Multiple Effect Evaporation System

- Heat exchanger with vertical tubes (called *Calendria*)

The calendria has a number of tubes which provide the required heat transfer surface. The liquor to be heated (generally to be concentrated) is circulated through the tubes while steam or vapour from previous effect condenses is on the outside of the tubes.

- Separation space for separating vapour from liquid
- A suitable demister is also provided to separate particles of the liquor from the vapour.
- Arrangement for incoming and outgoing liquid
- A circulation pump for the circulation of liquid through the tubes; which is usually located near the lower end of the tube bundle. Liquor from the flash chamber goes to the suction of the pump (in a flooded suction manner). The heated liquor leaving the top of the tubes flashes into vapour which leaves the unit from the vapour exit line.
- Inlet for steam/vapour from previous stage which is used as heating medium
- Control valves for flow of liquid, steam
- Steam traps for removal of condensate
- Sampling points for condensate from each effect
- Collection header for condensate
- Instrumentation for monitoring pressures, temperatures, flow rates

6.16.2   Observations to Be Made During Operation

- Flow rates, and temperatures of incoming (dilute) and outgoing (concentrated) solution
- Concentration of dissolved and suspended solid content of incoming and outgoing liquids
- Level of liquid in the separation space (chamber)
- Steam (vapour) pressure and temperature used for heating in every stage
- Operating pressure of each stage
- Current drawn by motor driving the circulating pump
- Quantity and analysis of condensate from each stage
- The condensate from each effect is to be analysed for pH, free acidity (if any), conductivity (depends on carryover of droplets of liquid). This is to confirm whether the condensate can be recycled to boiler or needs further treatment before reuse. *Careful operation of the MEE can enable reuse of the condensate for boiler feed, for process use...*
- *Separation/precipitation of dissolved salts are to be avoided during operation as they can choke the tubes. Operating temperatures must be chosen according to the solubility of the salts at different temperatures in the solvent.*

6.16.3   Maintenance Procedure

Isolation and Cleaning
- Stop feeding the fresh incoming dilute solution
- stop heating the unit.
- Close discharge (exit) valve for concentrated solution.
- Break the vacuum inside by opening vent carefully
- Drain out the inside liquid and transfer to process tanks.
- Isolate from all sides [incoming and outgoing vapour and liquid connections] The heat transfer (*calendria*) tubes shall be cleaned to remove all deposits of salts and then pressure tested @1.5 times maximum pressure of incoming steam.
- All joints with tube sheets and the projection of tubes beyond tube sheets must be checked during maintenance jobs; and should be as per OEM advice/original drawings.
- Clean the Calendria tubes by brush; check for any leaking tube and replace it.
- Inspect the tube sheet and attend.
- Clean the shell side of Calendria.

6.16.4   Items to Be Inspected and Attended

- Check the parts of circulation pump for corrosion, wearing out [shaft, impeller, gland, wearing rings] coupling with motor,
- Electric motor for pump: inspect cooling fan, fins on external shell, junction box, resistance between windings, and with casing, bearings etc. of the.
- Clean view glasses and light glasses.
- Examine the vapour –liquid separator for any corrosion or erosion. Take out the demister pad/candles and immerse in clean water followed by dilute alkali (NaOH) solution. [Do not use lime water as this can form deposits of insoluble calcium salts]. Again wash gently with fresh water.
- Do not use high pressure jets of water on demister pads.
- Clean and inspect thoroughly the flash chamber and vapour separator. Attend to any damaged portion of protective lining (fibreglass/rubber) and test at 5000 volts to make sure the lining has been properly attended.
- Check pressure taps; calibrate pressure gauges and vacuum gauges.
- Inspect thermo wells and thermocouples; dial thermometers.
- Inspect and overhaul the Steam Pressure Reducing/Controlling System components such as strainers, safety valve. Reset them at desired pressure if required.
- Open, inspect, and clean the heat exchanger section tubes gently by brush or as per advice from OEM or designer of the unit.
- Repair/replace the tubes as per OEM advice..
- Inspect, clean and repair the condensate recovery lines.
- Open, inspect and overhaul steam traps on individual stages (special type of Steam Traps are fitted to the MEE since it operates at progressively lower pressures)
- Carry out pressure testing of Calendria Tubes and Main flash vessel before restarting the unit.

6.17   Water Treatment Plants

Water is used for various purposes in chemical industries. The raw water shall be treated before use in the plant. The treatment shall be according to the quality specified by plant designers for different purpose in the process and utility sections of the industries.

- Process water (for making solutions), for cleaning process vessels
- Soft water for feeding to low pressure boilers and make up to cooling towers.
- Demineralised water for feeding to high pressure (more than 20 kg/Cm^2) boilers
- Very pure, distilled water for manufacture of medicines and laboratory work.
- Waste water (if found non-corrosive and suitable) can be used for toilets, floor washing and fire fighting.

- Steam Condensate from the heating of process units is generally fully recovered and reused as feed water for boilers. *It can be reused for process plant after treating for any free acidity, dissolved impurities, if any..*

Facilities required for use of water in the process plant generally consist of raw water storage, treatment plants of various designs and storages for treated water. All these facilities need to be maintained in proper working conditions *even when the production units are not working* because water is required for removing residual acids, for washing and thorough cleaning of process units, for making alkali and salt solutions etc. for undertaking maintenance work and plant start up afterwards.

6.17.1   Main Components of a Typical Water Treatment Plant

The main components of a typical water treatment plant are given below. All of them may not be required for the plant.(depending on quality of raw water):

- Raw water storage tanks
- Transfer pumps for raw water and treated water
- Settling Tanks.
- Sludge settler
- Water softeners with (Sodium exchange) resins in RL (rubber lined) vessels.
- De-mineraliser (DM) Plant with Cation and Anion exchange resins.
- Air Blowers for Degassing Towers.
- Alum and Polyelectrolyte dissolvers and Dosing Systems for Common Salt (NaCl), Alkali (NaOH), Acid (HCl)
- Pressure Sand Filter and Active Carbon Filter
- Instrumentation [TDS, DO, pH; Conductivity Meters]
- Reverse Osmosis Units with high pressure pumps and special membranes.
- Ultra filtration Units.
- Chlorination systems, Flocculator,
- Special piping made from CPVC, Stainless Steel, PP, HDPE, GI, CI etc. as per application.

6.17.2   Observations to Be Made During Operation

- Levels in all tanks.
- Working of all pumps, dissolvers, mixers.
- Analysis of raw and treated water at every stage. (TDS, pH, DO Concentration, turbidity etc.)
- Analysis of reject and back wash waters from DM plant, Soft water plant
- Analysis of Cooling Tower Bleed water
- Waste water from Process Reactors & other units.

6.17.3   *Maintenance*

Build up sufficient stock of treated water for all required purpose beforehand.

Backwash is to be given to the PSF and ACF when the pressure drop in the units build up to such an extent that the output of treated water is just a little more of requirement. This should be preferably be done in day shifts.

For complete cleaning-

Keep a stock of fresh clean sand ready for the Sand Filter which should be sufficient to replace the entire quantity to be taken out for cleaning. This will save considerable time for restart of the unit. Similarly, a sufficient stock of fresh activated carbon may be kept ready if possible. This can save time for cleaning and regeneration of the exhausted activated carbon (which will be taken out from the ACF.)

- Check the support plates for fixing distribution nozzles.
- Examine and replace the damaged water spray nozzles if any.
- Confirm that the valves in water inlet and exit lines as well as for flow reversal *are not leaking;* and can be operated easily.
- Examine and attend to any weak supports, ladder for climbing up the unit,
- Take samples from the sand and activated carbon taken out and analyse in laboratory.
- Check the internal lining/anti-corrosive painting and the shell for any weak spots.
- Resins: take out the entire resins and wash them by filtered water and then by treated water. Check support plates, distribution nozzles, and rubber lining inside the vessel.
- Reactivate them by dil. Acid and alkali and test in lab. Replace by new resins *up to original quantity* if the activity cannot be restored satisfactorily. Preserve samples of new resins for future reference.

6.18   Sludge Settler

Used for raw water treatment after addition of flocculating chemicals and thereafter allowing settling of sludge. The clarified water over flows into the launder at the periphery and is sent for further treatment. The sludge settles in the bottom of the conical portion. It is then taken out by sludge transfer pump installed externally. A recycle line is provided from discharge side of this pump to its suction side to prevent choking of the suction. In certain plants a compressed air line is provided to dislodge the sludge from the suction (chamber). The slow moving (rotating) rake mechanism serves to slowly push the settled sludge towards the suction "chamber" of the pump. Arrangement is available to lift up the rake mechanism and make it "free" if it gets stuck up in the settled sludge. Sufficient stock of treated water should be built up before stopping the settling and cleaning the unit.

6.19   Air Pollution Control System

Gases exiting from chemical industries contain various types of pollutants like dust (from grinding/pulverising operations) acidic gases (SO_2, HCl) and acid mist particles from reactors and absorbers; un-burnt fuel and carbon particles, fine ash. The common devices used for controlling their escape to the atmosphere are Impingement separators, Cyclone separators, Venturi Scrubbers, Packed Towers, Demisters, Electro Static Precipitators etc.

A study is made for each plant to estimate the likely pollutant load i.e. amount and concentration in gases; and quantity of gases to be treated; while taking into account the operating temperatures, pressures, moisture content are for designing and selecting a suitable Air Pollution control system which has to meet statutory regulations with latest amendments.

6.19.1   Observation on Individual APC Units

- Pressure drop in each unit; as well as actual pressure at inlet and exit.
- Temperature of gases and external surface of shell/enclosure.
- Operation of dust removal mechanism (e.g-rotary valve at bottom of cyclone, Dry ESP).
- Flow rate, pressure, temperature and analysis of gases at venturi, packed tower etc.
- Operation of scrubbing liquor pumps (check standby pump also)
- flow of scrubbing liquid through all units like venturi, packed tower.
- pH and flow of the scrubbing liquor at inlet of scrubbing units
- Alkali solution tank, valves, working of alkali addition pump to scrubbing units
- Circulation tanks for Venturi and Packed tower
- Confirm smooth operation of all valves (at pump inlet/outlet, sampling line, drain valves for circulation etc.)
- Working of pump for addition of 10—15% alkali solution to scrubbing liquor in the circulation tanks.
- Current drawn by circulation pumps and flow rate of purged (spent) liquor to ETP.
- Confirm smooth operation of all valves (at pump inlet/outlet, sampling line, drain valves for circulation etc.)
- Working of level indicator in all tanks.
- Connection of pressure taps to manometers/clean all pressure taps.
- Check operation of venturi throat valve and indication of flap.
- A set of removable nozzles shall be available for scrubbing liquor inlet to all units.
- Scrubbing liquor inlet pipes to all units shall be clean.

- Internal nozzles of liquor inlet shall be clean.
- Tower packing shall be clean and filled up as much as possible (up to about 150–200 mms below spray nozzles.
- Demister mesh pads must be securely fixed in spray/packed tower, polishing tower.
- pH controller: Dip the sensor in acidic solution (synthetic material made in a drum) and observe working of alkali addition valve regularly at defined frequency.
- Drain lines to effluent treatment (filter and recycle) and solid sludge to drying bed
- Arrangement to add make up water/recycled treated effluent
- *Gas analysis at inlet and exit of individual unit shall be done regularly at defined frequency determine scrubbing efficiency of each unit to pinpoint which unit needs immediate attention [if the analysis of exit gases from chimney indicate that the system is not able to meet pollution control norms set by Statutory Authorities.]*

6.19.2   Cyclones

These are provided to remove dust particles from gaseous streams. Incoming gases enter tangentially and form a vortex. The dust particles tend to settle down in the lower conical portion and cleaned gases leave from the central exit duet. The dimensions of the unit and the location of exit duct are designed on basis of gas volumes, dust content, particle size distribution, separation efficiency designed etc. Since there will be dust particles having size less the size considered for design, some particles will escape with exit gases. More than one cyclone are sometimes arranged in series to improve the system efficiency for separation of dust.

The settled dust can be removed from the lower (drain) nozzle by periodically opening the blind or by continuous operation of the Rotary Air Lock Valve. The exit duct is generally bolted on the top and is removable. Two/three exit ducts of different lengths are kept available.

Observations during operation
- Quantity of dust removed every hour and every shift.
- Screen analysis of the dust.
- Dust concentration in incoming and outgoing gases.
- Pressure drop across the unit.
- Condition of thermal insulation (if heat loss is to be minimised for downstream processing).
- Operation of Rotary Air Lock Valve; its drive mechanism and adjustment of rotary speed (if desired).

- The dust coming out shall be collected in trolleys and cooled by water spray (if very hot dust is coming out). The water vapours shall be sucked through a hood provided on the area and an ID fan.

Maintenance
- Isolate from upstream and downstream gas sides. Evacuate all residual gases by exhaust fans.
- Remove all accumulated (if any) dust from inside.
- Check condition of the shell & length of the exit duct. Check the thickness and attend weak spots.
- Provide new exit duct if the existing one has corroded or reduced in length/some of the incoming gases are directly going out through corroded, leaking parts.
- Carry out pressure test on the unit @1.5–2.0 times maximum operating gas pressure.
- Maintenance of drive mechanism shall be carried out as per schedule advised by OEM.
- Open removable exit pipe and clean thoroughly. Observe length inside and compare with original. This piece may be changed by another one if the performance during operation was not satisfactory. Also change any leaking ducts outside the cyclone
- Rotary Valve at bottom:- check condition of internal moving parts, drive chain/belt and motor. The height/length of discharge chute may be adjusted to ensure the discharged dust/ash should fall in trolley placed below. Inspect vanes of Rotary Airlock Valve and replace if corroded.
- Lubricate the (chain/belt) drive, gear box by using oil of recommended grade only.
- Examine support legs of the Cyclone and foundations.[whether the protective paint is alright or has peeled off].
- Check gas entry and exit dusts also.
- Check condition of dust collection (hand/trolleys/bigger trolleys which are towed away after getting filled up. In certain systems the ash is conveyed by steel belts.

6.19.3 *Impingement Separators*

- Most of the observations and maintenance activities are similar to that of Cyclones. The observation windows, cleaning manholes shall be opened to remove all (accumulated) dust and the impingement plates shall be thoroughly inspected for any corrosion or leaking spots. Check baffle plates for any holes

Important
- All residual gases are to be evacuated before opening any window. Flush out by injecting fresh air.

6.19.4 Bag Filters

Monitor the pressure drop and dust concentration in inlet and exit gases. Temperature of gases at inlet should not exceed the limit specified by OEM as the bags can get damaged at higher temperature (generally not more than 230–240 °C) Here incoming gases must be cooled by suitable means (atmospheric cooling duct) The dust deposited on the bags is shaken/dislodged by short blast of compressed air controlled by timer circuits. The duration and frequency of the blasts is adjusted to remove the accumulated dust.

It is important to monitor the moisture content in the incoming gases. Since deposition of acidic condensate (and acidic gases) can corrode the bags (hence choose the MOC of bags which can withstand such conditions)

Maintenance
Isolate the unit and commission standby unit if the plant is to be kept running. Vent out gases by connecting to an exhauster. Test for presence of any toxic, dangerous gases before opening the cleaning and maintenance manholes. Check all bags individually; change damaged ones.

Check working of timer for periodic back pressure of compressed air to clean the bags.

6.19.5 Ventury scrubber

Check condition of Spray nozzles at top. Adjustable throat Valve [damper, flap, shaft, operating mechanism], Pressure point connections at gas inlet and exit. The entire throat valve assembly can be changed by a new one if necessary. Check inside rubber lining (if provided).

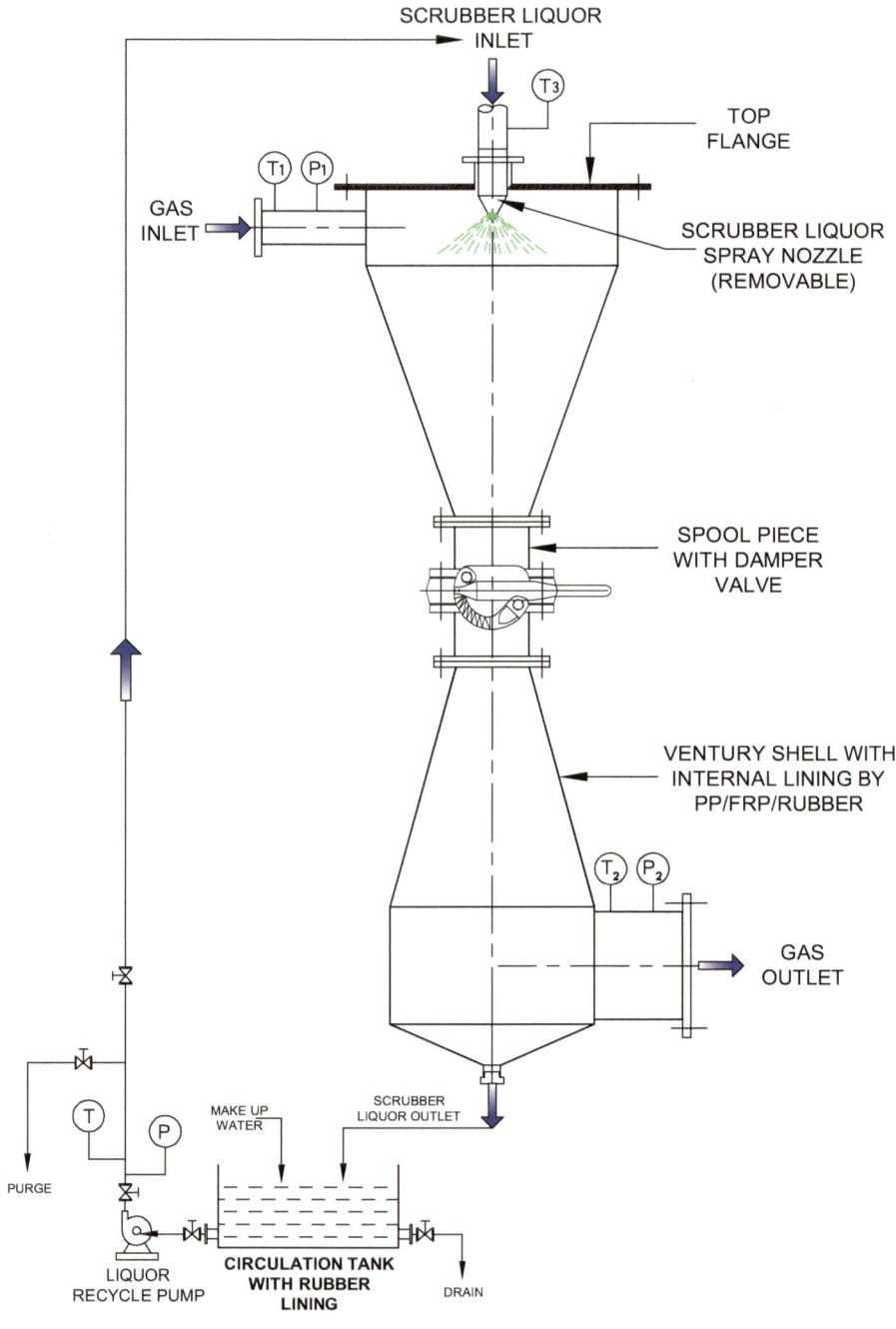

SCRUBBER LIQUOR
INLET

TOP
FLANGE

GAS
INLET

SCRUBBER LIQUOR
SPRAY NOZZLE
(REMOVABLE)

SPOOL PIECE
WITH DAMPER
VALVE

VENTURY SHELL WITH
INTERNAL LINING BY
PP/FRP/RUBBER

GAS
OUTLET

PURGE

MAKE UP
WATER

SCRUBBER
LIQUOR OUTLET

LIQUOR
RECYCLE PUMP

CIRCULATION TANK
WITH RUBBER
LINING

DRAIN

VENTURY SCRUBBER

Figure for Venturi Scrubber

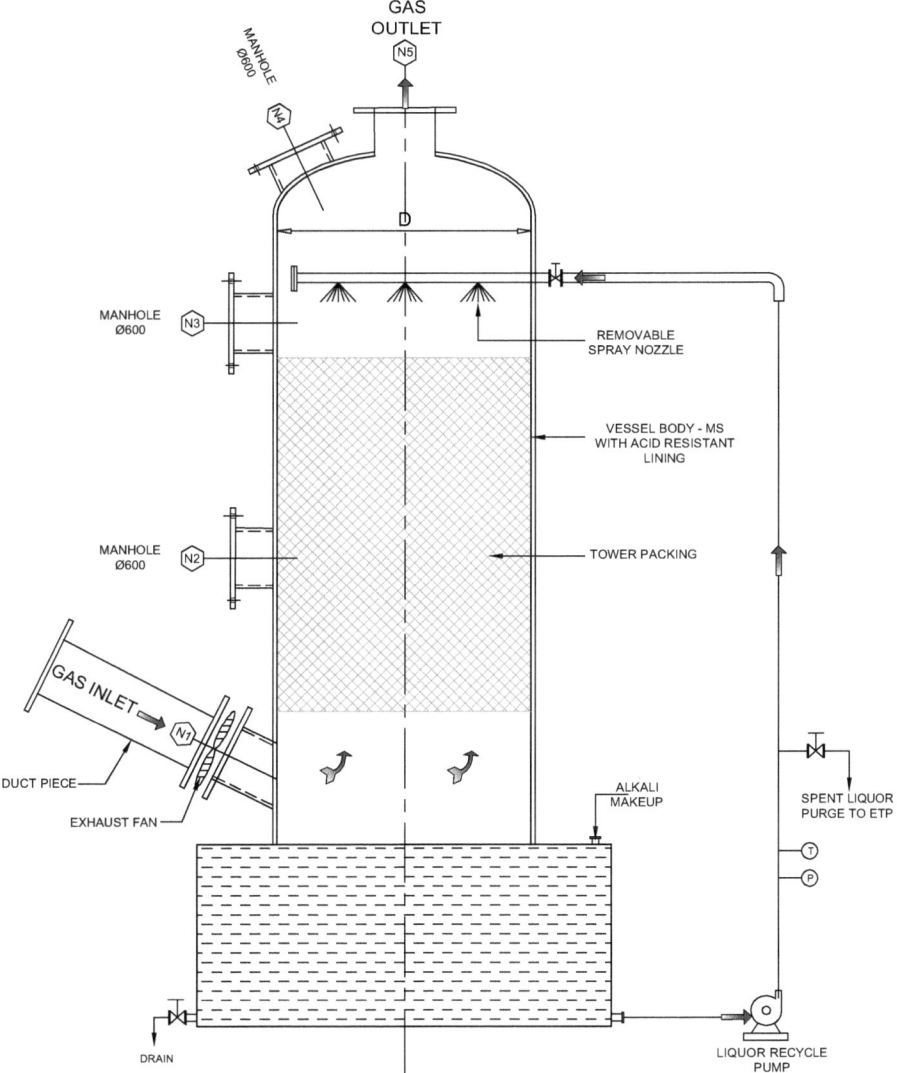

PACKED SCRUBBING TOWER

Figure for Packed Scrubbing Tower

6.19.6   Packed Scrubbing Tower

- Tower packings – Clean and replace damaged pieces. Remove one duct piece from gas inlet line and provide exhaust fan. It will suck fresh air from top and persons working inside will find it convenient.

- Check support for packings.
- Demister supports and pads. If the pads are tied together by stainless steel flats or wires then these are to be checked. Pads are to be washed gently by water (not by pressure jet) or by dilute alkali/other chemicals as per OEM advice.

6.20   Induced Draft Fans (Rubber Lined)

- Visual inspection of internals and spark test to check integrity of all rubber lines/ epoxy coated parts or vinyl ester coated parts.
- Dynamic Balancing of impeller at operating speed.
- Remove all ash deposits and wash internally.
- Check shaft, bearings, conditions of impeller vanes etc.
- Check foundation bolts, alignment with drive motor and VFD control for motor

6.21   Electro Static Precipitators (ESP) – Dry/Wet Type

These are used to remove micron sized particles (from gas streams) which are very difficult to remove by Venturi and Packed Tower Scrubbers. These particles are charged by high voltage electrodes. The dust particles get charged and are removed by collecting electrodes.

- Clean ash deposits on tubes and electrodes.
- Check electrical connections and insulators, supports for electrodes.
- Alignment of electrodes.
- Gas inlet nozzle, Gas inlet distributor plate.
- Wash water/spray nozzles at top of the system.
- During operation, TDS and suspended ash particle concentration in water shall not be allowed to exceed 200–250 rpm or the limits specified by **OEM** [Original Equipment Manufacturer]
- Earth connection for body – check continuity and attend.
- Check damage to any insulated portion and connection of high voltage lines, since the units operate at high DC Voltage.
- Rubber lined portions to be checked by spark test @ 10,000 Volts or as per OEM advice
- Pressure points taps.
- Drain valve and drain lines.
- Electronic Circuits/Power Controller for electrodes.
- Check instrumentation,
- Alignment of discharge and collecting electrodes
- Get technical assistance from OEM if required.

Chapter 7
Maintenance of Common Machinery, Process Units and Equipments

7.1 Stationary or Static Equipments

7.1.1 Chimney

Check and attend the following for maintenance: The plant should be stopped and blind should be put in gas inlet connection to chimney.

- Civil foundation, any corrosion of foundation nuts and bolts.
- Side wall thickness by Ultrasonic tester and acid resistant brick lining inside.
- Open inspection **window** to observe internal condition of Conical portion (at lower base section)
- *The chimneys generally have acid resistant brick lining in lower conical/lower 20–30% portion (approximately); followed by rubber lined upper pieces.*
- Make sure there are no toxic/dangerous gases or corrosive condensate inside before entering for inspection. Confirm by gas detector.
- No inflammable/explosive gas shall be present if cutting or welding is to be done inside or on the outer shell. Blow in air to remove such gases and test by gas detectors. [Hand held instruments as well as chemical test papers/ solutions]
- Open drain valve at bottom to remove acidic condensate from the chimney
- Open inspection **manhole** (after draining out any accumulated liquor inside) and check inside lining of bricks, rubber lining.
- Lightening arrestor from top to bottom.
- Aviation warning lamps.
- Work platform and gas sampling points
- Any tilting/inclination of the chimney itself
- Guy ropes (any corrosion/damage) and fixing with ground anchors.
- External paint (apply fresh coat of corrosion resistant paint).
- Monkey ladder with protective cage.

© Springer Nature Switzerland AG 2019

K. R. Golwalkar, *Integrated Maintenance and Energy Management in the Chemical Industries*, https://doi.org/10.1007/978-3-030-32526-8_7

- Pressure tap and thermowell at inlet.
- Provide sacrificial electrode if required for protection against corrosion
- Height shall be **as per latest norms** prescribed by Statutory Authorities.
- Check and attend the specially designed foundation bolts and their grouting in the respective pockets (in case of self supported chimneys).
- For other chimneys: inspect the guy wire ropes tied at different levels of the chimney to the well grouted strong anchors.
- Protective tubes of polythene with an internal grease coating are put oven the wire ropes to prevent rusting.

7.1.2   *Steel Structural Members:* **Regularly Inspect and Attend Any Weak Spots**

- Weight of all units – empty/full as per present operating conditions and for future expansion of production facilities. Compare with original design values/permitted weight.
- Vibratory load due to compressor, grinder, agitated vessels.
- All members (connecting steel angles, flats) and any exposed steel rods coming out from concretised floors or walls.
- Provide additional I – beams supports immediately for extra load due to any reason.
- Remove vibrating machine and install elsewhere if possible.
- Do not stock bags of raw materials on the floor.
- If there are more process units, do not run them together.
- Operate at lower level of load.
- Arrest all spillages, inject waterproofing compounds.
- Arrange external cementing to cover exposed steel rods.
- Provide rust protective paint/phosphating/anti corrosive zinc rich paint.
- Provide additional reinforcing flats, I beams, angle irons for any weak portions. Provide new structural members at the earliest.
- Get the complete structure examined by licensed structural engineer and Statutory inspection agencies.
- Provide additional separate supports to gas ducts, liquid pipes, valves, ladders, walkways to minimise load on existing structural members.
- Grouted foundation bolts.
- Any rusting observed.
- Vibrations of structural supports specially when a vehicle is passing near by.
- Damage to welded parts OR at riveted or bolted joints.
- Any dislocation of cross members.

- Peeling off of paint.
- Any spillage/dripping liquid falling on supports and corroding them.

7.1.3 Elevated Storage Reservoirs (ESR)

- The capacity is generally designed to be sufficient to meet emergency requirement of critical units like scrubbing system, boiler feed, water cooled electrodes, cooling jackets of process units (which handle volatile dangerous liquids) for 30 min. to 1 hr. or more if required. The purpose is to provide water till DG sets are started and come on load or other arrangements are activated to provide water supply.

Observe during running and attend wherever required
- Support columns
- Level indicator – a continuous flow from overflow pipe confirms ESR is full.
- Strainer at water inlet (clean every 15 days or earlier. Fit properly if displaced. Change the screen if it is damaged)
- Working of all valves. (remove rust if present. Lubricate where required)
- Outlet valve should open when power outage occurs (check electrical circuit every 15 days and attend immediately. Check if water supply can reach all important units when this valve opens- take trial)

7.1.4 Vertical Storage Tanks

Observations during running of the plant
- I-Beams for supporting at bottom.
- Level indicator – external pointer (which is operated by an internal float) moving against a scale, transparent external tube).
- Whether the actual level checked by dip stick is correctly being indicated by the instruments?
- Sampling nozzle
- Walkway should not rest on roof anywhere.
- Overflow nozzles of adjacent tanks may be interconnected (if they are handling same liquid) to prevent spillage due to excess filling when receiving supply from tankers or from production units.
- Condition of dyke walled enclosure; collection of any liquid taking place
- Condition of civil foundation.

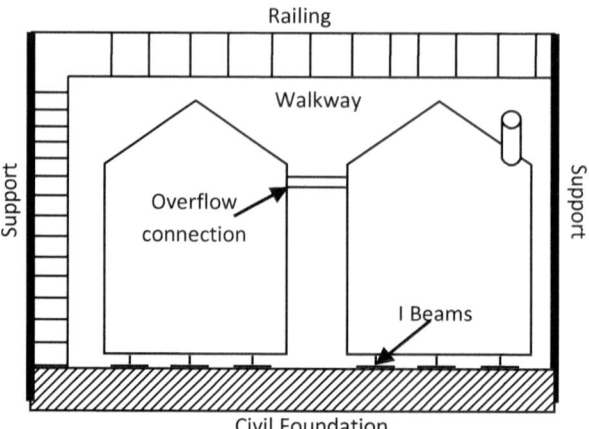

- Initial and final level every shift/day.
- Condition of supports for connecting piping.
- Temperature of shell.
- Pressure pH/acidity of fluid inside.
- Flow rates of incoming/outgoing fluid.
- Any leak of fluid from sides or bottom (check for leak from bottom plates)
- Working of valves for incoming and outgoing fluid stream.
- Release of vapour from vents.
- Ladders and Work platform, specially on top of the tanks.
- Moisture trap/Breather on Vent Lines.

Maintenance of tanks
- Empty out the contents by transferring to another tank/point of use and bring down the level to minimum.
- Flush out residual vapours by air/N_2. Connect the vents to scrubbing system through exhaust ID Fan.
- Open cleaning manhole after confirming no dangerous or vapour is remaining inside.
- Check wall thickness of side walls and bottom plates by Ultrasonic tester
- Acid resistant tile lining (two layers of 40 mm thick tiles with staggered joints at bottom; and a single layer on side wall up to a height of 500–1000 mm can be provided as an additional precaution)
- Clean portion below storage tank, bottom plate (between I beams) and inspect if any liquid is oozing out.
- Condition of breather and moisture absorber media inside. Change the absorbing media if necessary. No rain water, external dust should be able to enter the tank
- Use small pumps/buckets to take out remaining liquor and sludge from the manhole.

- Open bottom manhole and clean deposits of sludge by special anti-sparking shovels [aluminium or wooden spoons] Do not use sparking tools
- Confirm by gas detectors that it is safe to enter inside. Provide portable breathing apparatus to persons entering the tank. Arrange continuous supply of fresh air.
- Check earth connection for continuity.
- Collect oily sludge in drums and incinerate. Acidic sludge is to be taken out carefully, neutralised and send the sludge to authorised agency for land filling.
- Check thickness of walls, bottom plates, protective lining at $90°–180° – 270°–360°$ at least for 3–4 cross-sections which are 1.0 – metre apart or 25%–50%–75%–90% of height.
- Provide new patches at weak spots.
- Cut off entire weak plate if weak spots are more than 200 mm × 200 mm (for tanks operating at atmospheric pressure.)
- Check thickness of roof plates and reinforcements (flats, rings)
- Gussets at all nozzles.
- In case of any weakening of exit nozzles change the nozzle and provide new gussets.
- Check operation of all valves.
- In certain designs, an internal valve is provided to block the flow from outlet nozzle if a leak occurs between exit valve and storage tank. This internal valve is provided with a long operating stem and can be operated from top of the storage tank.

7.1.5   Horizontal Cylindrical Storage Tanks

- Disconnect (isolate) from all sides.
- Empty out completely.
- Wash to remove all residual material. The wash liquor should be sent to ETP.
- Measure thickness of tanks at various points,
- Confirm it is safe to enter inside; and also the smooth operation of all valves.
- Condition of breather and moisture absorber media inside. Change the absorbing media if necessary No rain water, external dust should be able to enter
- earth connection for continuity.
- condition of pad plates, bottom support plates and vent holes.
- water spray nozzles for storage tanks, which are used for storing volatile liquids.
- working of smoke detector and gas detectors at important locations around the tank.
- Calibrate pressure gauge, temperature gauge, safety valves.
- Confirm working of pressure release vent.
- External shed/protection from direct sunlight
- Check condition of external shed/protection from direct sunlight

7.1.6   Weigh Bridges

Observation and maintenance
- Clean top plate and remove rust.
- Ramp for incoming vehicles.
- Remove all dust, sludge, water from underground pit.
- Use dewatering pump.
- Check mechanical linkages underside of platform.
- Calibrate load cells.
- Place standard known weights and observe the accuracy of readings.
- Repair any damaged plates, reinforcing ribs, linkages.
- Check connectivity to computers.
- Obtain approval for calibration from statutory authority.

7.1.7   Ducts and Piping

Observation during operation

- Any leak from pipe or flange joints.
- Damage to support.
- Condition of external insulation
- Condition of expansion bellow.
- Pressure drop in pipeline (specially when they are long).

Visual checks for maintenance and repairs
- All valves and drain points in the pipeline.
- Condition of supports (specially rolling supports and any clamps provided on the pipeline)
- Any corrosion due to spillage of liquids from top.
- Any cracks in welded joints or escape of vapours.
- Operation of vent valves.
- Earth connections for pipelines and ducts carrying inflammable material.

7.1.8   Metering Tanks

- These are provided to ensure supply of correct quantity of products to the clients.
- The tanks can be calibrated in terms of capacity as 10MT, 12MT, 15MT, 20MT etc. by provision of exit nozzles corresponding to these capacities.
- The tanks will have overflow connections (with valves) at these capacities and will be connected by return lines to the main supply tank.

- The transfer pumps are run (generally alternatively) to fill the metering tank as per delivery of products to be made to clients.
- The valves at corresponding nozzles are kept open.
- Tankers are positioned below the metering tanks.
- After filling up the metering tank upto the required level (which can be confirmed by observing the flow in sight glass in return line) the transfer pump is stopped; and the tanker is filled up.
- Only correct quantity is filled up every time the metering tank is filled upto the mark for ultimately emptying into the tanker.
- Calibration of main supply tank shall be checked and confirmed that it corresponds correctly with overflow marks on metering tank.
- Metering tanks, shall be provided with a shed over them, or an insulation pad of mineral wool with a cladding to minimise effect of ambient temperature.
- Dyke walls shall be built around main storage tank.
- The metering tanks shall be installed on support columns in such a way that the drain (outlet) valve is at least one meter above filling point (inlet valve) of the tankers.
- Work platform shall be built for convenient positioning of the tankers and operation of valves.
- Clean at least once in three years, inspect thoroughly and repair weak spots

7.1.8.1 Advantages

- No spillage of product during filling of tankers as overfilling is avoided.
- Only correct quantity is delivered to client.
- A single inter connecting contact in return line can switch off the transfer pump as soon as the metering tank is filled up to required level.
- *Tankers of known standardised capacity (10–12–15–20 MT) shall be filled up and weighed on calibrated weigh bridge to confirm that the system delivers correct quantity every time.*

7.2 Electrical Items

A well designed and maintained Electrical System is essential for the safe and smooth working of chemical plants since many of them run continuously, and any sudden interruptions can result in development of dangerous situation as well as environmental pollution. The following guidelines will be useful for the electrical equipments to be used in the plant.

The considerations shall be for working out the requirement of power for erection, commissioning, operation, emergency loads, and efficient use of power generated in-house through heat recovery for captive use or export to external users.

7.2.1   Erection and Commissioning Activities (Including Site Fabrication)

- Site Fabrication of big process vessels and storage tanks, ducts, pipelines, structural supports, work platforms is generally done at site. Many welding sets run simultaneously during such activities
- Mechanical Trials of individual equipments/sections of plant- AC three phase
- Dry Run of Plant (most of the units are run as in case of normal production run of the plant—but without actually feeding raw materials)
- Project Commissioning
- Lighting load—AC single phase

7.2.2   Power Required for Operation

- Present plant operation at rated capacity-all mechanical drives, electrolytic cells, heating furnaces, lighting *(normal running load and total installed load which includes standby units)*, process control arrangements and automation devices
- For future capacity expansion/diversification

Maximum Demand MD
Permission for this is to be carefully applied for since any violation of the maximum demand even once in a day is generally heavily penalised by external power supply grid authorities. The MD should be minimised.

Other Power needs and balance with own generation
- Emergency power requirement for boiler feed water pumps, scrubbing systems of Air Pollution Control, fire fighting pumps, cooling water for reactor jackets and condensers handling volatile materials, lighting at important places …
- Balancing own generation by exporting excess power/obtain power from external source if own generation is not enough.
- Arranging automatic starting of Diesel generator sets and resuming emergency power supply to prevent mishap, environmental pollution (due to release of gaseous pollutants)
- System design to ensure (almost) unity Power Factor.
- Proper distribution of electrical load shall be arranged in all phases (for balanced load on transformers)

7.2.3   Equipments and Facilities to Be Installed

Isolation switchgear and Metering Panel with provision of instruments for measuring draw of power, KW,KVA, KWH, current drawn and maximum demand for power during every 30 minutes (with warning alarm) and safety arrangements as instructed by statutory authorities

Regular observations to be made

7.2.3.1   Metering Panel

- Incoming HT voltage, current, frequency (Hz)
- Any sparking at incoming power lines, visual damage to insulators, poles
- Total power consumption KWH per hour and per day
- Running load...KW, Current, Voltage, and maximum demand MD for power every 30 minutes. This must be monitored very carefully specially while taking trial of standby units (when the process is running normally) as it adds to the total load.

7.2.3.2   Step Down Transformer(s)

For reducing High Tension (11 KV/6.6 KV voltage at incoming power supply to Low Tension voltage 415–430 volts for supply to Motor Control Centres
In certain countries the supply is at 33 KV at input side; or HT motors are used which operate at 6.6 KV in special cases.

- present and future expected total electrical load for running the process units and machinery corresponding to the product mix planned
- any separate emergency power requirement
- if any arrangement will be required to adjust the voltage on LT side

Rectifiers for meeting Direct Current requirements
- Rectifiers are required for DC power supply to electrolysis plants at a voltage and current as per design of the series and parallel arrangements of the electrolytic cells.
- Other DC power required for UPS, instruments, controls etc.

7.2.3.3   Transformers and Rectifiers

- Running load in KW, Current, Voltage, at Primary and Secondary side
- Any humming or chattering sound (louder than usual)
- vapours coming out from breather or vent on conservator tank
- working of cooling fan
- temperature of oil and external surfaces
- any peeling off paint
- condition of moisture absorber in the breather
- deposit of dust on the equipment
- corrosion of power contacts
- covers for junction boxes
- spillage or oozing of oil from the body of the unit

Maintenance activities for Transformers
- Isolate from primary and secondary sides.

- Check resistance of primary and secondary winding between themselves; and with the body of the transformer.
- Check resistance of transformer body with earth.
- Take out samples of oil and analyse for (i) moisture, (ii) acidity (expressed as mg KOH per kg of oil), (iii) dielectric strength, (iv) suspended particles (which can be due to oxidised oil or debris from partly oxidised insulating material which is provided on windings. These suspended particles can settle on insulation and can reduce dissipation of heat (which can cause overheating of insulation and winding itself)
- Moisture and free acidity can corrode the insulation and internals. Check against standard given by OEM
- The dielectric strength should be at least 50 KV between two standard electrodes when the oil sample is tested by a standard test apparatus.
  Replace the oil if necessary (generally after 8–10 years of operation)
  High dielectric strength indicates good quality of oil (very low moisture)
- Check the working of cooling fan and temperature indicator.
- Operation of drain valve for oil and the connecting piping to oil collection chamber
- Check the silica gel (or $CaCl_2$ which may be used as drying agent) in the breather and regenerate it by drying in a current of hot dry air. Replace if it cannot be regenerated.
- Check connection to oil conservator tank and its vent line.
- Examine Bucholz Relay
- Carry out filteration of oil to restore its properties if the sample is not satisfactory.
- (i) On-line filteration can be done by a portable unit consisting of an oil circulation pump, oil heater and filteration cartridge/assembly for sludge removal.
  The heater will remove moisture also.
- (ii) Centrifuging–this method is difficult to use on load (when transformer is ON) and hence the production plant units will have to be stopped.
- Choice of the method depends on extent of impurities in oil, permissible down time of production unit, availability of fresh stock of oil. The method is to be decided by production engineer and electrical engineer.
- Test the filtered oil for moisture, suspended particles and dielectric strength and compare with specifications given by OEM before reusing. If the filtered oil is unsatisfactory, it can be disposed off through authorised agency (for incineration).

Maintenance activities for Rectifiers
- Check the resistance of primary and secondary winding, contacts with incoming HT cables and outgoing LT bus bars.
- Check the working of cooling fan and temperature indicator
- Examine external surface of the body for any hot spots (paint peeled off)
- Working of voltage adjustment taps at primary side
- All connections of bus-bars on secondary side (since it carries heavy current)

- Check oil samples and carry out filteration as above.
- The contacts at output side must be thoroughly cleaned since very heavy currents flow for electrolysis plants

7.2.3.4   Other Facilities

To be designed, selected and installed.

- Motors of various types as per need in the plant (three phase induction, DC, flame-proof, high safety standard...LT and HT type)
- Bus bars and power cables for transmission of large power;
- Starters, relays, fuses and safety interconnections
- Lighting arrangements with supply from different feeders (to prevent complete darkness in plant in case a feeder develops some fault)
- Diesel generators for supply of emergency power to keep the key equipments running during outage of external grid.
- Captive Power generation through Steam Turbine/Gas Turbine driven generators

- Uninterrupted power supply arrangement through battery backup.

7.2.3.5   Heating Furnaces

- Single phase or three phase
- Number of furnaces installed and those which are run simulteneously
- their power requirements—minimum and maximum for each furnace
- whether continuous or intermittent operation
- power input variation to be done in small changes, or large changes
- whether the changes are to be carried out *on-load* or *off-load*

7.2.3.6   Electrical Motors

Check points

- HP (horse power) rating of all connected and installed motors, operating speeds RPM their running times hours/day frame types, input power required as single/three phase and voltage (LT or HT), whether installation will be indoors or outdoors – Check Index of Protection Provided as per BIS standards. IP-65 is the preferable protection index as the construction does not allow ingress of dust, moisture in to the motor. (i) in dusty areas...(Rock Phosphate Grinding in single superphosphate plants), (ii) high temperature area...(near furnaces/burners), (iii) exposure to chemical fumes...(near reactors, process vessels),
- Use only flame proof in presence of inflammable vapours

Electrical Motors: Observations
- current drawn by the motor,
- any abnormal vibrations
- axial movement
- any loosening of foundation bolts
- any abnormal noise or smell from the motor
- Whether the cooling fins are clean
- any damage and fitting of cooling fan,
- temperature of external surfaces, any peeling off paint,
- deposit of dust on the motor,
- corrosion of contacts or covers for terminal (junction) boxes,
- oozing of oil from the bearings,

Inspection for Maintenance
Visual and external inspection for Mechanical and Electrical Tests

- axial movement (should not exceed 1.5–2.0 mm),
- check fitting of coupling half to shaft and alignment with driven machine
- greasing of bearings
- any loosening of foundation bolts or misalignment of motor with driven unit
- Apply very low voltage to the stator windings and determine the resistance of the winding itself. This is to be done for a few seconds only
- Resistance of winding with motor body, (This should be very high in order to prevent current leak to body)
- Resistance of earth with motor body, This should be very low to prevent electrical shocks
- Resistance of the individual rotor windings should also to be determined

Rewinding the motor
- This is to be done when old windings have damaged, insulation has come off
- New winding is provided with proper insulation and then inserted in the slots. Gentle warming by low power electric bulbs or keeping in a warm room is practiced and vacuum may also be used to remove moisture.
- Check the insulation of windings and resistance of winding with the body
- Check the bearings and the free movement of rotor,
- Clean the cooling fan and dust or oily deposit on the external surface
- On starting, check the current in each phase. It should be same in all phases; and should not exceed the value on name plate.

7.2.4   Maintenance of Other facilities

(i) Bus bars and power cables

Inspections and electrical measurement:

- Check corrosion of bus-bars, supports, fasteners at joints (clean surfaces and apply light coat of anti rust jelly)

- any damage of connectors or cable supports due to exposure to heat from furnaces, hot process units or ducts, direct contact with steam lines, or spillage of corrosive chemicals falling on the cables (hence select route of cables accordingly)
- broken portion or scratches due to sharp objects, external cover of cables and reinforcing wires
- check electrical continuity of individual wires (phases)
- resistance of the entire length from supply end to receiving end of power
- check resistance of conductors among themselves and with earth

(ii) <u>Starters, relays, fuses and interconnections</u>…… Check condition of fuses, contactors and any corrosion or dust deposit inside. Confirm rating of fuses should be as per protection required to connected electrical motors and other loads. Too high a rating can burnt out the downstream units as the power supply will not be cut off even in case of a fault.

(iii) <u>Lighting arrangements</u>….check contacts, external glass enclosures and internal reflecting surfaces, sealing rope for flame proof lighting.

Do not provide bulbs of high power rating in the vicinity of reactors or process vessels where inflammable vapours are present.

7.2.5   Diesel Generators

For supply of power during outage of external grid.

Capacity of the DG Generators: as required for emergency loads. Hierarchy of such loads to be decided by Senior Plant engineers in consultation with experienced operation staff.

Should have arrangement to start automatically in case of outage of external grid.

There should be auto load sharing arrangement among DG sets if more of them are installed and a single set cannot supply the entire required power.

Regularly operate once a week for a short time (30 min—one hour).

Regular servicing and inspection to be done by Service engineer from OEM. Annual maintenance contract is desirable.

Clean/replace air filter as per advice.

Check generator, electrical switches, battery for starting the set, exhaust silencer, acoustic enclosure, fuel storage tank, fuel oil pumps, coupling of engine with generator etc.

7.2.6   Captive Power Generation

Through Steam Turbine/Gas Turbine driven generators (details are elsewhere in this book)

7.2.7   Uninterrupted Power Supply UPS

Arrangement through battery backup.

Should have sufficient capacity to handle emergency process controllers, computers, safety devices, vents, lighting and preventing development of any dangerous situation.

The batteries must be kept properly charged always.

7.3   Rotary Equipments

Production engineers shall finalise the duty conditions, place of installation, whether continuous or intermittent working, material to be handled, safety features required etc. for the Rotary equipments in the process plant. These shall be discussed with maintenance engineers and detailed specifications shall be worked out before requesting for procurement.

Some typical considerations are given below. These can be modified/more points added as per need for the production plant, cost of equipment, delivery period, Quality Assurance Plan, erection and commissioning assistance from vendor, performance guarantees, quick availability of spare parts, annual maintenance contract offered….

7.3.1   Typical Rotary Equipments

- compressors, fans and blowers (centrifugal, lobe type)
- pumps of various types (vertical submerged, external horizontal, metering…)
- gas and steam turbines (condensing type, back pressure type etc)

To be used for
- transfer of fluids (for further processing)
- controlled feeding to process vessels, boilers
- compression of gases and vapours
- evacuation of process vessel, operating the process plant below atmospheric pressure (i.e. under suction)
- generation of power

Typical duty conditions to be specified
- operating pressures (during normal run) and maximum pressure
- volumes of fluids to be handled (normal and maximum rate of transfer,
- properties of materials to be handled-density, viscosity, melting and boiling points, inflammable/toxic nature, explosive, corrosive nature
- operating conditions—temperatures, presence of suspended solids, pH/free acidity

- whether operation will be continuous or intermittent
- whether two or more units will be run simultaneously

7.3.2   Features Required

- relief valves – in-built device/external device – *vendor may indicate*
- materials of construction for ensuring long life (for wetted parts)
- with heating jacket/with cooling jacket
- shaft sealing by simple glands/water cooled glands/mechanical seals
- for single stage/multiple stage machine….(for centrifugal machines like blowers, pumps)
- provision of inter-stage cooling
- dampening vessels on delivery side
- support legs, cleaning manholes, nozzles for lubrication and draining out internal accumulated fluids, vents for gas/vapour release,
- internal/external protective lining for corrosion control

7.3.3   Type of Speed Control Required

Purpose

 (i) to achieve intended process function (mixing, dissolving, reaction, calcinations...)
 (ii) to save power (per unit of product)
(iii) to ensure long life of moving parts (agitators, bearings)

Method for speed control

- V—belt drives operated by choosing pulleys of different diameters
- Positive Infinitely Variable drives (mechanical system)
- Electronic speed control (variable frequency drive)
- Hydraulic motors for providing: (a) high initial torque (b) reversal of direction of rotation if required

Type of drive
- directly coupled
- through gear box
- (i) *parallel axes/perpendicular axes* (ii) *speed reduction/speed increase*
- V- Belt drive
- through spur gear
- through helical gear
- through herringbone gear
- by electric motors
- by steam turbines

- by hydraulic motors (these can provide high torque required at start to overcome jamming of agitator in case of solidification or very viscous liquids in the vessel)
- The choice of drive depends on the load, speed desired, frequency of starting and stopping.

7.3.4   Types of Relief Mechanisms

- internal/external relief valves
- spring loaded valves
- rupture discs
- *(all above devices can be designed for release to atmosphere/recycle back to suction side)*
- soft flexible distance piece in discharge line
- warning device for high electrical load
- instantaneous tripping of drive motor

7.3.5   Installation in Plant

- equipments installed indoor in shed (should be protected from chemical vapours, acidic fumes, spillage of corrosive fluids from pipes).
- equipments installed outdoor (may get affected by dust, rain,)
- horizontal or vertical position
- at ground level or at a height
- inside process vessel (agitators, submerged pumps)
- outside process vessel (external transfer pumps)
- near control room (easily visible)
- away from control room (remote areas)

After successful erection and commissioning in the process plant the operating personnel must run the equipments as per Standard Operating Procedures; and never exceed limits on the equipments as given in instruction manuals from OEM.

7.3.6   Fans, Blowers, Compressors, Exhausters

These are used in many chemical industries for handling gaseous raw materials, supply of combustion air, sucking out/transfer of gaseous reaction products from the plants.

Fans are generally used for handling air for ventilation, for combustion air preheating and subsequently for supply to furnaces.

They are used to handle/deliver higher volume at less pressure; typically as induced draft fan for cooling tower

Observations to be made during running

any abnormal noise, vibrations, current drawn by motor, *apparently* less delivery as one may judge from exit gas volume from chimney (not accurate though)

Blowers
Centrifugal..

Single stage type units deliver high volume at less pressure while multi stage machines can deliver higher volumes at higher discharge pressure.

These are used in process plants for supply of process air, movement of gaseous reaction products through process units.

Can handle corrosive gases if proper material of construction is used, at temperatures above ambient temperature also. (example – ID fans in thermal power plants, in chemical plants after scrubbers used for acidic gases).

They are driven by direct coupling with motor or by belt drive or by fluid coupling etc.

Observations to be made during running
- discharge pressure (whether steady or fluctuating), abnormal noise, vibrations, current drawn by motor,
- any apparent problem in the operation of process plant (change in process parameters such as temperature, composition, pressures which can be related to change in flow rates of gases)
- flow rates of gases by suitable flow meters in gas/air lines.
- delivery of air/gases by visual observation of gases from chimney (not accurate though)

Positive displacement (twin lobe and three lobe type)
- discharge volume is more or less constant in-spite of higher pressure drop in process units downstream, (could be due to deposit of dust, rust or chemical salts.)
- generally used to deliver comparatively lower volume at medium pressure
- can be used as exhausters also to suck out gases from process units
- are driven by direct coupling, by belt drive, fluid coupling, or through gear box.
- dampening air vessel can be provided on discharge side to reduce pulsations of pressure (three lobe type have less pulsations)

Observations to be made during running
- delivery of air/gases by flow meter
- discharge pressure
- fluctuations in discharge pressure (should be less in case of three lobe type)
- any abnormal noise, vibrations, current drawn by motor
- condition of flexible piece/expansion bellow on discharge side
- any apparent problem in the operation of process plant (change in process parameters such as temperature, composition, pressures which can be related to change in flow rates of gases)

Compressors
- used for handling/compressing chemical gases like Ammonia, SO_2, refrigeration plants, Nitrogen, Oxygen at higher pressure
- They deliver comparatively lower volume at high pressure

Observations to be made during running
- any abnormal noise, vibrations of compressor,
- high current drawn by motor,
- Frequent blowing off safety valve
- Gas leak from pipes or gaskets
- condition of belt drive, lubricating oil pump, cooling system
- flexible piece/expansion bellow on discharge side if provided

7.3.6.1   Maintenance Checks and Activities

The maintenance engineer must always check the following:
Visual check:

- condition of foundation bolts and base frame, coupling halves, silencers/filters on suction and discharge side
- condition of belt drive and pulleys
- any deposit of dust/rust on blower and motor external parts of body
- **Alignment**
- Alignment of motor, gear box and the driven unit must be checked very carefully to prevent damage to bearings, drive shafts, coupling halves etc.
- Centre lines of the shafts should be exactly in line. They should not be parallel, or at an angle (skew). Also check individual tooth of the gears
- This can be checked by Dial gauge.
- Laser alignment kit shall also be used for accurate measurement.
- The reading shall be compared with maximum permissible limit given by OEM and corrective action taken immediately.
- Misalignment can also cause shaft seal failure.
- deposit of dust and rust on internals of the machines (clean by kerosene or recommended solvent by OEM)
- Check timing gears (any wearing out or mismatch)
- Any scratches on lobes
- Condition of flow control vanes and operating levers and their locking devices..
- Check setting of internal/external pressure release valve.
- Cooling system for lubricant
- adequacy and grade of lubricant to be added (as specified by OEM)
- Condition of bearings and lubricants
- Axial play of shaft, internal clearances and bearing clearance should not be more than permissible range given by OEM.
- Radial play of shaft–should not be more than permissible range given by OEM (higher play indicates wearing out of bearings)

- *Take out sample of oil and check. Replace carbonised, viscous, oxidised lubricant by new one recommended by OEM. The amount should also be as per level marked on gauge glass.*
- Check trip setting of relays, fuses and alarms for high load on drive motor
- Typical safety interconnection is the electrical interconnection of blower with boiler water level, signal from photo electric flame sensor (alarm for flame extinguished condition) to trip oil supply pump to firing system
- **General**
- Make a list of equipments, the sequence and time for which they will be stopped
- Make a detailed list of jobs which will be taken up on each unit – e.g. lubrication of bearings, changing impeller, repair to coupling, replacement of drive belts etc.
- Inform production engineer, other technical personnel (instrument engineer, boiler operators etc.) and other maintenance sections who may also take up some work (electrical checking, attend leaks);
- Store keeper to provide necessary spares
- Marketing shall check stock of finished product (if their production is likely to get affected)
- Estimate idle time of the machine; spare parts, lubrication arrangement required
- Obtain Work Permit for the day (fresh permit for every day)
- Remove fuses from electrical mains
- Isolate from inlet and exit connections; put blinds in gas and liquid lines (except cooling water lines which may be required)
- Observe dust and chemical rust on outside
- Clean the unit/machine thoroughly from outside and inside
- Check bearings, impeller, casing, wearing rings, shaft seals (any leak of oil/vapour/fluid being handled), gland packing and their cooling arrangement
- Opening of seal and replacing by new one…follow instructions from OEM specially for mechanical seals.
- Check for any corrosion of internal parts like shaft, impeller, shaft key, wearing ring
- Nozzles and pipes which connect to vents, safety valves should be cleaned of any dusty material deposited inside.

7.3.7 Reciprocating Compressor

- Measure vibration in horizontal, vertical and axial directions and whether there is any tendency for increase

  Visually check for excessive shaft movement

- Any abnormal rise in temperature or noise
- Open inspection windows, and then the casing/volute etc.
- Check condition of gland packing, wearing rings
- Check axial and radial movement of shafts.

- Inspect and replace strainers in suction lines [to prevent erosive particles, tramp material entering the reciprocating machine. It can damage cylinders and piston]
- Take samples of lubricant/oil and examine (under microscope)/by suitable means for presence of metallic particles *which indicates which part is getting eroded*
- Calibration and Setting of all Safety Mechanisms (Internal and external Safety Valve, rupture disc, trip setting on electrical relay of starters for motors)
- Check and attend high pressure alarm

  – High temperature/low oil level alarms for gear box

- Connecting rods, movement of cams.
- Condition of all spring loaded parts, foundation bolts, dampening pot at discharge side.
- Expansion Bellows on discharge side, expansion joints, gaskets
- Internal clearances among lobes; impeller and volute, wearing rings ; lobes and casing
- ***Standard Feeler gauge/strips may be used as per advice from OEM.***
- Shaft Keys on coupling, Cotter pins, Bibi coupling springs etc.
- Replace damaged shaft sleeves, lobes, bearings as per advice from OEM.
- ***Do not use re-conditioned spares it OEM does not allow***

Rotary Machines/Units
Installed directly on/inside a process unit [example – high speed agitator]

- Material of Construction of critical/wetted part.
- Water cooled gland/air cooled gland/other means for cooling the shaft seals or other parts of the machine.
- Continuously operated or run intermittently;
- Provision of Safety devices [tripping system, high temperature/high pressure alarms.]

Observations during operation
- Suction and Discharge sides – pressure, flow rates, temperatures of fluid being handled.
- Current drawn by drive motor.
- Steam pressure, temperature at inlet and exit of driving turbine. (if electrical motor is not used)
- Vibrations in the machine itself or in discharge side piping [check piping layout and flitting provided – Non return valve, Pressure (flow control valves, Dampening pot, supports and clamps provided for piping.
- Abnormal noise or temperature rise of bearings;
- Any axial movement/radial movement in horizontal and vertical direction [perpendicular to axis] taking place?
- Examination of lubricating oil at periodic intervals as recommended by OEM [will indicate wearing off shafts/bearings/other parts]
- Leakages from shaft seals, water cooled parts or body of the machine itself.

7.4   Some Auxiliary Equipments

- Keywords: Effluent treatment plant, dissolvers, rotary screens, ball mills, ribbon blenders, oil firing system, pumps
- Abstract: These are very important auxiliary equipment required for the smooth operation of chemical plants. They are used for treating and feeding of raw materials to the plant; for heating some process units, and for treatment of effluents.

7.4.1   Typical Effluent Treatment Plant

- These are designed on the basis of maximum pollution load i.e. for maximum concentration of pollutants like heavy metals, organics, BOD, COD, low pH, suspended and dissolved solids etc.
- Incoming effluent is collected in a equalisation tank with a grit removal chamber, either inside the tank or separate.
- Submerged transfer pump is provided in the equalisation tank with a recycle line.
- It serves to provide slow mixing of the effluent and to make it uniform.
- The treated effluent from active carbon filter outlet can be further treated by passing through ultra filtration system and reverse osmosis plant for further purification if desired by the plant engineers before recycling to the plant.

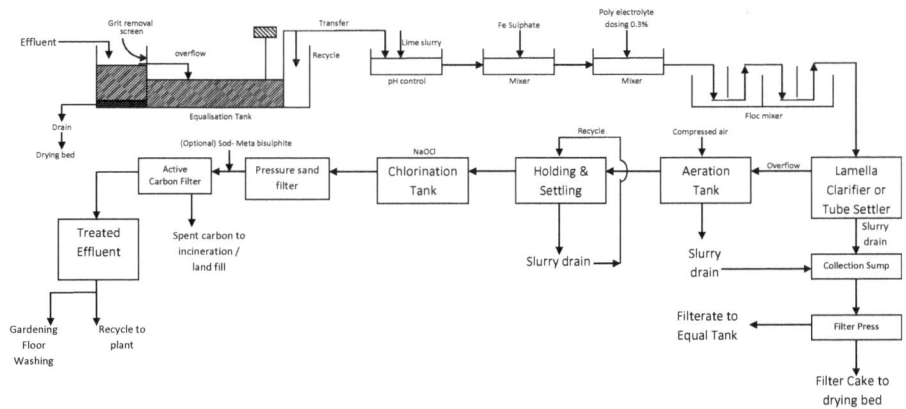

Figure for a Typical Effluent Treatment Plant

7.4.1.1   Main Equipments

These are typical units. More process units are provided if the effluents are containing difficult to treat pollutants or flow rate is varying. (The above drawing is indicative and does not show all the units stated below)

- Equalisation Tank (this should have sufficient capacity to hold sudden high volume of effluents due to some problem in the process plant)
- pH adjustment for effluent
- Alkali, Lime, alum, Fe sulphate, flocculant dissolvers and dosing units (tanks, dosing pumps, mixers etc.)
- Settling tanks, (tube settlers, Lamella clarifiers)
- Flocculators
- Polyelectrolyte doser
- Lamella clarifier or tube settler.
- Air compressor and aerators
- Filter press
- Pressure sand filter.
- Active carbon filter.
- Transfer pumps.
- Filteration system (Filter Press/Sand Filter)
- Effluent Transfer pumps
- Electrical installations, Emergency power supply [DG sets]; Plant lighting
- Storage Tank for treated effluent.
- Instrumentation for incoming and treated effluent pH, temperature, TDS, dissolved oxygen, flow rates, levels in various tanks and dissolvers
- Drying Beds for sludge taken out from Filter Press, clarifiers etc.

Observations during operations
- Flow rates and analysis of incoming effluent.
- Levels in Equalisation Tank, Settling tank, dissolvers, aerators etc.
- The treated effluent may be recycled to an earlier appropriate stage if it is not able to meet statutory requirement.
- More quantity of effluent may be stored in equalisation tank for incoming effluent if it becomes necessary to reduce the through put temporarily or maintenance activities are in progress on downstream units.

7.4.1.2   Maintenance of Following Items Is Required

- Regular removal of settled sludge from tanks, settlers.
- Examine and replace tubes/lamella plates.
- Examine and replace agitators in dissolvers if necessary.
- Examine and replace control valves for dosing chemicals if necessary.
- Examine flow distributor nozzles, tube sheet, internal rubber linings.
- Screen for grit removal.
- Transfer pump.
- Mixers for chemicals.
- Plates for lamella clarifier.
- Tubes for tube settler.
- Filter cloth for filter press.
- Spares for air compressor.
- Spares for air spraying assemblies in aeration tank.

- Spares for sludge transfer pump.
- Active carbon (replacement may be required 2 or 3 times in a year depending on the effluent quality. Check different samples of ACF in own lab to confirm suitable grade.
- Fresh sand (replacement may be required) 2 or 3 times in a year.
- Valves in liquid limes, aeration lines and chemical addition lines.
- VFD control spares for air compressor
- Quantity and grade of sand and Activated Carbon in the respective PSF and ACF should be as per original design of ETP. The sand and AC particles should be big enough so that they do not choke up slots in liquid distributor nozzles.
- During cleaning of the PSF, ACF the nozzles shall be taken out cleaned and fitted back on distributor sheet.
- Check valves in Back Wash lines to confirm proper operation and that they are not leaking.

7.4.2 Dissolvers

These are process vessels, in MS, SS, or PP/FRP lined construction and provided with agitators (turbine, anchor, propeller type) and baffles on sides.

The agitator blades are either welded or bolted to main shaft. The latter design enables easy removal and replacement by another set of blades.

Similarly, baffles which are fitted on side or internal walls can be removed and replaced by another type [with different holes, width, length, slots]

Observations to be made during operation:
- Level of liquid (solution) in the vessels.
- Rate of addition of solvent and solute to the dissolver if it is a continuously operating unit.
- Analysis of solution.
- Flow rate of solution from exit.
- Temperature of the solution inside.
- Pressure of steam if it is used for heating
- Escape of any vapours or toxic fumes from shaft seal or gland packing. These should be removed by an exhaust fan and passed through a scrubber before venting to atmosphere.
- Flow of cooling water through water cooled gland.
- Current drawn by agitator motor and its speed of rotation.
- Direction of rotation—should be as per original design
- Working of the external liquor circulation pump if provided for faster rate of dissolution.
- If external cooling arrangement is provided – Flow of cooling medium such as water, chilled water or brine.
- Temperature of cooling medium at inlet and outlet of dissolver. This is very important in case of solids which evolve heat during solution and can cause

release of toxic vapours or abnormal temp rise. The rate of addition of solid and speed of agitator must be controlled.

Maintenance
Precautions
Isolate (stop) supply of solvent and solid/solute. Stop supply of heating steam. Put blind in the process stream connections and steam line. Stop cooling water/chilled water supply if there is a chance of solidification of solution and consequent jamming of the agitator. Drain out the unit completely and flush by water/appropriate solvent. Evacuate all residual vapours inside the unit by an exhauster vacuum pump. Pass the vapours through a scrubber.

Remove fuses of agitator motor; and de-couple from motor if some internal inspection or repair work is to be done.

- Slowly take out the agitator assembly from the dissolver if necessary.
- Replace damaged/badly corroded agitator blades or bolts by new ones and provide lock nuts/cotter pins to prevent loosening while in operation. Repair internal lining by cutting off damaged portion and fixing new sheets. Test rubber lining/Fibre glass lining by electrical spark test at 5000 volts. *Use higher voltage only if mutually agreed between vendor and plant engineers for the type of lining and the thickness of rubber layer provided.*

Note
 (i) It will be found useful to provide strainers in fluid lines to remove hard solid particles or sharp pieces of suspended impurities which can damage the lining.
(ii) Flouride content in liquids can corrode glass lining in acidic conditions.

7.4.3 Melters, Ball Mills, Ribbon Blenders

a. **Properties of materials being handled**

(Selection, operation and maintenance of the above units depend on following properties:)

- Density (bulk)
- Explosive/inflammable/corrosive nature.
- Moisture content
- Lump size.
- Particle size as available/as desired.
- Solubility in the solvent.
- Presence of hard or sharp particles.

b. **Required operating capacity of unit**
c. **Operating cycles/hours per day**
d. **Location for installation in plant.**

Maintenance:
Check the following and attend wherever necessary (for all units).

- Feeding system screens and feeder controls: Too big lumps or too high rate of feeding can overload the motors and damage the agitators.
- Exhaust hood and exhaust fan for removal of dust/vapours from the unit.
- Hood, support, ducts to be coated with rust resistant paint (acidic vapour resistant) or FRP lining. Exhaust fan also should have similar coating.
- Charging chute and cover should also have an exhaust hood and fan (This is necessary because considerable vapours are evolved from melter if there is moisture in the solids).
- Discharge nozzles and drain valves.
- Check drive mechanism (v-belt, pulley) and speed control (VFD) for units.

7.4.3.1 Melter

Steam supply line and setting of safety valves, steam trap, external insulation and cladding, drain valve. Certain melters have continuous charging of solids at a controlled rate and overflow of molten material from exit nozzle. Check the screen at overflow nozzle to arrest escape of solid particles. Melting point, latent heat and specific heat of the solids

- Heating jacket/coil – should be pressure tested. If steam is used as the heating medium, a safety valve shall be provided in the steam line and should be set to a pressure less then the pressure at which it is tested (Steam pressure to be used depends on MP, latent heat etc.)

7.4.3.2 Ball Mill

- *observations to be made before start of the unit*
- analysis of feed material to be charged in
- condition of feed silo, feed regulator and chute
- free rotation of the mill,
- inspect belt drive system (pulley, belts, motor)

  - *Protective Guard on complete system is necessary.*
  - *Examine the condition and tension in each belt.*
  - *Examine the individual groove in driven and driver pulley*
  - *The belts will have to be replaced by a new set if more than 10% have worn off.*
  - *Check alignment of driver and driven pulley and adjust position of the drive motor.*

- condition of the balls (replace balls if old ones are damaged, by new ones. Obtain certificate from vendor for hardness, composition and size distribution

- check inner liner of the mill,
- cyclone, bag filter and ID fan for dust control

Observations to be made when the unit is running
- Smooth working of drive mechanism (belts, pulleys, motor, etc.)
- Arresting of dust by cyclone, bag filter, ID fan, whizzing fan.
- Screen analysis of product from ball mill.
- Check presence of any debris (particulate matter from balls).
- Any abnormal vibrations or noise.
- Power required by drive motor/current drawn

7.4.3.3   Ribbon Blenders

- Check condition of ribbons for forward and backward movement of material
- (thickness, smooth movement inside, clearance from sides)
- Condition of shaft and bearings
- Screen analysis of feed material
- Hardness of particles in feed
- Charging chute, feed control
- Vent valve
- Drive mechanism
- Speed control provided
- Proper grouting of foundation bolts

7.4.4   Oil Firing System

These are provided for firing of rotary kilns, pre-heating or regular operation of furnaces, or oil fired systems for heating thermic fluids or small packaged boilers.

7.4.4.1   Observations During Operation

- Level of oil in day tank
- Working of oil heater (in severe winter when oil becomes viscous)
- Working of oil pump and setting of its safety valve.
- Discharge pressure of oil pump.
- Discharge pressure of air blower for combustion of oil. Position of dampers in primary, secondary and tertiary airlines.
- Proper fixing of oil burner
- Condition of flame in furnace. It should not impinge directly on refractory lining or tubes for heating thermic fluid.

- Temperature of flue gases from the unit being heated.
- Presence of any soot particles in the flue gases.

7.4.4.2 Check the Following During Maintenance Jobs

- Cleaning the oil tank thoroughly. Check level indicator.
- Proper operation of the valves in oil lines, air blowers.
- Free movement of the shaft of oil pumps; alignment with motor.
- Earth connection of oil tank, pump, motor, oil heater.
- Temperature setting of thermostat provided for oil heater.
- Strainers in inlet line of oil tank and suction line of pumps. Change the strainer screen if damaged.
- Setting of safety valve on oil pump.
- Check cover on electrical contacts and cable connection box.
- Thermal insulation on hot oil lines.
- Working of dampers in air lines and ratio controller for oil supply.
- Clean nozzles in burner assembly. Partially choked nozzle will spray the oil in some undesired direction – and the flame may even impinge directly on refractory lining.
- Working of flame monitoring photo electric sensor.
- Manometer/pressure gauge for Air pressure.
- Interconnection of low air pressure with oil pump.
- Electrical spark plug for starting oil burner-change if the spark is not satisfactory or if the insulator is damaged.

7.4.5 Rotary Screen

Observation during operation:
- Screen analysis of feed material and output material.
- Smooth movement of the rotary screen.
- Observe speed control mechanism if provided.
- Working of exhaust hood, exhaust fan and bag filter.

Maintenance
- Check individual screen. Change if damaged.
- Drive mechanism and speed control.
- Lubrication of bearings of shaft.
- Feed silo.
- Rotary valve at a bottom of feed silo.
- Charging chute.
- Confirm (i) proper fixing of exhaust hood on the rotary screen (ii) working of exhaust fan (iii) cyclone separator for dust removal.

- Length of screens for separating out particles of different sizes should be such that separate heaps shall be formed if more than one screen is fitted for classifying the solid particles as per size desired.

7.4.6   Pumps for Process Liquids

Various of types of pumps are used in a chemical plant for transfer of liquids. These are selected on the basis of the capacity required, discharge head required, liquid to be handled (density, viscosity, presence of any suspended solids, temperature of operation), position of installation of the pumps, (as a submerged unit in a tank or installed outside).

The shaft sealing of the pumps are selected on the basis of inflammable or toxic nature of the liquid (conventional gland packing are used for water and mechanical seals are to be used to toxic, dangerous liquid).

The material of construction of the wetted parts (impeller, shaft, wearing ring etc.) are chosen on the basis of liquid properties and corrosive nature.

Observation during operation:
- Check the foundation and levelling of the pump.
- Alignment of the pump with drive motor/turbine.
- Strainers in suction line (when the liquid contains suspended solids)
- It is advisable to have a net positive suction head when the liquid is volatile and is to be handled at higher temperature.
- Smooth working of valves in suction and discharge line.
- Proper operation of the non-return valve in the discharge line.
- Proper steam flow and working of the steam trap when the pump is to be kept hot by steam. This is required when the liquid may become very viscous or even solidify at lower or even ambient temperature
- Discharge pressure of the pump.
- Current drawn by the motor.
- Any abnormal noise or vibrations

Maintenance
- Commission the standby pump for running the plant
- Isolate the pump to be taken out for maintenance from suction and discharge side
- Check the alignment of the pump with the motor and loosening of the foundations bolts.
- Check the gland packing/mechanical seal.
- Open the pump body and inspect the internals. Care should be taken to use only the lifting lugs.
- Check the clearances between impeller and body, any damage to wearing ring etc.
- **Gear Pump**: Check the clearances between gearwheels and body and any internal safety release valve.
- Check strainer screen and replace if damaged

- Further checking will depend on the type of pump in use.
- Follow strictly the instructions from OEM in addition to above.

  Please see Chap 9 Sec 9.3 for more information.

7.5 Pressure Vessels

- Many chemical industries use vessels which operate under pressure.
- (for manufacture of ammonia, liquid sulphur trioxide, Nitrogen and Oxygen gases, processing of petroleum, for recovery of waste heat by boilers to generate steam).
- These must be designed, fabricated, erected, commissioned, operated and maintained properly to prevent any mishap in the manufacturing plants.
- Therefore the operating conditions, materials which will be handled, their intended use, location for installation (inside or outside covered shed) etc. must be correctly informed to the designers, fabricators and all other concerned persons before procuring them to avoid mishaps during use in the plant.
- Approval must be taken from local Statutory Authorities in the country for the design, fabrication, material of construction, welding processes, etc. before commissioning
- **Important steps related to these and transport, erection, operation as well as maintenance of pressure vessels are given below**. It will be useful to look into the following description for a better understanding of the matter.
- **Suggestions from readers for improvement of the written matter are welcome**.

7.5.1 Classification of Pressure Vessels: Fired and Unfired

Examples of fired pressure vessels:

- Which are heated by fuels like coal, furnace oil, propane, LPG, ...
- Those which have external heating jackets/internal steam coils;
- Those heated by process gases (Waste Heat Recovery Systems)

Examples of unfired pressure vessels:
  Any closed metal container that is not directly in contact with some heating source but is subjected to internal pressure (where no vapour or steam is generated by heating).

- Condensers (with incoming vapours at high pressure)
- Compressed air receivers
- Inert gas cylinders,
- Vessels carrying Dissolved Acetylene,
- CO_2 cylinders for fire extinguishers
- Those which may carry toxic chemicals

Unfired pressure vessels having capacity of more than1000 Litre should be sub-jected to periodic thorough examination for ensuring safety of personnel and plant units.

Unintended pressure vessels in chemical plants
- Vessels which can get pressurised due to (i) failure or insufficient supply of cooling medium; (ii) runaway reactions due to overfeeding of reactants (iii) high speed of agitators instead of slow agitation (e.g. Vaporisers for Liquid SO_3 *heated by warm water)*
- Equipments where some exothermic reaction may occur due to ingress of moisture or some foreign material and the system gets pressurised unknowingly-*SO_3 condensers, 65% Oleum coolers-which need special designs*
- Chemical Storage tanks which can get pressurised due to high ambient temperatures in summer or because of proximity to units operating at high temperature.
- Furnaces with water cooled electrodes or jackets. Water can enter the hot furnaces if the electrodes leak inside and pressurise the unit due to generation of steam.

7.5.2   Some Examples of Pressure Vessels

Production of Crude Petroleum Oil
- Vessels (which operate under high, medium and low pressures) for separation of crude oil and gas at oil wells.
- Heater treaters for treating crude oil before processing,
- Desalters for removal of salt,
- Induced gas floatation units for separation of effluent from fine oil droplets

Chemical Process Industries
- Caustic soda & chlorine plants—compression and storage of chlorine and hydrogen in cylinders
- Effluent treatment plants having dissolved air floatation units for particulate matter removal
- Air drying plants and receivers for supply of compressed air

Pressure vessels in Sulphuric acid and oleum plants
- Steam jacketed Pressure leaf filters—for filtration of liquid sulphur,
- Gas Heated/Steam heated Oleum Boiler for generation of SO_3 vapours,
- Waste Heat Recovery Systems for generation of steam,
- Process Reactors and condensers for Liquid SO_2 production plants,
- 65% oleum coolers, liquid SO_3 condensers and storage tanks
- Carbon di-Sulphide storage tanks use water pressure for transporting CS_2 to the point of consumption.

7.5.3 Considerations for Procurement of Pressure Vessels

Specify Process requirements

Specify the intended use: As reactor, condenser, heat exchanger, boiler, air receiver.

Specify duty cycles and conditions:

 (i) whether it is to be used intermittently or continuously
(ii) frequency of pressurisation/depressurisation during use of the vessel

• Physical and chemical properties of materials to be handled by the vessel

Boiling points, density, viscosity, erosive, corrosive, toxic, inflammable explosive nature.

• Operating pressures and temperatures – normal, minimum and maximum

Chances of any sudden rise of pressure and/or temperatures during operation (due to runaway reaction, failure of cooling water flow.)

• Composition of materials to be handled: presence of suspended solids; or dissolved solids which **may choke** the nozzles for safety valves, exit for materials, pressure taps.
• Location in plant and position in which it will be installed:

 (i) location (indoor/outdoor), ground floor/at a height
(ii) position: horizontal/vertical

Volumetric Capacity required (including some empty space to be kept at top)

• Will it be fitted with an agitator/without agitator?
• Will it be a fired or an unfired vessel? Will it be heated or cooled—internally or externally?
• **Details of arrangement which shall be provided for heating/cooling should be informed to all concerned (designer, fabricator, plant operator, maintenance team, statutory inspector who will finally give approval...)**
• Nozzles required for incoming and outgoing process materials, safety vent, safety valves, pressure and temperature indicators, drain valve, inspection and cleaning manholes, entry of agitator shaft, observation nozzles...
• Nozzles are also required for level indicator, sight and light glasses, sampling points, inlet of reactants and exit of products, draining out the vessel.
• Design pressure—it is the maximum pressure to which the vessel can be exposed (used for design)
• Maximum allowable operating pressure MAOP—it is the maximum pressure which the weakest component can withstand. Never operate above this pressure.
• MAOP is less than MAWP (maximum allowable working pressure)
• Test pressure at which the vessel is actually tested.

- *The test pressure is always considerably higher than the operating pressure to give a margin for safety.* **Typically the vessel may be tested at 1.5 times the maximum allowable working pressure or as per instructions from statutory inspector**
- As a matter of abundant precaution do not exceed the lowest value from above while running the plant.

Mountings and other arrangements required
- Safety valves and rupture discs
- Pressure and temperature gauges, gear box and motor for agitator,
- **Externally connected pipes and ducts should be supported independently to prevent mechanical stresses (due to bending loads) on the nozzle.**
- Manholes for inspection, cleaning and maintenance
- Lifting lugs and support legs
- External heating/cooling jacket or limpet coils
- Drain valves, Vent valve and Pressure taps
- Support legs and lifting lugs

Shaft seals
The designer should select a suitable shaft seals for agitators as per operating conditions, and properties of material inside (inflammable, explosive or toxic nature).

Some of the options are – conventional glands, water cooled glands (for lower pressure only) and mechanical seals.

Wall thickness
- This shall be calculated as per maximum internal/external pressure. Designer shall also consider maximum operating weight of the vessel (when full of material inside), load due to external fittings, valves, connected piping, any shock loads from connected compressors, wind loads etc…
  Thereafter safety margin allowances are to be added to the calculated value for thickness of the walls of the vessel. These additional allowances (*as safety margins*) are for corrosion, severity of operating conditions, and whether dangerous/inflammable fluids are to be handled.
  Reinforcement as gussets at nozzles, pad plates can also be provided externally

7.5.4   Further Considerations for Design and Fabrication

- Various design and fabrication codes like ASME Sec.VIII Div.I and II, TEMA, ASTM Codes for materials, IBR-1950 with amendments, and codes should be generally followed. If there is any deviation from the standards/codes it must be with a written mutual understanding between the fabricator and the production and maintenance engineer of purchaser.
- Approval to be obtained for the design and fabrication drawing, welder's qualification from statutory authority before placing order for fabrication to Vendor

- Orientation drawings for all required nozzles should be approved by the purchaser or his consultant before commencing fabrication. Supporting pad plates (and vent holes for pad plates) and lifting lugs position (with their details) shall also be checked and confirmed.
- Radiography (100%) for all welded joints should be insisted upon specially if dangerous materials are to be handled.
- Welding rods used for fabrication: statutory approval shall be taken.

However, care should be taken to see that no Statutory Regulation is violated.

- Corrosion allowance actually provided shall never be less than the (design) value as indicated in the drawing finally released for fabrication.
- External Insulation and painting shall be provided only after completing all tests at vendor's shops and repeating the tests at site after completing erection.
- External enclosures for isolation of the vessel may be erected at site. The vessel shall be installed at a place in the plant where minimum number of persons work.
- Connecting pipelines should be independently fixed and supported separately so that no stresses are created on nozzles of the pressure vessels due to expansion and contraction of the pipes due to temperature changes or any other reason.

7.5.5   Materials of Construction (MOC)

Test certificates for Materials of Construction must be provided by the vendor (fabricator). Samples of MOC can be tested independently by purchaser also to confirm the quality *in approved laboratories.*

- Test Certificate for tensile strength, corrosion resistance at operating conditions against chemicals handled, composition of MOC used and bills of procurement for material etc. shall be obtained from fabricator
- . The pressure vessel should be fabricated from standard size sheets available in market as virgin material (fabrication of pressure vessels shall not be done from recovered material (used for some other vessels earlier) or from ship breaking yards.
- Purchaser may suggest the MOC suitable for the process materials (reactants and products to be handled) and as per the operating conditions of maximum pressure and temperature likely during use in the plant.
  Test the MOC and the vessel for high impact resistance, especially for vessels used in low temperatures

  Following are some of the common MOC used. However the suitability for the intended use in the operating conditions (consider most severe conditions) must be approved by the designer, fabricator, statutory inspection authorities for pressure vessels before commencing fabrication.

- Carbon steels: IS 2002 Grade A, ASTM A 560, SA-516, A-179
- Carbon Manganese steels,

- Carbon Molybdenum steels,
- Chromium Molybdenum steels,
- Chromium Molybdenum Vanadium steels,
- Chromium Nickel Stainless steels
- Duplex steels, Super Duplex steels

Ordinary Mild Steel (e.g. I.S.2062 or its international equivalent) *shall not be used.*

- **Limitations of Materials of Constructions**
- Carbon Steel for Non corrosive service: up to 500°C
- Stainless Steels (*18–25% Cr; 8–20% Ni 0.05–0.2% C*): Corrosive conditions, up to 650°C
- Clad steels (Carbon steels *with cladding of Stainless steel*)
- Nickel alloys: Monel for marine/organic service
- Inconel: for dairy/food processing service
- Carbon Molybdenum steel: up to 550°C
- Chromium Molybdenum steel: up to 650°C

7.5.6   Heads for pressure vessels

Approval for their design, welding joints, electrodes to be used must be obtained from statutory authority before placing order for fabrication

- Flat head: For low pressures up to 5 kg/cm^2
- Torispherical head for pressures up to 15 kg/cm^2
- Ellipsoidal head generally for pressures 15–25 kg/cm^2
- Hemispherical generally used above 25 kg/cm^2

7.5.7   Checking Fabrication Activities

- Fabrication to start only after the fabrication drawings are approved and Quality Assurance Plan is accepted
  In case special electrodes are to be used as per fabrication drawing preserve the electrodes stubs to confirm correct type of electrodes have been used.
  Observe the settings on welding machines. (preferably use Direct Current for welding) Job is to be done by qualified welder with valid certificate on the date of welding.

Inspections during fabrication activities

Stage Inspection—storage facilities for electrodes, plates, bought out components,, thickness and composition of plates, Cutting diagrams and cut plates, weld groove preparation, orientation of nozzles

- Assembly of Vessel *with only tack welding* prior to final welding, root run of welding
- Final Visual Inspection—Overall dimensions, Pad plates and vent holes, lifting lugs, safety vents/valves, Rupture disc provision, drain points, sample points, Thickness of all flanges and the holes drilled, gussets for nozzles, Ovality check
- Final hydraulic test must be carried out in presence of purchaser's representative and shall be as per statutory requirements

To observe during visit to Vendor's shop
- Nozzle *orientations* for connecting to other process units and machinery *must match the finalised layout of plant.*
- *Important: load bearing capacity of selected site must be checked before erecting the vessel.*
- *No residual water in crevices or nozzles or drain points to remain inside—which can react with acidic gases to form corrosive products inside the vessel and reduce its life*

Transportation and Erection at site
- Transporter should obtain instructions for loading at vendor's shop and unloading at site;
- erection and commissioning manuals

   Do not use slings on any nozzle for lifting. Use only lifting lugs.
   Close all openings by blinds protecting the flange faces before transporting to site by using suitable gasket,

- Site preparation—civil foundations, base frames inspection, and installation precautions, internal lining–*if required*–at site. Corrosion resistant paint must be applied to bottom sides before placing on base frame.
- .Supply separately Safety valves, Rupture Discs, Pressure gauge, thermocouples with Test Certificates

7.5.8 Documents to Be Available to Plant Engineers

- Documents for Transporter
- Order copy giving intended use/service conditions,
- G A and Approved Fabrication Drawings with inlet and exit nozzles, their orientations, nozzles for safety valves, vents, level and temperature indicators, light and sight glass, drain point, pad plates for reinforcements, details of all welded joints, lifting lugs, support lugs, *electrodes used (test certificates for compositions) used*

- **Post Weld Heat Treatment (Stress relieving**)–Heating and Cooling curves records
- Radiography records, Hydro-test records,
- Test and Guarantee Certificates for bought components–Safety valves. Sight Glass, Level indicators, Light Glass, Rupture discs
- Quality Assurance Plan verifications
- Minimum three sets of Manuals for Erection and Commissioning to be given by vendor to site engineer

7.5.9 Quality Assurance Plan

To be approved by Production, Maintenance engineers and Purchase officers before placing order.

Vendor to specify which codes and standards will be followed during fabrication. Some of the well known codes are:
- ASME Sec I: Boiler and Pressure Vessel Code
- ASME Section VIII: Rules for Construction of Pressure Vessels.
- Pressure Equipment Directive (PED)
- Japanese Industrial Standard (JIS)
- German Pressure Vessel Codes (DIN)
- Indian Boiler Regulations IBR 1950 *with latest amendments*
- IS 2825–1969 for Unfired Pressure Vessels (these are Indian codes.)

Following Indian Standards or their Equivalent international standards shall be considered for selection and procurement of the pressure vessel.

IS 2825–1969 for Unfired Pressure Vessels
- **Class I Vessels**: for carrying lethal/toxic substances – Full radiography is mandatory.
  Only IS 2002 Gr A (boiler quality steel) is allowed for construction of these vessels.
- **Class II Vessels:** Many vessels in Chemical Process Industries are classified as per this Class. *All longitudinal and circumferential joints to be radiographed.*
- **Class III Vessels: Maximum** Pressure 7.5 Kg/cm^2, Maximum temperature 250 °C. These vessels are for light duties only and are n*ot to be used below zero °C.*

Stage inspection and Final inspection programme must be offered.
Final test must be carried out in presence of purchaser's representative and shall be as per statutory requirements. It is better to test the vessel once again after erection at site as per standard procedures.

Documents to be obtained for
- Safety margin considered and corrosion allowance added
- All materials used for fabrication shall be tested and certified by approved Test labs/procured from reputed vendors (including electrodes, fluxes). *Preserve test coupon.*

- Traceability of material used (stampings on plates, original purchase bills for bought out items, stores records for receipt of incoming materials and issued for fabrication.)
- Welding process followed and records of setting of current and DC voltage on transformer
- Welding electrodes used
- Records of Post Weld Heat Treatment done for stress relieving with heating and cooling curves
- Stage inspections and final inspection offered. Equivalent International Standards shall be used.

Records of hydraulic pressure tests carried out at Vendor's works. Pressure gauges used for these tests must be calibrated and certified by approved Testing Agency.

7.5.10 Statutory Documents: To Be Made Available to Purchaser by Vendor

These are required in many countries for obtaining permission to erect and hydro test (and also for future reference/records of the purchaser).

- Order copy giving intended use/service conditions,
- GA and approved fabrication drawings with details of all nozzles and their orientations,
- Post Weld Heat Treatment records
- Radiography records, Hydro-test records
- Test and guarantee certificates,for bought components: Safety valves, sight glass, level indicators, light glass, rupture discs
- Quality Assurance Plan verifications

7.5.11 Safety Valves

- These are generally bought out items. Vendor should furnish test certificates with markings on them for identification and the date of test for the safety valves.
- Purchaser shall confirm the values for design, operating/working and test pressures from design engineer and Statutory Authorities (Factory Inspector, Govt Inspection Agency) while providing the safety valves.
- Should open in case of increase beyond set pressure and prevent further pressure rise: the valves shall be suitably sized
- Close by itself after pressure is released
- Shall provide tight shut-off during normal operations (to prevent loss of inside materials through a leaking valve)

- At least two safety valves are to be provided to release the excess pressure
- Reinforce the nozzles for mounting the safety valve in consideration of chances for bending moment during blow off.
- These nozzles should be always kept clean (remove any deposit of solids during regular inspection)
- *Setting of safety valves shall be done as directed by Statutory Authorities*
- **Other pressure release systems (design and construction** shall be approved by Statutory Authorities)

  – To be provided as close to the pressure vessel as possible
  – Provide locking arrangement so that unauthorised change of setting for release pressure shall not be possible

- Should function automatically (provision of motorised drive or compressed air shall not be necessary for pressure release)
- Provide good lighting and easy approach for vent valves and pressure release valves so that they can be opened manually *if they do not open automatically during dangerous situations.*

  – Should be easy to maintain
  – Weather proof construction

Some examples of pressure release system
- Seal Pots on vent lines with tall vents

  – Fusible plugs on body. These shall melt in case of high temperature and can extinguish fire below

    • Rupture Discs—need replacement
    • Adequately sized vents with release points away from work/inflammable storage areas

7.5.12 Pre-commissioning Checks

- Confirm details of order copy should be same as per name plate)
- Check foundation bolts, lock nuts, nozzle orientations, proper gaskets in the connections with all piping,
- Provision of calibrated pressure and temperature gauges,
- Setting of all safety devices—Safety valves, alarms for high/low temperatures, pressures etc. and rating of rupture discs
- Smooth working of all valves and hydro test with fittings
- Heating and cooling arrangements—as applicable
- Working of alternative (emergency) cooling arrangements
- Provision of mechanical seals, speed control for agitator drives
- Vibrations at full speed shall be within acceptable limits
- External thermal insulation

7.5.13   Operation of Pressure Vessels

- Prepare a Standard operating procedure with important instructions:
- Never exceed the maximum allowable operating pressure
- Set the Safety valves as per statutory rules/less than this pressure.
- Never exceed design operating temperature—do not shift to areas of high temp/ near hot equipments unless HAZOP is done
- Operate at 75–80% of maximum allowable pressure only. As a matter of abundant precaution the safety valves may be set to open accordingly
- Provide warning lamps, hooters and tripping systems if pressure increases to dangerous values.
- Do not use for different fluids than specified.
- Do not change colour code
- Do not disturb the *supports—check them regularly*
- Pressurise and de-pressurise slowly as per OEM instructions only—never suddenly open/close valves
- Drain out accumulated condensates (if any) periodically
- **Check setting of safety valves and safety interlocks regularly**—on scheduled dates in presence of Factory Inspector, Safety Officer and Representative of workers' union.
- Check thickness of vessel by ultrasonic instrument periodically
- Generally keep about 15–20% empty space at top
- **Extend operation beyond permitted date only with written Statutory Approval**
- Do not use beyond end of service life.
- Monitor physical and operating conditions—take immediate corrective actions if deviations are found more than permissible limits.

7.5.14   Additional precautions

- Provide multiple safety devices—safety vents and valves, Rupture Discs, high temperature and pressure to disconnect power to fuel feed pumps (for preventing overheating or pressurizing).

  - Should be used for handling only *that material* for which it is designed and constructed.
  - Installation must be on properly designed and constructed foundation
  - Vent holes at reinforcing pad plates must be kept clean
  - Protection must be provided from (i) any spillages from nearby ducts or pipes carrying corrosive fluids (ii) harmful effects due to surrounding such as dust, corrosive vapours, sun, high temperature units nearby,
  - Do's and don'ts recommended by OEM and Statutory Authorities must be strictly complied with.

- Calibration of all instruments used for pressure, temperature, flow rates, pH for handling (filling, emptying) materials in to the vessel and for operation of the pressure vessel must be checked regularly.
- Provide big sized dials (with red areas to indicate danger zone for operation) which can be read easily from a distance.

• Operating range and frequency of operation for increasing and decreasing pressure should be as per recommendation by designer and within limits given by OEM. Never exceed limits of allowable rate of rise/reduction of pressure and temperature.

Avoid
• Repeated rapid pressurisation and depressurisation
• storing/entry of corrosive materials;
• wrong method of lifting;
• interruption of cooling water/medium supply,

- excess/wrong sequence of feeding of raw materials can lead to run away reactions
- Reaction material feeding system must be provided with VFD speed control/metering arrangement/screw feeders
- Provide safety release valve on feed pipes if there is chance of excess feed even after VFD control.
- Cooling jacket shall always remain full—even when supply of cooling water fails due to any reason
- provide power by DG set to run the cooling water supply pump.,
- Elevated Storage Reservoir should provide cooling water for temperature control in case of emergency.

7.5.15   Maintenance of Pressure Vessels

Take the vessel out of service as per procedure given by OEM. Isolate from incoming and outgoing process streams and heating arrangements. Insert blinds in the connecting pipes. Only closing of valves may not serve the purpose as they may pass some streams.

Remove fuses from drive motors for pumps, agitators etc. to prevent start up by mistake if there is a chance of a mishap.

Never depressurise suddenly. The pressure is to be reduced slowly (generally at a rate not more than 0.5–1.0 kg/cm^2 per minute or lower as per OEM instructions) by venting out the pressurised contents to another process unit or through a suitable scrubbing system.

When the pressure is reduced completely, the residual liquid or any accumulated condensate is to be drained out and collected in suitable containers. It may be reused or treated in the effluent treatment plant before disposal.

Now flush out any remaining vapours or gases by compressed dry air or nitrogen whichever is suitable. The gases are to be vented out as above.

Check the gases being vented out by hand held instruments or suitable chemical detectors to confirm that all dangerous material has been removed.

Provide appropriate personal safety devices to the technicians. Confirm that suitable arrangements for fire fighting, personal breathing apparatus and rescue systems are in place.

Get authorised Work Permit from responsible senior plant engineer and checked by safety officer. Fresh permit is to be obtained every day before starting maintenance work

The inspection/cleaning manholes are to be carefully opened and the vessel is to be thoroughly cleaned from inside after neutralising the residual material in crevices.

Check the vessel and all nozzles for any cracks, leaks, weak gussets visually first. Now check it by hydraulic pressure at 1.5–2.0 times the working pressure for about 4–8 hours (or as per instructions from statutory authority). Examine thoroughly if any drop in pressure is noticed. Check thickness by ultrasonic tester also.

Any weak spot, minor leak, weak gussets, leaking portion is to be repaired by cleaning and welding. Post Weld Heat Treatment is to be carried out.

However, the vessel may be derated by the inspecting statutory authority after this. They may permit use at lesser pressure or for handling safer fluids which are not dangerous.

It should never be used as per original operating conditions if derated.

A new vessel shall be procured and the old one may be discarded or used for storing water, dilute alkali etc.

Note

- **Post weld heat treatment PWHT is** *is mandatory* when vessel is > 38 mm thick. **It should be asked for**.
- *To relieve stresses introduced during fabrication*
- *To improve corrosion resistance*
- *To improve strength and toughness*
- *No changes or corrections involving cutting/welding are permitted after PWHT*.
- **Vendor shall have proper facilities for PWHT as**:
- A furnace with heating systems with c*alibrated thermocouples*
- *Automatic temperature control*
- *Proper Loading and Unloading facilities*
- *NDT facilities*
- Inspection after PWHT must be done for welded joints, any deformation of nozzles

- **Radiography of welded joints is a must when** lethal fluid is handled, when the vessel will be used below minus 10 °C as per Indian standards. Follow equivalent international standards. **Such precautionary steps are essential for better life of the vessel and minimizing breakdowns**.

7.6 Civil Works and Their Maintenance

Civil Works are to be carried out in Chemical Plants for equipment foundations, control rooms, internal roads, storages for raw material and finished products, water tanks, sewers etc.

7.6.1 Preventive Steps for Civil Structures and Foundations

Design of these facilities shall be done after considerations of:
Load bearing capacity of land at site. This should be checked at all four corners and a few more (selected) places in the plant layout area. These could be around the places where heavy process units and machinery are to be erected. These are very important steps for expansion of plant capacity as additional units may be erected.

7.6.2 Safe Load Bearing Capacity of Soil

There are several methods available for testing. However expert advice must be obtained.
The safe load bearing capacity of soil must be established before constructing civil structures specially for concentrated loads like process reactors, boilers, distillation columns, overhead condensers, elevated storage reservoirs etc.
Type of soil (rocky, sandy, with clay) and water table in nearby area, location of water bodies, rivers, highway with heavy traffic, railway, shall be taken in to account while designing the footing. Its own weight should also be considered.

7.6.3 Alternative Plant Layouts

Two or three different alternative plans for installations of the process units, machinery, control room etc. shall be kept ready since some area may be found unsuitable to erect heavy units (due to less load bearing capacity). *It will not be possible to erect heavy units and their accessories at such spots.* The layout may need major changes in such cases in order to address such condition.
The other alternative plans for the layout can be considered in such case (since it will be difficult, costly and time wasting to search another site or changing the process units).

7.6.4   Civil Structures and Foundations

Estimate the weight of each process unit, machinery, and all necessary equipment (when full of materials and fitted with required accessories like gear box, motors), work platforms, railings ladders, safety vents and valves, connected gas ducts and liquid piping etc.

Consider other weights which may also get added due to some amount of raw materials stored near to the process unit *on the work platform itself*, and weight of the operating personnel.

The civil and structural design may be done with an adequate safety margin for above loads, plus wind loads and an additional margin to withstand an earthquake up to 7.0–8.0 on Reichter scale (if more safety is desired in case the site is earthquake prone)

The design incorporating safety margins (and with steel reinforcing members) is to be obtained from **licensed Structural Engineers** and constructed under strict supervision.

All load bearing members which may be subjected to corrosive conditions during plant operations should be covered with corrosion resistant coating.

These will include columns and beams specially for agitated vessels. Special attention is to be given to them when planning for expansion of plant capacity, diversification to more products or when some additional equipments are proposed to be added.

The steel bars, schedules for bar bending, underground footing for the foundations and composition of **cement-stone-sand mixture** shall be approved by experienced licensed civil engineers and architects.

Adequate ventilation, escape routes, wide stair cases, glare-proof lighting must be available in the working area.

All construction activities including curing of concrete mix must be supervised by experienced construction engineers.

No process unit or machinery should be placed on the foundations, beams, floors etc. till these have been cured and have acquired sufficient strength (as per tests to be carried out by civil engineers before granting clearance for the same).

Elevated Storage Reservoir support columns should be designed with extra margin for load since this will be a water storage tank with thousands of liters of water at a height. It should also be **Certified by statutorily licensed Structural Engineers**.

7.6.5   Other Civil Works

- Provide strong internal roads for movement of heavy vehicles and process units like heat exchangers, process reactors.
- Construct Storm water drains with slope towards effluent treatment plant,
- Cooling tower basins must be water proof.

- Raw water storages, Soft water and De-mineralised water plants. ETP equalisation tanks, settling and treatment tanks should also be of sound leak proof construction
- Civil foundations of all pumps, process units, compressors must be at least 300 mms above finished ground level
- .Provide dyke walls around storage tanks, and all those areas where such chemicals are handled
- Provide acid/alkali resistant tile lining to all such foundations and dyke walls.
- Provide Rust resistant, corrosion resistant coatings on all structural members, columns and beams, roof trusses and covers,
- Electrical Installations–drain pits and kerb walls around transformers and rectifiers shall be checked and approved by electrical engineers and local statutory authorities.
- Provide lightening arresters for storage tanks to be used for inflammable materials

7.6.6   Observations and Maintenance

- Check any tilt or cracks in the foundations or any exposed steel rods.
- Check any water logging in the premises, any damage to walls of buildings, leaking water taps, drains, and choking of storm water drains.
- Check any corrosion of reinforcing steel members or any ingress of corrosive liquids in the beams, columns in chemical process areas.
- In case of seepage make slurry of cement with water proofing agents and inject under pressure at selected spots. Check by injecting a strong water soluble dye. Do not take into use till it is set and fully cured again
- Repair all acid and alkali resistant lining on foundations, support beams. Provide new tiles (minimum 40 mm thick) with acid resistant cement. Cure by brushing with dilute acid while having good ventilation fans at work place.
- Reduce operating load on beams immediately or provide additional support columns if some tilt is observed. Shift the heavy load to another location at the earliest and reconstruct the beams. Obtain expert advice.
- Use only high strength composition of cement, clean washed metal (stone pieces) and graded clean washed sand.
- **Use potable water for construction and curing. Do not use sea water, waste water, acidic or chemical effluents for curing as it can weaken the structure.**

Chapter 8
Tools and Facilities for In-House Maintenance

8.1 Arranging Facilities for In-House Maintenance

It is necessary to keep a close watch on all units related to:

- **Safety matters**: all units operating at high pressure, temperature and handling dangerous materials, fire fighting, emergency power supply, piping etc.
- **Effluent Treatment Plant and Air Pollution Control units**
- **All plant units** to be operated for Production and maintaining product quality
- **Process units and plant machinery** due for major maintenance or for replacement
- **Jobs to be taken up** during annual shutdown for overhaul

Regular condition monitoring, preventive maintenance, predictive maintenance and timely repairs of such equipments must be carried out by coordination with production department in close consultation with marketing and stores departments before taking up such jobs.

8.1.1 Maintenance Tools and Facilities

Some of the essential tools and facilities to be always available for Mechanical and electrical maintenance trades are

- Sufficient sets of oxy-acetylene gas cutting sets and brazing sets.
- Welding transformers (AC and DC based) and special warm, dry storage facilities for welding rods
- Personal Protective appliances for welders, fitters

© Springer Nature Switzerland AG 2019
K. R. Golwalkar, *Integrated Maintenance and Energy Management in the Chemical Industries*, https://doi.org/10.1007/978-3-030-32526-8_8

- *Welding sets shall be equipped with sufficiently long insulated cables (including earth cables).* **Existing process piping or vessels shall not be used for earth connection.**
- Movable tripod stands along with chain pulley blocks of various capacities
- Separate sets of chain pulley blocks (1-2-5-10-20 MT capacity),
- Number of link chains of sufficient lengths
- Wire ropes, nylon ropes,
- Sets of spanners (fixed/ring/box type), Allen-key sets,
- pipe wrenches, rubber hammers, wooden hammers
- Tools such as bearing heaters, bearing and coupling pullers, torque tightening machines,
- feeler gauges of different types for checking clearances in pumps/blowers
- Portable folding type scaffolding sets
- Light folding type aluminium ladders
- Bench vices in adequate numbers
- Tripod stands with chain pulley blocks
- Dial gauges, Laser alignment test kits
- Belt tensioners, screw extractors,
- Manual hydraulic test unit (water tank, HP pump, standardised pressure gauge) shall be very useful for in-house testing of pressure vessels, pipelines.
- Lapping arrangement for steam traps and grinding of valve seats

8.1.2   Dynamic Balancing Machine

It is advisable to have such machine(s) in-house, if there are a large number of rotating equipments. This machine is costly but it will be found useful and it can warn about the sudden break down of rotating machines.

- Tong testers (**clip-on-meters**), hand held testers, **portable dryers** for electrical work.
- Transformer oil filtration and drying unit where a number of electrical transformers are installed and it rains frequently at site.
- Rewinding of electric motors and their careful baking in a temperature controlled furnace.
- Resting of resistance of winding by megger, hand held **Multi meter** instrument
- Crimping tool, cable lug filling and soldering torches, hand-held dust blowers (for cleaning electrical mains),
- **fuse extractors from mains**
- Exhaust fans and air circulating fans (pedestal/wall mounted type)
- 24-volt lighting system (flame-proof type, if necessary)
- Instrumentation repair tools (as per advice from manufacturer)

8.1.3   Maintenance Tools and Facilities for Process Units

- Personal Protective appliances for all persons for working inside closed vessels.
- Gas detectors (portable) for working in closed vessels.
- Portable breathing apparatus
- Ultrasonic thickness testers for checking shells of storage tanks and process units.
- Vibration analysers for blowers, compressors, crushers and agitators or if abnormal vibrations and abnormal rubbing noise are noticed.
- Hand-held (portable) infrared temperature detectors for detecting heat loss for from furnace shell or gas ducts, temperatures of bearings (can indicate need for immediate cleaning and lubrication), and electrical bus bars (can indicate loose contacts);
- Thermo graphic imaging cameras
- Dial thermometers, standardised temperature measuring instruments.
- Gas analysers and detectors for detecting malfunctions of scrubbers/process reactors.
- Orsat apparatus/electronic analysers (for gas composition analysis)
- Tools for digging out hard sludge and clinkers from melters, dissolvers, furnaces, storage tanks (non-sparking tools shall be used in presence of inflammable vapours/deposits). Wooden scrapers and spades can also be used
- Rotary and vibratory screens with various different screens for screening catalyst mass, castable refractory, filtering media in hot gas filters, sand from water filters
- Collecting bags, containers for the dust and screened material
- The screens shall be motorised and covered by a hood along the length. An exhaust fan shall be attached to the hood to suck out the fine dust. Discharge side of the fan should be connected to dust collectors (cyclone, bag filter) before discharging to atmosphere.
- Special scraping and collecting tools/wire brushes for removal of packing from absorption towers, scrubbing systems
- Exhauster for slowly sucking out dust, particulate matter from process vessels like tanks, converters, shell side of heat exchangers, before they are opened for cleaning
- Lifting ropes, nylon slings and pulleys for removal of internals (demister candles, support grids, separating plates, demister pads) from absorption towers, reactors, scrubbers etc.
- Washing tanks of sufficient capacity for careful cleaning of tower packings like raschig rings, demister candles; and tanks for preparing solutions of cleaning/neutralising chemicals.
- Rubber, plastic, HDPE, Teflon hose pipes with external braiding by stainless steel; and couplers to transfer corrosive chemicals
- Circulating systems for cleaning shell sides of boilers, heat exchangers by inhibited acid, and collecting the waste liquor for neutralising subsequently
- Ventilation Fans and blowers.

- A stock of sealing compounds to arrest leakages from various piping, flanges, etc. shall always be available. These should be compatible with the fluid inside
- Sufficient numbers of empty drums and bags of 50, 200 and higher capacity for storing the washed material prior to filling back in respective process unit
- Special multi pronged wire brushes to clean support grids of catalyst, filter media, sand and active carbon filters
- Scrapers for cleaning grooves in plates of filter press, PHE,

8.1.4 Refractory, Rubber, and Other Linings

- Scrapers, cutters (for removal of damaged portion of old lining) and for laying new lining. Special fixing wooden blocks, pipes with smooth plastic ends are useful to keep holding the lining for the initial set.
- Curing facilities (steam supply, oil firing, small portable stoves etc) are to be used thereafter for baking.
- Acid resistant lining is to be cured by brushing with dilute acid. Exhaust fans shall be available for work inside a vessel.
- Quick setting cements mixture, wooden plugs, rubber plugs, lead plugs and quick setting chemical resistant pastes for emergency use

8.1.5 Equipments for Fabrication

Management should procure drilling machine, plate bending machine, lathes, grinders, shapers, band saws if more repair jobs are to be taken up or minor fabrication is to be done in-house.

These can be used for some site fabrication work of ducts or small tanks for water etc. in-house and hence can be useful for future expansion work also. More facilities may be procured and installed if funds are available and required by plant engineers.

Pressure Vessels shall not be fabricated in-house unless statutory permission is taken.

8.2 Instrumentation Control

8.2.1 Introduction

Proper Instrumentation is necessary to efficiently achieve the main aims of the chemical plant which are:

- Safety of human beings, process units and machines
- Environmental Pollution Control,

- Minimise energy consumption—Conserve and recover energy
- Ensure quality of products
- Guidance to minimise cost of production
- Rate of production to meet market demands
- Research and development

Safety detection and warning of unsafe conditions like (i) high pressure or temperature (ii) high or low levels in process tanks or cooling/chilled/boiler feed water tanks (iii) extinguishing of flame in oil fired units (iv), detection of toxic gas concentration in closed vessel.

Environmental Pollution Control detection of pollutants in stack gases, analysis of effluent.

Quality of Products meeting specs given by clients.

Minimise energy consumption by better process control.

Monitor energy recovery Waste Heat Recovery System (Vacuum in Condenser for Steam Turbine working of cooling tower, DM feed water analysis and conductivity, Dissolved Oxygen content.

Smooth and efficient operation to minimise cost of production
This depends on monitoring the process, treatment of raw materials to remove harmful contaminants which can reduce catalyst activity, affect process units and product quality. Analytical instruments are essential for checking treatment of raw material and quality of other inputs for the process. Hence timely maintenance of instruments is essential.

- **Developing a new technology with own Research**: Very careful commissioning is required. Accurate measurement of process parameters for (i) ensuring safety of personnel and machines, (ii) preventing loss of material (due to escape of unconverted/uncondensed materials)(iii) minimising generation of effluents is essential

8.2.2 Parameters Generally to Be Measured and Controlled

Pressure, flow, temperature, levels in process vessels, pH, solution strengths, gas concentrations, moisture contents, humidity, consumption of power and fuel..

Typical plant equipments which need such measurements and controls
Raw material storage tanks and feed systems for process, water treatment, effluent treatment plant, reactors, furnaces, condensers, exit gas analysis after air pollution control,

8.2.3 Procurement of Instruments

- It is easier to operate the process when the right instruments are procured with sufficient spares. Provide full details of the process fluid, desired characteristics of accuracy of measurement, quick response to change in process parameter, wide range for indicating change in values, easy to install and repair, minimum effect of nearby objects, data transfer connections, easy availability of spares as well as position of installation, line size, desired end connections(−flanged/ screwed ends).
- Maintain spares for two years smooth operation.
- Real time connectivity to Computers in Control room
- **Types of instruments required**—Level indicators, thermocouples, pressure transmitters, flow rate controllers, pH controllers, analysers of exit gases and effluents
- **Multi functional**: such instruments can indicate, record, control the process parameters and warn about development of dangerous situations

Some common instruments
- Thermocouples, pressure gauges, dial thermometer, level indicators, conductivity meters

8.2.4 Some Special Instruments

- Electromagnetic Flow meters (Temperature, viscosity and density do not affect the readings of flow measurement. Generally no maintenance is required for the flow sensor since it is a simple straight tube with no moving parts) It can be available in wide selection of corrosion resistant lining
- Follower Magnet level indicator—Magnet inside stainless steel tube(for oleum tanks)
- Internal graduated scale in liquid SO_3 tank—with sight glass and light glass
- Toxic gas detectors (H_2S, CO, Chlorine, ..)
- Inflammable gas/Smoke detectors
- Limit switches –Proximity switches
- Polluting gas detectors—High volume air sampler with dust collector, suction air blower and timer sampler
- CO, CO_2 detectors for exit flue gases—for combustion efficiency
- On-line detectors for acidic gases exiting from chimney
- Steam Flow meters
- Opacity of Clear liquids
- Dual gauge (pressure and vacuum) *for SO_3 condenser*
- Conductivity indicators for boiler water

8.2.5   Typical Working Principle of Some Instruments

- Thermocouples—generation of current
- Pressure gauges—deformation/displacement of mechanical linkage
- Contactless Flow meters – generation of induced current in external coil
- Rotameters—rise of float in tapered (glass) tube.
- Strength indicators—measurement of conductivity and comparing with standard
- Optical pyrometer—Flame monitor-Photoelectric cell

8.2.6   Procure Necessary Accessories

- Thermo wells suitable for process conditions
- Connecting cables, cable trays and selector switches
- Air drying plant with accessories (air filter, compressor, motor, air drier columns, air receiver (with safety valve, pressure regulator, auto on/off switch, moisture drain..)
- Sample Cooler/Conditioner –filter
- Weather protection covers/enclosures
- Wetted parts of instruments should be of suitable MOC
- Bypass lines—isolation valves——work platforms and railing
- Mounting nozzles—siphon legs/coils for steam pressure gauges
- **Calibration test bench**, standard weights, gas cylinders of known purity to make synthetic gas samples, substances with known melting and boiling points

8.2.7   Controlling Instruments

These are based on accurate measurement and comparison with desired values/set points at output (in case of automatic controls) followed by corrective actions.

Means of Control: Control Valves, Solenoid valves, mechanical, electrical or pneumatic actuators.

Feed back control: control action is initiated on input side when a deviation from set point value of process parameter is detected on output side. It works on the principle of detection of error from set point and may take longer time to change if the volume of process unit is large.

Feed forward control can be possible when the effect of control action at input side can be reasonably estimated on the exit side value of the process parameter.

Choice of the control depends on the through put rate, tolerance limit of deviation from set point, accuracy of sensors, time required for control action and knowledge of effect of control action on process.

8.2.8 Precautions to Be Taken During Installation of Instruments

- Avoid dusty location, where rain water can fall directly
- proximity to surfaces and ducts at high temperature,
- There shall be no direct impingement of flame on sensor;
- Strong magnetic fields nearby can disturb working of electromagnetic instrument
- Bus-bars carrying heavy currents—can have induced magnetic fields around them
- Vibrations from heavy machinery nearby (crushers and grinders) can disturb set points
- Avoid inaccessible locations—repairs are difficult at great height if safe working platforms ladders and railings are not provided
- Improper lighting on dials/probes/connectors make repairs difficult
- Logical increase/decrease in controlled parameter not in accordance with control knob movement
- Non—linear dials; very small dials/pointers of very short length not reaching up to graduations on measuring scales can result in erroneous reading by operators.
- Dripping/leaking acid or alkali; proximity to steam lines can damage connecting cables of instruments and wrong indication may be seen in control room

8.2.9 Auxiliary Facilities and Equipments to Be Kept Ready

- Calibration test benches
- Solutions of known strengths/pH
- Test papers for NH_3, H_2S, SO_2...Cl_2, pH papers, universal indicator solutions
- Convex mirrors to "see "the reflection (amount) solid in a deep tank which is at a height
- Metering tanks for sale of liquid products; To be provided with a sight glass and overflow line back in to main supply tank (to prevent excess filling of tankers)
- Overflow alarm for tanks
- Small day tanks in plant to prevent excess feed (the tank will get emptied)
- Metering pumps instead of centrifugal pumps to prevent excess feed
- Gravity flow through rotameters only—bypass lines only if necessary
- Vents and Safety seals with distant release of gas through alkali seal pots
- Variable Frequency drives with RPM indicator for shaft speed
- High temp alarms

8.2.10   Some Common Reasons for Wrong Indications/ Malfunctioning of Instruments

- Choked pressure taps; damaged sensors, coating of dust on probes, MOC of thermo wells not being a good conductor of heat, wrong/far away location of sensors, flame impingement on sensors, vibrations in equipments/pipelines, gas or acidic liquid leaking and falling on sensors or connecting cables
- Lower thermal conductivity of thermo well—(ceramic)
- Improper positioning of thermo well/sensor..(should be at interface of catalyst bed and quartz bed)
- Damaged sensor––/ compensating cable is cut/contacts are corroded
- Choked pressure taps due to dust
- Parameter value more than maximum of instrument range
- Long sampling line—(delay time.. = volume of line/rate of draw of sample)— this can delay control action
- Sample cooler not provided—(for hot boiler water; hot depleted oleum)
- Hot surfaces near probe—
- Flame impinging directly on sensor for temperature
- Dial is not linear—difficult to estimate—pointer stuck up—heavy vibrations nearby
- Poor lighting on dial—dial is too small—many dials on same panel may cause confusion during emergency or panic situation
- Electronic circuit faulty
- Leaked diaphragm in pressure gauges

8.2.11   Corrective Actions

- If a particular instrument shows faulty readings—first check from list below.
- 1. Repair—sensor – connecting cable-instrument component--
- 2. Recalibrate the instrument
- 3. Replace by another properly calibrated instrument and confirm process parameter readings Then adjust process feed/controls accordingly.
- 4. Check another property—
- *Example from sulphuric acid plant*
- *If the temp of furnace is suspected to be wrongly indicated then check the SO_2%, glow inside, production rate, rise in converter bed temperatures, colour of gas leaking from vent points provided for this purpose*

8.2.12 Maintenance of Instrumentation on Regular Basis

• Clean the instrumentation probes, or pockets in which they are installed for correct indication.
• Clean the gas sampling lines or provide suitable filters upstream of probes to arrest dust particles, acid droplets in them.
• Avoid long sampling tapping lines since these may get choked or can reduce response time of the instruments (while a quick response is essential for immediate warning about dangerous situation)
• Check Gas Sample Conditioners for cooling and removing acidic condensates,
• Pressure Points – Remove choking by compressed air/ by cleaning rods.
• Clean electrical contacts for quick response to dangerous situations
• Provide thermo wells of correct length and proper Material of Construction [Ceramic tubes, Stainless Steel, Heat Resistant Steel) which is suitable for the location in process unit, gas duct, furnaces
• **Connecting cables** from sensors to measuring instrument, recorders to be checked every week
• *Avoid a path which is very near strong magnetic fields, high temperature units (furnaces) or is directly below strong acid/alkali pipes.*
• Float operated level indictor – clean the guide tubes of any deposit inside.
• Conductivity probe for low level in boiler is a device which can be more reliable for safety of boiler.
• Adequate lighting (flood lights) for level indicators, gauge glasses, dial thermometers, operating valves, pressure gauges must be in good working condition.
• Air drying plant-regenerate the dryer columns. Check drying media and replace if necessary. Check the air receiver controls, safety valves, drain valves, trip system, moisture removal trap etc. so that controllers and other instruments will have assured supply of dry air as per limits of dew point specified by OEM of instruments.
• Check mechanical links of actuators and lubricate as per OEM advice.
• Check electrical contacts and circuits of electrical actuators for controllers and attend.
• Special tools for maintenance for instruments shall be procured as per advice from manufacturer. The maintenance shall be carried out under expert supervision; and calibration and working to be checked after the job is over.

8.2.13   *Some Typical Thermocouples*

Types	Conducting wires	Temp. range °C	Remark
B	Platinum-30%/ Rhodium	1300–1600	Generally not used below 600°
E	NiCr/Constantan	0–850	Better accuracy
J	Iron/Constantan	0–750 (see remark)	Gets oxidised above 550 °C
K	Ni-Cr/Ni-Al Chromel-Alumel	0–1260	Gets oxidised above 750 °C, hence recalibration required above this temperature
R	Platium-13%/ Rhodium	850–1400	Stable

8.2.14   *Selection of the Thermocouple*

The selection of the thermocouple depends on process conditions, corrosive atmosphere, maximum temperature which may reach occasionally. Platinum based thermocouples are costly.

Chapter 9
Trial Runs and Restarts of Equipment After Maintenance Work

The production engineers shall carry out the condition monitoring of all equipments and note any shortcomings in the performance. The required improvement in performance of the equipments shall be worked out for achieving the required safety, product mix and energy efficiency. They shall discuss this with the maintenance engineers.

9.1 Analysis and Planning for Maintenance

Plant engineers and experienced technical personnel shall look in to the likely (or more probable) causes by identification of problem areas which can result in disturbance in performance. This can be done on the basis of operating and maintenance history of the equipments; and their own knowledge and experience of maintenance.

- Root Cause Analysis of the deterioration in performance shall be carried out to find out the real reason.
- Plan the activities to restore the required improvement in the performance of individual machinery or the process units in the plant.
- These activities can be planned by preparing a list of activities. The improvement expected, the cost required and the time required for executing them should be carefully studied. Required inventory of spares, tools and other necessary inputs shall be arranged before taking up such activities.
- Compare the planned activities with original design and OEM instructions also.
- Procure spares from OEM (reconditioned spares to be used only if original spares are not available)

© Springer Nature Switzerland AG 2019
K. R. Golwalkar, *Integrated Maintenance and Energy Management in the Chemical Industries*, https://doi.org/10.1007/978-3-030-32526-8_9

9.2 Mechanical Trials

- There will be different procedures for the process units and machineries. These shall be studied and any doubt shall be clarified from senior engineers or the vendors. Some general precautions are:
- Check proper fitting of foundation bolts of equipments
- Gaskets of appropriate material and thickness to be provided in flanged joints. Provide covers on flanged joints to prevent spray of liquids in case of a leak
- Confirm power rating, direction of rotation, trip settings, and couplings of the motors for each driven unit for which trials are to be taken.
- Check free movement of rotating shafts by hand/wrench for bigger unit; quantity of grease/suitable lubricant in the bearings, gear boxes of rotating machines.
- Safety valves shall never be set at more than working pressure. **No safety device or interconnection shall be bypassed.**
- It will found to be useful to confirm working of safety valve/vent, and alarms to take care of any dangerous or undesirable condition in all process units and machines of the plant beforehand. Confirm setting on electrical tripping relays to stop the concerned unit in case of high pressure, high electrical current drawn by motors or as per recommendation of OEM.

9.3 Pumps Used in the Plant

9.3.1 Vertical Submerged Pumps

These are *generally used for liquid service in the process plants* and do not require priming. Confirm there are separate support frames for the pumps. The vibrations due to operation of the pumps should not disturb the protective lining of the tanks on which the pumps are mounted. External arrangement shall be a permanently available installation for taking out the pump for maintenance.

The process liquid drips for some time when the pump is being taken out from the tank. Collect the liquid and put back in the tank. Thoroughly wash the pump (if the liquid is acidic/alkaline) and then neutralise the wash liquid.

Inspect all internals and body (impeller nut, shaft sleeve, wearing ring,) delivery side piping etc. Check balancing channels (holes) on impeller and clean them. Change the worn out parts by getting spares from OEM. Keep the pump in a horizontal position and support it on wooden blocks to prevent stress on it/distortion of shaft.

Examine the non-return valve and stop valve on delivery side. (properly working NRV can prevent water hammer when the pump is stopped)

Check setting of electrical trip to prevent overloading the motor.

Confirm direction of rotation and return liquid from process tank/flow rate to receiving tank by actual measurement. Either drop in level of supply side tank or rise in level in given time (30 min) may be measured against current drawn by motor and discharge pressure of the pump.

Provide anti- vibration pads or shims to prevent vibrations.

9.3.2   Horizontal Pumps Installed Outside

Open body and inspect internals (shaft, impeller, wearing ring, lock nut for impeller etc.) and replace if worn out. Provide new gland packing. Inspect bearings, coupling halves, foundation bolts etc. and attend. Follow maintenance manual/advice from manufacturer. Replace worn out components, mechanical seals etc by spares procured from original manufacturer (for pumps required to handle inflammable, corrosive, dangerous fluids).

9.4   Steam Jacketed/Heating Jacketed Pumps, Pipelines and Valves

Carry out pressure test by applying a pressure 1.5 times working pressure after the pump has been cleaned of all deposits of solidified/sticky material. Check internal clearances, wearing ring, bush bearing, condition of shaft and impeller.

Check for any steam leak also.

Store at a clean dust free place on wooden blocks.

9.4.1   Reciprocating Pumps

- Check gear box, mechanism for adjusting the stroke length; non return valves at the suction and discharge ports and the movement of piston.
- Check the pump capacity against the indicating scale by measuring the actual discharge.
- Replace the rupture disc on discharge line by a new one.
- Check dampening pot and connecting line on discharge side. The line should be clean always.
- For steam jacketed pump: check setting of safety valve and pressure reducing valve on steam supply line. Open and clean the steam trap.
- Check setting of electrical trip relay (in case the pump gets jammed due to any reason, it must trip immediately.)
- Open and clean the screen in strainer in the suction line. Change the screen by a new stainless steel screen if the holes in the old one have become bigger.

9.5   Metering Pumps

- These are generally gear pumps or reciprocating type positive displacement pumps.
- They are used to deliver the process liquid at a constant desired rate which can be adjusted on the indication dial/knob on the pump.

9.5.1   Check Points

- Calibration of supply tank.
- Strainer in suction line of metering pump
- A positive suction head is desirable, though these pumps can suck the liquid from a certain depth (consult OEM regarding limitations of the pump).
- Proper operation of valves in suction, discharge and bypass lines of the pump.
- Provision of dampening pot on discharge side.
- Working of pressure control valve, safety valve and steam traps if the piping and pump is steam jacketed to ensure proper flow of liquid/preventing of pump is steam jacketed to ensure proper flow of liquid/preventing of solidification of liquids like sulphur in the pipeline or jamming of the pump.
- All gaskets and bolts at flanged joints.
- Start the pumps at a capacity about 20% of maximum and open the valve in recirculation line (returning the flow into supply tank). Observe the flow and note the working/if there is any noise from the pump/current drawn by motor.
- Slowly raise the setting of discharge in steps of 10% each to 50-60-70-80% and run the pump for about 30 minutes at each setting.
- The volume of liquid delivered at different settings may be measured in another vessel/drums in 10–15 minutes to confirm proper working. If the volume does not correspond to the increased setting, the control knobs needs to be checked or the pump internals shall be examined again (some flow may be getting recycled to suction side due to improper closing of suction valve; or increased clearances in gears).
- The pump may be taken on service after satisfactory discharge capacity is confirmed, motor is not overloaded, and there are no abnormal vibrations or noise.

9.6   Corrosion and Erosion of Pumps in Chemical Plant

The pumps can get eroded and corroded due to suspended solids and free acidity in the liquids being handled.

- These are to be looked in to and addressed in order to increase life of the pumps
- Provide filter for suspended particles and neutralize free acidity

- Provide strainers (running and standby strainer in parallel if possible) at suction especially for metering pumps. Check any damage to grit removal screen
- Liquid circulation pumps in scrubbers, absorption towers – These can get corroded and eroded due to suspended solids, debris from tower packing materials, acidic conditions. Hence use only treated water for addition to the circulating liquor; glazed packings in towers to reduce suspended solids in circulating liquor and proper MOC for piping and pump internals. Provide pH controllers by adding alkali to scrubbing liquor.
- Use filters for process liquids (and cyclones/demister pads for gases)

9.7 Filter Press

Check the following before taking trial
- Complete assembly with filter clothes. No piece shall be torn or with passages of a larger opening than required.
- All drain valves should be closed
- Operating screw should be lubricated
- Pressure gauge on liquid inlet pipe should be calibrated. The pressure tap connection should be clean—there shall be no deposited sludge inside.
- Pressure release valve on liquid inlet pipe to be set below maximum permissible pressure to prevent damage to filter cloth.
- Clean basin below filter press.
- Check working of pumps for dirty liquid supply to filter press.
- Note the level of liquid in dirty liquid tank.
- Slowly start supply of clean water in inlet line. Open vent valve on exit line.
- Observe all gaskets, plates for any leak. Open drain valves one by one and observe flow of water.
- Close all valves and observe flow of water from exit line. Record the pressure drop in the unit when clean water is flowing through
- If there is no leak, stop water supply. Flush out residual water by compressed air if possible. Now start flow of dirty liquid which is to be filtered. Add filter aid to the dirty liquid feed tank if recommended by designer.
- Preserve sample of dirty liquid. Collect sample of filtered liquid and analyse purity (clarity). A quick test is to collect on a white sheet of paper or cotton cloth.
- Check pressure drop across the unit and flow rate of filtered liquid.
- Continue filteration and collect filtered liquid in separate tank
- Recycle to dirty liquid tank if the quality is not satisfactory.
- Keep a watch on the pressure drop across the unit as the filteration is continued.

9.8 PHE (Plate Heat Exchanger)

These are constructed by assembling a number of parallel plates with narrow grooves (passages) for flow of the liquids.

The liquids flow at high velocity when they pass through the grooves; and the heat transfer co-efficient increase as a result. It is necessary to prevent deposits in these grooves and hence strainers must be installed in the incoming liquid lines to the PHE.

- Openings are provided at upper and lower corners with special gaskets between the plates at the openings. When the assembly of the plates is tightened a continuous passage gets created for the incoming and outgoing [hot and cold fluids]
- High pressure drop indicates obstruction to the flow of liquid which could be due to some deposits in the grooves. Also check if the there is less cooling of hot fluid.
- The PHE may have to be backwashed in such situation (under expert supervision from OEM engineer) for removal of the deposits. If this does not improve the performance, then the assembly of the plates of the PHE may have to be opened, cleaned and reassembled. Special care is required for taking out the gaskets and fixing them back again. The adhesive chemical and new gaskets shall be procured from OEM only [Example; Concentrated 98% sulphuric acid is cooled from 70°C to 60°C by a PHE constructed from Hastelloy C- 276 plates and Viton gaskets.

9.8.1 Checking Before Trial Run

- Complete assembly (with recommended gaskets only), foundation bolts of the unit and connections to inlet and exit process fluids.
- Smooth operation of valves in inlet and exit process fluid lines
- Cooling water flow to be checked first. Test by N_2/Air (if permitted to check).
- Provide pressure and temperature gauges in fluid lines.
- Slowly start flow of cooling medium (generally water). Check if there is any leak
- Observe pressure drop and flow.
- Slowly start flow of hot fluid.
- Observe temperature and flow of both fluids. Compare with expected temperatures

Observations to be made during trial

- Pressures of fluids (both hot and cold) and the temperatures at inlet and outlet.
- Pressure drops on hot and cold fluid side as compared to the pressure drop when the PHE was initially commissioned and was clean.

9.9   Air Pollution Control System *(Venturi and Packed Tower)*

- Check smooth operation of venturi throat control valve.
- Variation of pressure drop on gas side in venturi as the valve is operated.
- Flow rate of scrubbing liquor through venturi.
- Observe through sight glass (with good lighting from light glass) the flow of scrubbing liquor and its distribution on the tray in packed tower.
- Pressure drop on gas side through the packed tower.
- Gas analysis at inlet and outlet of all units to check the removal of pollutants
- Presence of mist particles in exit gases of packed tower will indicate displacement of or damage to demister pads. Open the manhole and fix the pad properly/replace the pad.

9.10   Wet Electro Static Precipitator WESP

9.10.1   Checking Before Trial Run

- Gases with particulate matter (and some corrosive components) are admitted from lower side of this equipment. Internal parts of the **WESP** are rubber lined. Water is also sprayed from top to wash away the particulate matter from the unit. This has better performance than a dry ESP
- It operates at high voltage DC electric discharge between the electrode and collector tubes which are made of special alloy.

9.10.2   Maintenance Check

- Dirty gas inlet duct to the unit.
- Water fogging nozzle at inlet. Check the fine water spray in to these gases by fogging nozzles.
- Alignment of electrodes and collecting tubes
- Electrical connections to electrodes and collecting tubes
- Electrical power control circuit.
- Water spray from top to wash the tubes.
- Check the integrity of the rubber lining by spark test at 5000 volts.
- Supporting (hanging) arrangement for electrodes and fixing of collector tubes
- Remove any accumulated deposit
- Drain water from cooling tower water basin if TDS has reached 3000 ppm. Add fresh water

9.10.3 Wet ESP

Check
- Rubber lining inside. Cut off damaged portion, provide new rubber sheets (neoprene) 5 mm thick and test @ 5000 volts.
- Provide acid resistant coating on supports, top cover tec.
- Check safety valve and rupture disk nozzle at top.
- Check electrode assembly and alignment, and bus bars.

Trial Run
- Start water fogging spray in gas inlet line.
- Switch on DC power at lower DC voltage to start electrical discharge.
- Draw known volumes (through an aspirator), of dirty gas i.e. gas with suspended dust particles from the inlet gas line and cleaned gas from exit duct.
- This can be done by High Volume Air samplers for about 30 min. twice per shift, or over a continuous 24 hours period.
- Dust collector discs in collection containers are provided as below:

    (i) Dust particles above 10 microns.
    (ii) Dust particles between 10 microns to 5 microns.
    (iii) Dust particles smaller than 5 microns but bigger than 2.5 microns.

- Adjust fogging spray and power supply to arrest maximum particulate matter more than 2.5 micron, with steady power.
- Wash the collection tubes every 4–6 hours to remove accumulated particles.
- The unit may be bypassed for a few minutes or the plant run at lower gas flow rates if possible.
- Gas analysis at inlet and exit of the units to check removal of pollutants (acidic gases) and particulate matter

9.11 Oil Firing System

9.11.1 Check During Maintenance and Trials Thereafter

- I-Beam supports below oil tank.
- Thoroughly clean the tank. Flush out all deposits from cleaning nozzle and drain out
- Proper operation of the valves in oil lines, air blowers.
- Free movement of the shaft of oil pumps; alignment with motor.
- Earth connection of oil tank, pump, motor, oil heater.
- Temperature setting of thermostat provided for oil heater.
- Calibration of thermometer.

- Strainers in inlet line of oil tank and suction line of pumps. Change the strainer screen if damaged.
- Vent on oil heater shall be clean.
- Electrical contacts/cable connection box must be covered.
- Thermal insulation on hot oil lines.
- Vent on combustion chambers must be open.
- Working of dampers in air lines and ratio controller for oil supply.
- Clean nozzles in burner assembly. Partially choked nozzle will spray the oil in some undesired direction – and the flame may even impinge directly on refractory lining.
- Working of flame monitoring photo electric sensor. This is very important. It may be checked by masking the sensor to see if the signal is transmitted to stop oil pump.
- Air pressure: Air low is pressure to oil burner, due to any reason the oil pump must stop immediately.
- Electrical spark plug-change if the spark is not satisfactory or if the insulator is damaged.
- After checking above, first start air supply; and then preheated oil supply.
- Observe the condition of flame (should be steady, bright, without flickering)
- There shall be no unburnt oil/soot particles in exit gases.

9.12 Rotary Screens (for Separation)

Check
- Individual screen. Repair if holes have become bigger/Change if damaged too much.
- Drive mechanism and speed control.
- Bearings of shaft.
- Feed silo and charging chute.
- Rotary valve at a bottom of feed silo.
- Check the operation by feeding in a mixture of graded mass of grit/crushed refractory or a mixture of sand and pebbles.
- Confirm (i) proper fixing of exhaust hood on the rotary screen (ii) working of exhaust fan (iii) cyclone separator for dust removal.
- Length of screens for separating out particles of different sizes should be such that separate heaps shall be formed.

9.12.1 Trial Run of Rotary Screens

Check
- Screen analysis of feed mixture.

- Separate heaps formed when the unit is run for about 30 min. when a mixture of graded mass is fed in.
- Screen analysis of the separate heaps.
- Rate of throughput from the system.
- Removal of dust by exhaust fan.
- Check any jerks during rotary movement or jamming of drive chain.
- Provide anti-rust-coating on all supports dust collection hood and gas ducts.
- Check inclination of the screen for the desired screening rate and speed of rotation.

9.13   Condensers (Check After Maintenance)

- Flow of condensate and cooling fluid
- Rise in temperature of cold fluid.
- There should be no emission of vapours from the vent OR any build up of gas pressure at inlet.

9.14   Converter (Check After Maintenance)

- Fluid stream analysis at inlet and outlet—to check degree of conversion
- Pressure drop through the system.
- Less power consumption by the air blower (when feed is gaseous)
- Production rate of the plant.
- Temperature of fluid stream at inlet and outlet.

9.15   Ball Mill (Maintenance and Subsequent Check)

- Change the liner if damaged
- Add more balls to replace the damaged balls. The new balls shall be as per original specification.
- Product analysis at exit
- Production (throughput) rate
- Working of dust control system.
- Power consumption per unit output.
- Operation of belt drive and speed control (where provided)

9.16   ID Fans (Check During Maintenance and Trials Thereafter)

- Wash the internals (volute and impeller) by clean water till all sludge inside is removed completely.
- Drain out the wash liquor and send to effluent treatment plant.
- Check bearings for any damage.
- Check internal protective lining.
- Check lining on tips of impeller if it is run on high speed.
- There should be less vibrations, less noise and more suction created by the ID fan when all deposits have been removed.

9.17   Melter/Dissolver (Steam Heated with Agitator) During Maintenance and Trials Thereafter

- Working of pressure reducing valve and steam trap
- Setting of safety valve on steam inlet line.
- Rate of melting/preparation of solution. (there should be no solids in overflow at exit)
- Power consumption by agitator.
- Working of fume extraction system (bag filter/scrubber and ID fan).

9.18   Heat Exchanger (Check After Maintenance)

- Exit temperature of cold fluid.
- Exit temperature of hot fluid.
- Calculate the heat balance to check the heat transfer efficiency.
- Pressure drop on hot fluid and cold fluid sides. (whether any further cleaning needs to be done)

9.19   Hoists and Electrically Operated Travelling EOT Crane (Check After Maintenance)

- Confirm and check pin and hook, wire ropes. (no strand should be broken)
- Smooth movement while lifting and transferring loads.
- Current drawn by the motors.
- Noiseless operation
- No wobbling in vertical and horizontal movements.
- Proper winding of wire rope on the drum (without getting entangled).

9.20   Belt Conveyor, Bucket Elevator (Check After Maintenance)

- No spillage of any material during conveying
- Proper pick-up of material at feed end
- Proper discharge from chutes at delivery end
- No rubbing of buckets on the casing.
- Check operation of tripping switch by pulling a string along the length of the belt.

9.21   Screw Conveyor (Check After Maintenance)

- Smooth movement of the material through the unit.
- No rubbing action on the flights against the casing
- Load on the drive motor
- No jamming of the unit due to excessive feeding of material (check working of feed inlet and if there is any choke at discharge port)

9.22   Disposal of Waste from Various Units

- Metallic scrap: Clean and recycle usable components. Repair damaged components if possible. Sell scrap to manufacturer of alloy steel plants.
- Oily sludge: Take out carefully by using wooden shovels or non-sparking alloy shovels and collect in drums for use as fuel **if permitted** by statutory authorities for certain users or send to authorised party who operate incineration plants. Physical and Chemical analysis of sludge shall be done before disposal. Records should be available. Preserve samples for future reference.
- Waste catalyst: Regenerate if possible or return to manufacturer. High value catalyst shall be checked for recovery of costly metals like nickel, silver.
- Demister mesh pads. Reuse in plant by repairs the holding cage if possible. If the mesh itself is damaged then do not reuse it. Instead it can be used for primary chamber of ETP for arresting floating impurities.
- Toxic sludge or solids: Should be neutralised at site and then packed in leak proof drums, shall be sent to an authorised party who operates incineration plants. Physical and Chemical analysis of sludge shall be done before disposal. Records should be available. Preserve samples for future reference.
- Refractory waste, catalyst bed support (ceramic): Can be crushed, screened and cleaned by removing dust for reuse of this material as filter media in hot gas filters or as filler material during repairs to refractory lining (which are to be operated only at low temperatures not exceeding 600°C.) Manufacturer of refractory may buy back some of the waste material for reprocessing.

- Sulphur sludge from melters. This is an inflammable material which gives off SO_2 gas on burning. This can be used for production of liquid SO_2 by the special process developed by NEAT Services, Mumbai, India.
- Plastic waste and other combustible matter: This can be sent to cement plants and used in small quantities as supplementary fuel. These cement plants must operate at a temperature above 1200°C and should be equipped with multistage air pollution control units since the plastic waste can contain harmful compounds. *Permission from local statutory pollution control authorities is necessary for this.*

Chapter 10
Management Approach to Increasing Energy Efficiency

10.1 Erection

A considerable amount of site fabrication is required in many projects for large vessels like storage tanks, converters, distillation columns, absorption towers, structural supports, gas ducts of large diameter. These fabrication jobs need considerable energy, expenses and efforts.

The demand of power for erection work should be minimised till regular production activities are started because certain fixed charges are levied on the basis of application for power.

10.1.1 Mechanical Trials

Repeatedly stopping the plant during commissioning (**due to frequent interruptions caused by equipment or machinery problems**) delays stabilising the production run, can generate off-spec products, more effluents and also becomes necessary to reset all feed rates, control valves, air/gas flows, heating-cooling system flow rates.

Such interruptions can also waste fuels, starting materials, power and efforts.

Hence the mechanical trials of individual equipments/sub systems shall be carried out for detecting any faults in erection *before commissioning* of the plant.

To be checked before taking mechanical trials
Civil foundations and base frames, securely fixed foundation bolts of machinery

Orientation of nozzles for incoming and outgoing process materials (solid, liquid, and gases; reactants and products; heating and cooling utilities) of process units.

Convenient ladders, work platforms, railing and lighting should be ready for the operation and maintain the valves/dampers in pipeline and ducts.

© Springer Nature Switzerland AG 2019
K. R. Golwalkar, *Integrated Maintenance and Energy Management in the Chemical Industries*, https://doi.org/10.1007/978-3-030-32526-8_10

Convenient location with good lighting for indicating and controlling instruments to observe and attend them whenever necessary.

Confirm proper assembly and alignment of driven machines (pumps, compressors, blenders.) with respective drive motor, and direction of rotation of the motor shall be as marked on the machines as well as lubrication. This eliminates waste of materials, energy, time and efforts during commissioning.

Confirm setting and actual working of all safety devices and logical sequence of safety (electrical) interconnections before taking mechanical trials to prevent accidents. Any malfunctioning device shall be attended and set as per safe limits required.

It is advisable to take the trials by running the equipment slowly (initially below the rated capacity) when these are to check for any leak, initial electrical load, vibration for pumps, blowers, agitated/jacketed process vessels. Later on they can be run **but only under expert supervision**. at higher speeds to check discharge pressure, flow rates, electrical load wherever possible, and Keep a strict watch on all pressures, flow, temperatures, vibrations, any abnormal noise during the trial runs.

Mechanical trials of new units like pumps, blower, fan, refrigeration and heating system, feed of raw materials, etc. shall be carried out strictly as per procedures given in the manual supplied by OEM/as per commissioning engineer.

10.1.2   Trials of More Machines/Sub-Sections of the Plant

Further trials of equipments/machinery can be thus be taken one by one in sections where:

- erection has been completed
- internal fitting have been provided
- the unit/machine is not connected to other units and independent trial is possible
- the unit can be kept isolated from other units till they are ready for trials
- the performance observed during trials is found to be satisfactory and can be relied upon during next trials/plant commissioning

10.1.3   Dry Run of the Plant (Without Feeding Raw Materials)

- A dry run is necessary before starting the actual production runs of plant.
- It is advisable to check working of all units of Effluent Treatment Plant/Air Pollution Control units and keep available necessary chemicals (alkali, lime, alum, FeSO4.) and ash buckets, sand buckets before starting any further trials.
- It will found to be useful to confirm working of safety valve, vent, and alarm for development of any dangerous or abnormal/undesirable condition in all process

units and machines of the plant beforehand. Confirm setting on electrical tripping relays to stop the concerned machine in case of high pressure, high electrical current drawn by motors or as per recommendation of OEM.

- The capacity of boiler feed water pump shall be confirmed before heating of the boiler (for steam generation) is started.
- Filling of day tank (level indicators, overflow controllers), controllers for fuel oil pumps and burner, combustion air blowers, control dampers in air lines, for high and low temperature etc. shall be checked and attended for any malfunction.
- It will be found to be useful to check such additional points and take precautions before commissioning of plant (before production runs are commenced).
- This will minimise any further interruptions due to the maintenance required during start up and also the wastage of materials, energy, manpower as well.

10.1.4   Restart of the Plant After Major Stoppage/Annual Shut Down

It is done after major repairs/replacement of some unit or after annual shutdown for overhauling the plant. It may not be necessary to follow a very elaborate procedure for restart (similar to very first commissioning but the general precautions as above can save materials, power, water, fuel, energy and time. The instructions of senior engineers or recommendations of OEM should be followed.).

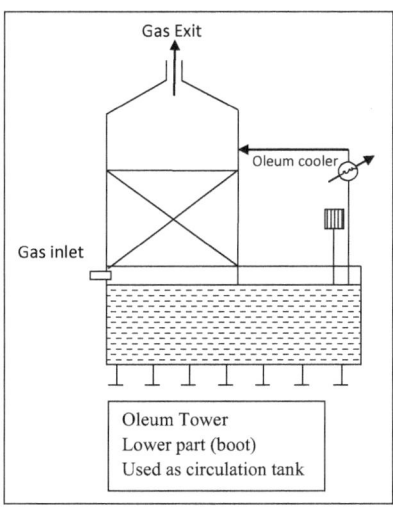

Oleum Tower
Lower part (boot)
Used as circulation tank

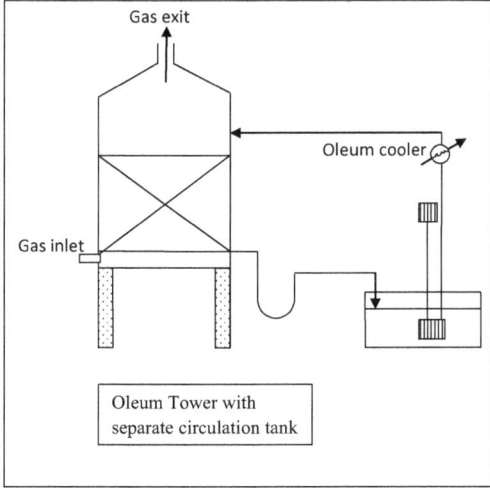

Oleum Tower with
separate circulation tank

10.2 Better Technology

10.2.1 Better Technology for Oleum

The design shown in figure above can save space, cost and maintenance for the separate circulation tank and exit piping from the oleum tower; and may operate with a little reduction in the circulation rate of the oleum.

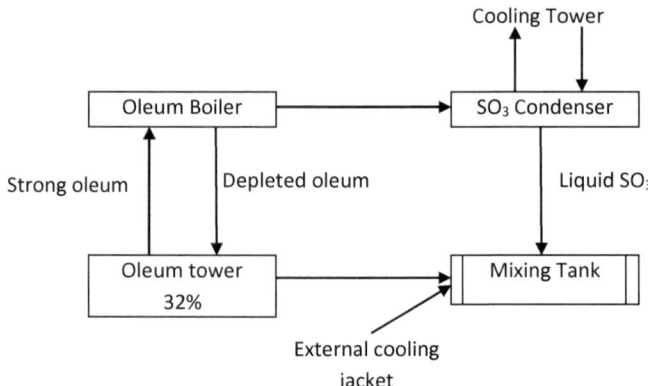

Production of 65% Oleum by batch process

65% oleum production by the batch process (please see figure above) is carried out by mixing measured quantities of liquid SO_3 with 32% oleum in an externally cooled mixing tank. It needs SO_3 condenser.

10.2.2 Continuous Process for 65% Oleum

SO_3 vapours generated by boiling 32% oleum are directly absorbed in a circulating stream of cooled 65% oleum (please see figure below). The strength is maintained by continuous addition of 32% oleum.

Product 65% oleum is transferred to storage tanks. This process does not require condenser for SO_3 vapours, nor a mixing. The 65% oleum cooler is sufficient for the purpose. It needs less space, less energy to operate and is safer.

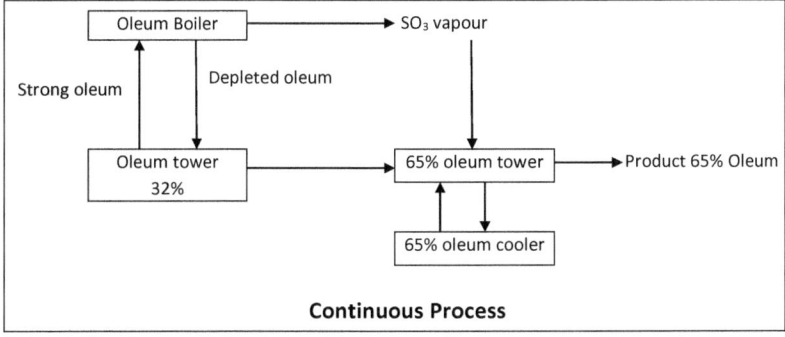

Continuous Process

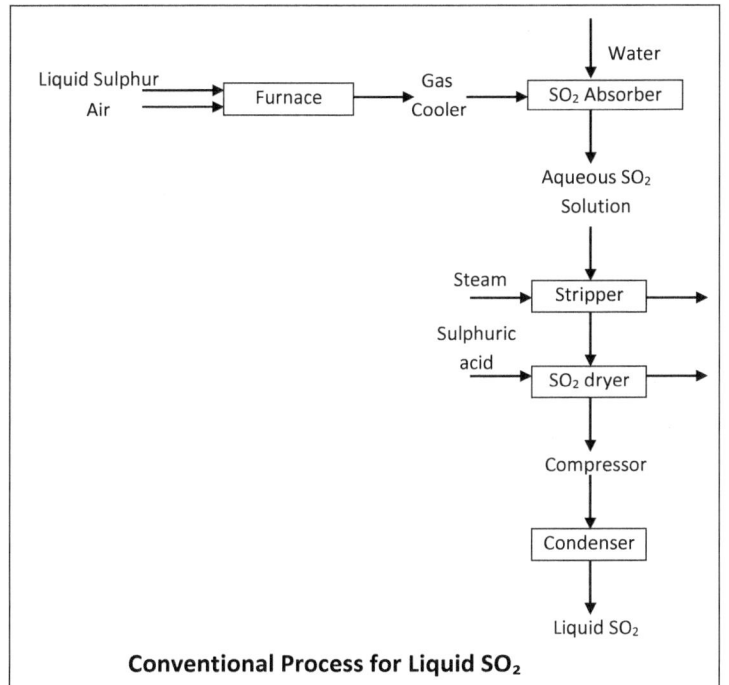

Conventional Process for Liquid SO₂

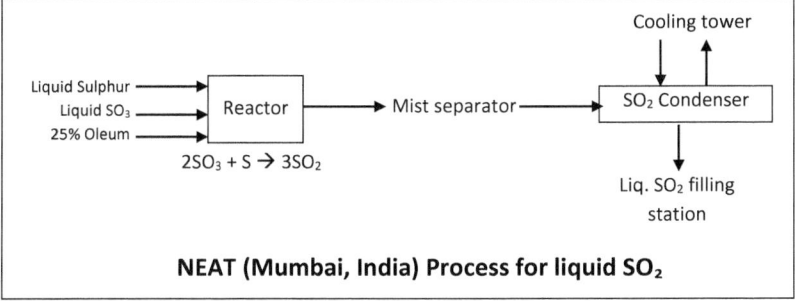

NEAT (Mumbai, India) Process for liquid SO₂

10.2.3   Better Process for Liquid SO₂

The conventional process for manufacture of liquid SO_2 involves many steps like burning of sulphur, cooling of the hot gases, absorption of SO_2 in water and subsequent stripping, drying, compression and liquefaction.

The process developed by NEAT Services, Mumbai (India) with some assistance from the author is a simple process. Liquid sulphur trioxide and sulphur are reacted in an oleum medium. Pure SO_2 is produced under pressure and can be easily liquefied. It does not need compression, refrigeration, stripping etc. No dilute sulphuric acid is produced. It needs less space, less energy and is easy to operate and maintain.

10.3   Better Plant Layouts: Some Examples

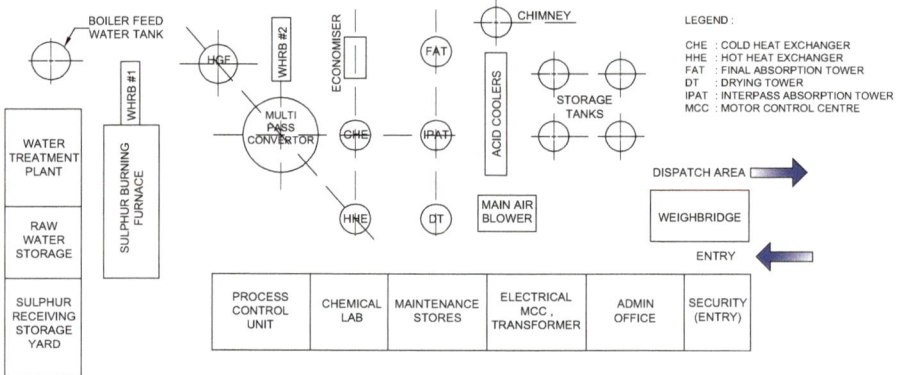

LAYOUT FOR SULPHURIC ACID PLANT

10.3.1   Layout for Sulphuric Acid Plant

Movement of materials is minimised.

It is easy to keep watch on entry of raw materials and dispatch of finished products.

Operation of important units in the plant can be conveniently watched and controlled from the Process Control Unit.

10.3.2   *Guidelines for Improving Plant Layout*

Sr. no.	Units in plant/premises	Possible measures
1.	Fuel tanks such as FO, diesel, propane, etc. and dangerous chemicals	Should be away from main process plant, working area and the place where high temperature units are operated
2.	Crusher, grinder, heavy duty compressor	They can cause vibration, noise, and dusty condition. These should be installed on ground floor but away from working area
3.	Oil firing/reheating unit/electrical heater	They should be located on ground floor where they can be seen easily
4.	Boiler and feed water pumps, maintenance store and electrical MCC	They should be located on ground floor where they can be seen easily (especially boiler water level and pressure gauge)
5.	Sulphur burning furnace of Sulphuric acid plant	It should be installed on ground floor where the inside flame can be easily observed from the sight nozzle
6.	Condenser in carbon di Sulphide CS_2 refinery	At a height for ease of draining into storage tanks (since it is no advisable to pump CS2)
7.	Day tanks for (i)fuel oil for burners (ii)chemicals which are to be added into reactor	Install at a height to use gravity flow. It saves the cost for pumping again
8.	Release point of vent on safety valves	At a height above roof of processing section and should be away from working area.
9.	Drain points, cleaning manholes of process vessel	Orient towards the sloping floor with acid/alkali resistant lining/or to join main drain leading to ETP
10.	Cooling tower	At a height/nearby unit having possibility of fire [water will be available for dousing/cooling the unit]
11.	Elevated water storage reservoir	Centralised location of plant so that water supply will be available to all unit easily
12.	Ammonia/Freon refrigeration unit	Near cooling tower

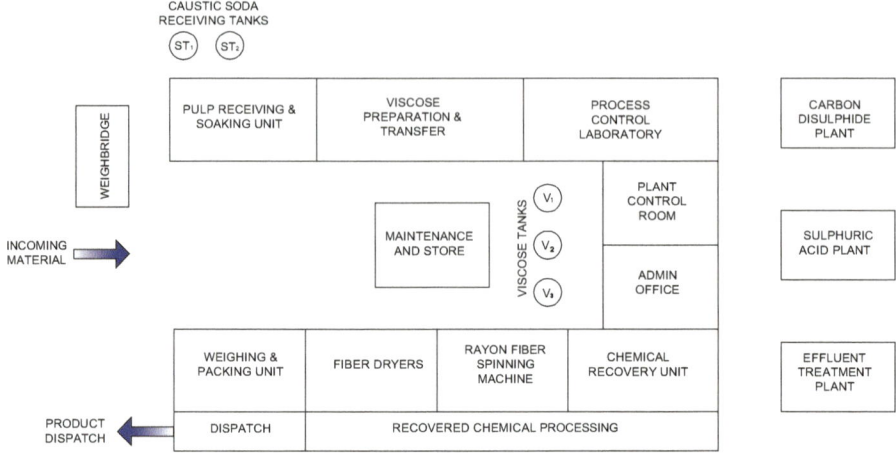

LAYOUT FOR VISCOSE RAYON PLANT

10.3.3   Layout for a Typical Viscose Rayon Plant

Movement of materials is minimised.

It is easy to keep watch on entry of raw materials and dispatch of finished products.

Operation of important units in the plant can be conveniently watched and controlled from the Main Plant Control Room. There are smaller process control units for individual plants in the premises.

Maintenance facilities and stores are easily accessible to all sections.

Corrosive chemical plants are located a little away from main viscose rayon plant.

10.4   Packaged Boilers

These are smaller capacity units with working pressure generally less than 10 kgs/cm^2 and steam generation up to 500 kgs/hr.

They are used to supply steam to start the initial plant operations (since the bigger / main boiler will generate steam after the plant is started). e.g. melting of sulphur for starting a sulphuric acid plant or evaporation of liquid ammonia to start nitric acid plant.

The packaged boilers can be used for heating jackets, tray dryers, stream tracing of process lines for batch processes and in smaller plants for which it will be costly to run bigger boilers continuously. They can be available as skid mounted unit which are ready to be commissioned in a short time.

Following shall be discussed with the vendors while procuring such boilers..

- **Required capacity** for the present and future.
- **Fuel to be used**–coal lumps, agro waste pellets, furnace oil, light diesel oil
- **Arrangement for Combustion of fuel**

  (i) Any pre-treatment of fuel required–crushing, drying, filtering, heating.
  (ii) Type of combustion chamber provided.
  (iii) feeding arrangement for fuel –whether manual or automatic.

Complete analysis of available fuel should be studied by production and maintenance engineers.

- **GA drawing** showing all auxiliary equipments and battery limit of supply by the vendor–(i) FD Fan, ID Fan, Air Pre-Heater, (ii) Electro Static Precipitator (optional), (iii)Scrubbing System for Air Pollution Control.,(iv) Supporting combustion by Oil Firing/LDO/HSD. (v) Safety Interlocks provided for safe operation (vi) Quality (analysis) of feed water required and whether vendor will supply water treatment facilities, feed water pumps, economiser (vii) civil foundation drawings, (viii) chimney for flue gases as per rules by pollution control authorities (ix) safety valves, and all necessary instruments for safe operation etc. (x) foundation bolts to be supplied by vendor
- **Generating steam from a cold start and** quickly meeting additional demand

  – What will be the maximum steam generating capacity?

- **Purpose for procurement** –initial plant start up, process heating, steady / intermittent requirement and whether it will be run continuously or intermittently?
- **Controls provided** for Safety, Feed water level control, Automatic blow down (as per water quality inside).
- Space required to install the boiler with all accessories
- Space to be kept around for Cleaning and Maintenance for the boiler
- Time required to come on full load from a Cold Start
- Time required to completely stop from full ON -LOAD condition
- MOC of Shell—Technical/trade name of Alloy Steel with Code no/Composition and Mechanical Strength properties. Get samples for testing if possible. (ii). Vendor and Country of origin
- MOC of Tubes, and Tube sheets--- Details of Composition and Mechanical properties. Vendor to give samples for testing
- Post Weld Heat Treatment to be carried out by vendor
- Instrumentation provided for safe and smooth operation
- All necessary Statutory certificates and clearances to be obtained by the vendor.
- Test certificates shall also be provided by the vendor for tubes. Shell, all components and bought out items
- Power requirement to operate the boiler (feed water pump, fuel pump. Air blower etc.)
- (*Cost of the System for generating steam at rated output shall be estimated*)
- Any other important feature which is necessary. (ash removal mechanism/soot blowing),

- thermal insulation and aluminium cladding to be provided by the vendor
- refractory lining for combustion chamber to be provided by the vendor..
- base plate for oil pump, air blower to be provided by vendor
- Civil Foundation required. (base plate, foundation bolts to be supplied by vendor)
- Time required for the completion of the project (from placing order to handing over)
- Essential spares for smooth operation of 2 years
- Any other item necessary for smooth erection and commissioning (but not mentioned in the above)

10.5   Heating Systems

Various sources of heat are used for heating systems in chemical process plants. Some of these sources are electrical power, firing of different fuels, heating by hot gases, coal firing or by heat transfer oils (thermic fluids) etc. Process plant engineers should consider the following factors while selecting a heating system:-

- heat load (amount of heat required as kcal)
- rate of heating required (as kcal/hr)
- whether the process plant is to be run continuously or intermittently
- operating temperatures to be reached in order to initiate the reactions
- temperature of process unit to be maintained for running the plant
- accuracy desired for temperature control
- safety in use of a particular method
- cost of equipments required for the heating system
- operation and maintenance costs for the heating system
- Some typical observations to be made during their operation and activities which are necessary for properly maintaining these systems are given below

10.5.1   Electrical Heating

Common Application – Manufacture of Aluminium metal, Calcium Carbide, Carbon Disulphide

Typical features of the System:
- Power supply can be Single phase/Three phase
- Heat is generated *directly inside* the unit being heated and hence efficiency is more. Typically one KWH of power input can generate 860 Kcal of heat. Most of this is generally available for heating the material inside the unit when the losses from the unit are minimised by high temperature resistant refractory bricks followed by insulation bricks.

- High temperature can be achieved (1000–1500 °C) and controlled accurately
- Quick heating from a cold start is possible due to and high rate of heating
- Operating cost depends on cost of power (it is generally higher than other methods)
- *This can be reduced if water cooled electrodes can be replaced by graphite electrodes.*
- Minimum contamination of reaction material since no ash, flue gases are introduced by the heating medium. This is a clean method for heating.
- Maintenance cost depends on cost of electrodes and fittings; repairs to refractory lining.

Some important Components of such system are:
- Transformer for power supply
- Circuit breakers and Isolation arrangements
- Process vessel/Reactor to be heated
- Oil/Air circuit Breakers
- Bus bars/Power Cables to supply electrical energy at appropriate voltage and current
- Electrodes (graphite /alloy steel)
- Arrangement to provide in new electrodes when old ones get consumed.
- Cooling system for the electrodes or their glands (water cooled) *if required.*
- Protective Refractory lining for the unit being heated.
- Feed nozzle (s)/feeding arrangements for incoming material to the process unit.
- Exit nozzle (s) for products
- Cleaning nozzles for waste generated, if any
- Manholes for internal inspection, cleaning and repairs
- Observation windows, Safety Vents/Rupture discs
- Ammeters, Voltmeter, KW, KWH meters
- Thermocouples and pressure gauges at appropriate locations
- Safety Valves, Vents and Rupture discs for release of any excess pressure (developed due to excessive heating)
- Ash handling system is not required

Some typical observations to be made during the operation
- Temperatures of the process unit (outer shell, exit materials and gases)
- Electrical parameters-power input to the system, voltage applied, current drawn by different phases *to keep track of power consumption in different phases*
- Power consumption per unit of output from the plant
- Any sparking or abnormal heating of shell near electrodes (indicates damage of electrodes, internal refractory lining or electrical insulation)
- Any abnormal rise of pressure or temperature in the system (may be due to excessive rate of heating)
- Condition of thermal insulation and cladding
- Condition of electrical insulation for electrodes, gland packing, bus bars, cables
- Any damage taking place due to proximity to vibrating machinery, dripping of rain water or process fluids on electrical components or on the shell itself

10.5.2   Heating by Steam

This requires a steady supply of steam at controlled pressure either from own steam generating boiler which can be a Gas fired, Furnace oil fired, Coal fired, or a Waste Heat Recovery Boiler *(which recovers heat from hot process gases).* Steam can be taken from exit nozzle of a back pressure type Steam Turbine as per overall system design for the plant.

Main Features
- Temperature of the material being heated can be controlled within 8–10 °C.
- Large heating surface is not required due to high coefficients of heat transfer when condensation of steam takes place.
- Generally used for heating up to 120–140 °C by using steam at moderate pressure.
- However, Steam at a much higher pressure is required for heating beyond 180 °C. This can increase cost of the system since the components have to withstand high pressure.
- **Main Components of the system**

   All these are covered by applicable Statutory Regulations for Pressure Vessels.

- Boiler with accessories, water treatment plant, feed water and steam piping, Steam Pressure Regulating Station with Main Isolation valves, Control Valves, Safety Valves, Pressure gauge
- Pressure and temperature measurement instruments (local and remote indication as per need)
- Jacketed Vessels/Internal Heating coils.
- thermal insulation on hot surfaces of process vessels, steam pipes
- equipment construction to be looked in to: external jacket, limpet coils, internal heating coils
- Steam traps and Condensate recovery system.
- Water Treatment Plant for Boiler feed water.

Common Application
- Evaporation of dilute solutions (which may be affected if dried at higher temperature), melting of solid sulphur, drying of wet materials, wet textile or rayon fibers (at lower temperatures)

   – The temperature of material being heated can be controlled within 8–10 °C.
   – Large heating surface is not required due to high coefficients of heat transfer from condensing steam.
   – Generally used for heating up to 120–140 °C. by using steam at moderate pressure.
   – Steam at a much higher pressure is required for heating beyond 180 C.

Factors for selection of system components
- Total heat required- amount of material to be handled, Sensible and latent heat, drying load for material with more moisture content
- Rate of heating required – continuous melting, evaporation, drying
- Temperature to be reached—for initiating reactions, for melting, for boiling
- Only maintaining process units at required temperature—pipelines, tanks (jackets or with tracing)-to compensate heat losses only
- *Whether the condensate is to be removed immediately OR is to be allowed to cool further*
- Steam pressure at inlet of Process Vessels
- Layout of Condensate recovery line
- Desired to recover heat from condensate also (*possibility of preheating of incoming material by outgoing condensate)*
- Thermostatic Traps—for removal of air; for further cooling of condensate in the jacket itself. Heat recovery is more but rate of heating is less.
- Thermodynamic Traps—fast removal of condensate (fast heating rate)

Some typical observations to be made during operation:
Steam pressure in main supply line and at inlet nozzle of the process unit being heated (confirm setting of safety valves), operating temperature and pressure in the process unit being heated, flow of condensate from steam trap and its recovery, leak of steam from gaskets / pipe threads, any abnormal noise or vibrations of the heating jacket or coils, condition of thermal insulation and cladding (any ingress of moisture through cracks)
Dripping of rain water or process fluids on system components

Maintenance
- Pressure Reducing Station:-isolation valves, safety and control valve, pressure gauges, pilot line and by pass line, steam traps.
- All supporting columns, clamps on steam lines,. Pressure testing of steam coils, jackets for process vessels shall be carried out @ 1.5 times the maximum working pressure.
- Attend thermal insulation pads and cladding
- Open and clean internals, actuator links of bucket type traps and discs of thermodynamic traps.

10.5.3   Oil Fired Heating System

Common Applications
- Oil fired furnaces, small packaged boilers or heating systems for thermic heat transfer fluids, rotary kilns, incinerators for waste..(where direct contact with the flue gases is allowed).
- Quick heating from a cold start is possible due to high rate of heating.

Typical features of the System

- High temperature can be achieved (1000–1200 °C) and controlled within +/−10 °C
- Quick start of the system and high rate of heating is possible.
- Contamination of reaction material is minimum since no ash or flue gases are introduced by the heating medium.
- This is generally a clean method *when no spillage of fuel oil is allowed to take place.*
- Operating cost depends on cost of oil, and operating temperature in the system. (flue gases will exit at high temperature from the system if operating temperature is high. It is advisable to have equipment like air pre-heater to recover heat from flue gases)
- Cost of maintenance depends on cost of spares for oil burners, air blowers, repairs to refractory lining.
- Easy to handle and feed controlled quantity of fuel oil

Main facilities and components required for using the system

Fuel Oil Storage tanks with accessories, Oil transfer and feeding Pumps, Day Tanks located near point of use, air Blowers [for atomising and combustion air], oil burners, Photo Electric (Flame Monitoring) cells, view glasses, oil and air flow control valves,, safety vent and rupture disc on oil combustion chamber, air pre-heater..

- Ash handling system is generally not required.
- Accumulated sludge from oil storage tanks should be removed once a year or every 2 years.
- Air Pollution Control equipments (these are necessary if fuel oils have higher sulphur content.)

Observations and Safety Issues During Operation

- When a fresh supply of fuel oil is being received- Check net calorific value, viscosity at ambient and operating temperatures, pour point, density, boiling point; ash, moisture sulphur content etc. This should be as per specifications in Purchase Order
- Check level of oil in day tanks every hour
- Check pressures in oil and air piping
- Ensure proper setting of flow rate on oil feeding pump, and pressure of atomising air
- Monitor oil consumption rate, and cumulative quantity consumed every shift/day.
- Combustion air blower- Current drawn by drive motor, discharge pressure of blower and setting of control dampers in Primary and Secondary air lines.
- Flame should be steady during the oil firing. A flickering flame indicates choking in system, improper combustion; need for adjustment of oil flow, air flow...
- Check presence of carbon particles in flue gases (indicates inefficient burning)

- Colour of smoke from chimney. (Black smoke indicates un-burnt oil)
- Temperature rise of the unit being heated. Compare with desired values.
- Condition of thermal insulation and cladding
- Do not allow any dripping of rain water or process fluids on electrical components
- Steady oil flow through the unit.
- Electrical contacts of heating coil and working of thermostat
- Earth connection of heater body, and whether it is becoming excessively hot.
- Oil Temperature at exit to be confirmed by another standard (mercury) thermometer so that excessive heating of fuel oil does not take place.
- Pressure in oil line and spray of oil in the furnace (flame should be steady and well spread). It should not impinge on the walls of the furnace.
- Pressure of air at inlet of burner; position of dampers in primary, secondary and tertiary air supply lines
- Colour and length of flame. Dull yellow/orange and long flame indicates insufficient air supply. The flame should be colourless (with bluish edges) or very bright indicating better combustion. Adjust pressure and flow of oil to the burner.

  - Check presence of soot particles in flue gases by letting a small amount of gas to impinge on a white asbestos piece.
  - Fire fighting arrangements must be readily available at oil storages and day tanks. Dyke walls must be constructed around storage tanks to prevent spread of oil from any leak or spillage. Day tanks must have suitable enclosures around them. They should not be located near any hot units or ducts.
  - Advisable to have arrangement for cooling water spray on storages to keep them cool
  - All oil pipes must be pressure tested before commissioning and at regular intervals

Maintenance
- Check the internal parts (including pressure relief valves) of oil pumps and air compressors and confirm that they will be able to develop sufficient discharge pressure for atomising the fuel oil. Clean spray nozzles of oil burners regularly. Replace the nozzles if the holes have become bigger and atomising of oil is not satisfactory.
- Check and attend any leak from oil lines.
- Inspect strainers in oil lines and change damaged strainer plates.
- Inspect all electrical contacts—*confirm that there are no loose connections*
- Inspect refractory blocks and lining around the oil burners. Repair small cracks by castable refractory. Replace the blocks in case of a bigger damage. Adjust position of burner nozzles if the damage to refractory lining has taken place due to direct impingement of flame.
- Photoelectric cell – Clean the port. Check the wiring to confirm proper warning signal will be transmitted in case the oil flame gets extinguished.

10.5.4   Coal Fired Heating

Typical features of the System
- High temperature can be achieved (600–800 °C).
- Control of temperature within +/− 10 °C is difficult; and depends on rate of coal feeding. It can take more time to raise the temperature as compared to oil firing.
- It can take a few hours to reach desired temperature from cold start of the system. Fast rate of heating is generally difficult.
- Heating surfaces can get deposits of ash or particles of coal. This can reduce efficiency of heat transfer. Hence ash must be removed regularly. Check presence of un-burnt coal particles in the ash which indicates loss of coal (inefficient operation.)
- Operating cost depends on cost of oil (it can be higher than coal firing)
- Cost of maintenance depends on cost of spares for coal crusher/pulveriser, screw feeder, coal burners, air blowers, repairs to refractory lining.

Main Components
Coal handling – Coal Storage shed/yard, Weigh Bridge, Conveyor, Pulveriser, Screening System, Dust Control arrangement, Silos for storing pulverised coal, Screw feeder

Air Blowers for: Primary and Secondary Air supply

Combustion Air supply for better efficiency by completing the combustion; and to protect inner walls from local hot spots.

Air pre-heater for combustion air.

Air lines and flow control dampers

Air Pollution Control [Cyclone, ESP, Scrubbers etc.]

Ash removal and disposal arrangements

Rotary air lock valves below ash collection cyclones and bag filters, trolleys to shift the collected ash to disposal/reuse arrangement.

Refractory lined Furnace with exit duct and emergency vent for exit gases.

Temperature Control – It is difficult to accurately control temperatures of Process unit by coal fired furnace, say within +/− 7–8 °C.

Takes a long time to start and reach operating temperature.

Can generate hot flue gases up to 900–1000- °C for heating process units

Observation
- Monitor and control CO_2% in exit gases
- Temperatures of preheated combustion air and furnace exit gases
- Control on rate of coal feeding by screw feeder,
- Regular ash removal and presence of un-burnt coal particles in ash.

Ash deposit can occur on heating surface of process unit and this can reduce heat transfer efficiency; hence it must be removed regularly by soot blowers or other mechanism provided.

Maintenance
It can take a long time of 12–48 hours for the unit to cool down before internal inspection, cleaning and repairs of the brick lining of the unit (e.g. furnace, reactor) can be undertaken. Remove clinker deposits. Examine any damaged spots which indicate any un-even heating that has taken place during operation.

Check and repair firing grate

Rotary valves and screw feeders in the coal feeding system

Screen the circulating refractory material (if used for getting uniform temperature in the combustion furnace). Discard fines less than 4 mms diameter and replace by bigger particles. Confirm alumina content-which should not be less than 45%-in the material being added. *Do not use particles obtained from crushed insulating bricks.*

Air injection nozzles: clean them off any deposits or clinkers inside.

Safety issues
Following shall be carefully observed during operation and recommendations from OEM should be followed.

Operation of mechanised loading system, pulveriser, air blowers, combustion of pulverised coal, safe removal of hot ash from the system…

10.5.5 Fuel Gas Fired Heating (Natural Gas, LPG, Propane)

Main Components
- Main incoming gas supply line with accessories (pressure regulator, flame arrestor, pressure release vent, pilot burner and chimney), moisture trap and drain valves, gas flow meter with totalizer
- Safety systems having – lightening arresters, electrical earth connections, provision of fire fighting arrangements
- Smoke detectors in the vicinity of storages and gas lines, warning boards to restrict entry of vehicles with internal combustion engines
- Any other arrangements as per instructions of Factory inspector and Insurance company
- Combustion air fans
- Air flow controller for proper combustion
- Induced draft fan for the system being heated
- Flame proof motors, lighting and other electrical fittings *if required*
- Pressure indicator for incoming fuel gas and air
- Temperature of exit gases after combustion

This is generally a clean method

Observations to be made during operation
- Pressure in main incoming gas supply line,
- Gas flow Totaliser meter readings

- Combustion air fan: speed, condition of drive belts, current drawn by drive motor, discharge pressure, condition of filter on suction side
- Setting of control valves for air flow
- Flame colour, any flickering of flame
- Carbon particles in exit gases
- Induced draft fan
- Temperature and pressure of incoming Fuel gas, exit gases
- Working of pilot flame in Combustion chamber at lower side of Chimney for venting excess gas

Maintenance
- Check and attend all parts of combustion air fan and ID fan
- Check strainer screens and control valves in oil lines
- Ensure smooth operation of dampers in Air flow controller—clean pressure tapping and lubrication
- Consult manufacturer of Control Valves for fuel **before lubrication** as there can be fire in the valve if incorrect lubricant or procedures are followed
- Valves in vent line for releasing excess supply gas pressure

10.5.6 Heating by Hot Process Gases

This method is used in many chemical industries. It serves to control temperatures in downstream units as well as recover heat from the hot process gases. It thus saves cost of fuel in the plant. However, the process gases should not have corrosive properties. Temperature of the process gases in the system should not be allowed to fall up to the dew point if corrosive gases and moisture are present.

Main Components
- Incoming, outgoing and bypass ducts (for flow of hot process gases) connected to the process unit being heated
- Moisture traps/drain lines for the gas ducts
- Control valves in the gas ducts.
- Pressure and temperature indicators for the process units being heated and the downstream units. These are required to adjust the flow of gases through the unit/through bypass duct.
- Indicators for the flap position of the control valves wherever possible. The operating lever for the valve should have arrangement for locking in the desired position.
- Thermal insulation and cladding for all hot surfaces

Observations
- Setting of Control valves in the ducts.
- Pressure and temperature of the process units being heated and the downstream units
- Integrity of thermal insulation and cladding for all hot surfaces
- Operation of the process units being heated as well as downstream units.

Maintenance
- Thoroughly check the control valves in the gas ducts.
- Check accuracy of pressure gauges (or manometers) and thermocouples for temperature indication for the process units being heated and the downstream units.
- Check indicators for the flap position of the control valves wherever possible. The operating lever for the flap should indicate correct position. The arrangement for locking the lever should not be damaged.
- Thermal insulation and cladding for all hot surfaces

Safety issues
- Work platforms for control valves, thermocouples and pressure taps should be safe to go up and operate. Sufficient lighting should be available at the work platforms.
- Gas detectors must be installed gas ducts and flanged joints.
- Any leak of process gas must be immediately attended
- Thermal insulation and cladding on ducts and valves shall be thick enough to prevent the temperature of external surface rise beyond 50–55 °C.

10.5.7 Heating by Heat Transfer Oils (Hot Thermic Fluids)

This method is used for heating the process units to 280–290 °C without pressurising them. The heating jackets or the coils will have to be designed and constructed to withstand very high pressure if steam is used as a heating medium.

Main Components
Thermic fluid oil circulation tank and coils for heating circulating oil; circulation pumps (working + standby) with mechanical shaft seals, expansion tank, fuel (oil/gas/coal) firing system, piping and control valves, flame proof motors, thermal insulation and cladding.

All components having oil circulation should be capable of withstanding an operating pressure of at least 8 kgs/cm^2 and should be tested prior to use.

Ash handling system is required only if coal is fired to heat the thermic fluid.

Observations
- Level of fuel oil in day tank every hour
- Steady combustion of fuel oil/coal/gas
- Working of air-preheater (if provided), combustion air fan and Induced Draft fans
- Rise in temperature of circulating thermic oil when it passes through the heating coil.
- Pressure in thermic fluid lines at discharge of circulation pump, at inlet and exit of heating coil and process unit
- Working of circulation pumps and any leak from the shaft seal
- Occasional release of vapours from vent line on expansion tank (the vent line should not be choked)
- Condition of thermal insulation and cladding on oil circulation tank, piping from heating coil to process unit being heated; and return piping up to circulation tank.
- High temperature alarm and cut off arrangement of firing

Maintenance
- Level indicator and overflow alarm of fuel oil day tank
- Condition of volute and rotating parts of combustion air fan/induced draft fans
- Thermocouples for circulating thermic oil when it passes through the heating coil.
- Check pressure gauges and taps in thermic fluid lines at discharge of circulation pump, at inlet and exit of heating coil and process unit
- Circulation pumps and their shaft seals
- Vent line on expansion tank to be kept clean (should not be choked)
- Condition of thermal insulation and cladding on oil circulation tank, piping from heating coil to process unit being heated; and return piping up to circulation tank.
- High temperature alarm and cut off arrangement of firing

Safety issues
- Work platforms for control valves, thermocouples and pressure taps should be safe to go up and operate. Sufficient lighting should be available at the work platforms.
- All hot oil lines must be pressure tested regularly.
- Fire fighting arrangements must be available in the immediate vicinity.
- Any leak of hot oil gas must be immediately attended
- Thermal insulation and cladding on hot oil lines and flue gas ducts and valves shall be thick enough to prevent the temperature of external surface rise beyond 50–55 °C.

10.5.8   Format

Production and Maintenance Engineers shall compare the various methods of heating in the following format and select the most suitable one.

System Item	Electrical heating	Coal firing	Steam heated	Oil firing	Gas firing [Natural gas]	Thermic fluid	Process gas	Remarks
Main components of the system								
Process unit to be heated-(e.g. reactor, melter, dissolver)								
Material being handled -refer to properties of materials [MSDS]								
Thermal -melting and boiling points, specific heat, latent heat								
Chemical-toxic/ inflammable								
Input streams/ materials to be heated								
Temps. To be achieved and maintained;								
Accuracy of temp. Control required								
Heat input required								
Rate of heating required kcals/hour								
Total heat required for a batch of process								
Safety precautions, and interconnections								
Site infrastructure required for safe storage								
Other accessories required for unloading, handling, feeding								
Operating cost heat available kcal/per unit cost								
Heating cycle whether intermittent or continuous								

System Item	Electrical heating	Coal firing	Steam heated	Oil firing	Gas firing [Natural gas]	Thermic fluid	Process gas	Remarks
Items which may need more maintenance efforts and /or cost								
To be considered by plant engineers.								

10.6   Selection of Better Design of Process and Equipments

- Double Conversion Double Absorption (DCDA) process in place of Single Conversion Single Absorption for manufacturing of sulphuric acid. By operating at a higher concentration (10.5–11% SO_2) at the inlet of converter the gas volume being handled is reduced considerably as compared to SCSA process which used a lower (7.5–8.0%). This could reduce consumption of power by the air blower.
- Use of membrane process for manufacture of caustic soda in place of older mercury cell process could reduce power consumption by a large extent. Recently the Nafion membranes have further improved the efficiency and reduced power consumption further.
- Water cooled electrodes in electric furnaces have been replaced by graphite electrodes. The power consumption is much less when graphite electrodes are used. The furnace is also safer to operate since there is no chance of pressure build up due to any water leak inside.
- Plate heat exchangers in place of shell and tube coolers. The heat transfer coefficients are more and space required for installation is less. Cold fluid (generally water) can be preheated with an approach upto 5–6 °C temperature of incoming hot fluid. Heat is recovered as pre heated water and reused elsewhere in plant. Area of heat transfer can be increased easily by adding more plates if required (this is not possible in case of shell and take coolers).
- Disc-and-donut Heat exchangers (please refer below).
- Super phosphate plant: Acidic liquor is produced when gases are scrubbed in venturi and absorption towers. It contains H_2SiF_6 which is recycled to the reactor. It saves consumption of sulphuric acid; and the effort to produce diluted acid also.
- In certain locations a solution of alum is required. It is advantageous to sell them directly a solution of alum produced in the plant instead of trying to concentrate it and produce solid alum. This saves the energy required (steam) for evaporation.

DISC AND DO-NUT HEAT EXCHANGER

N-2 GAS EXIT SHELL SIDE

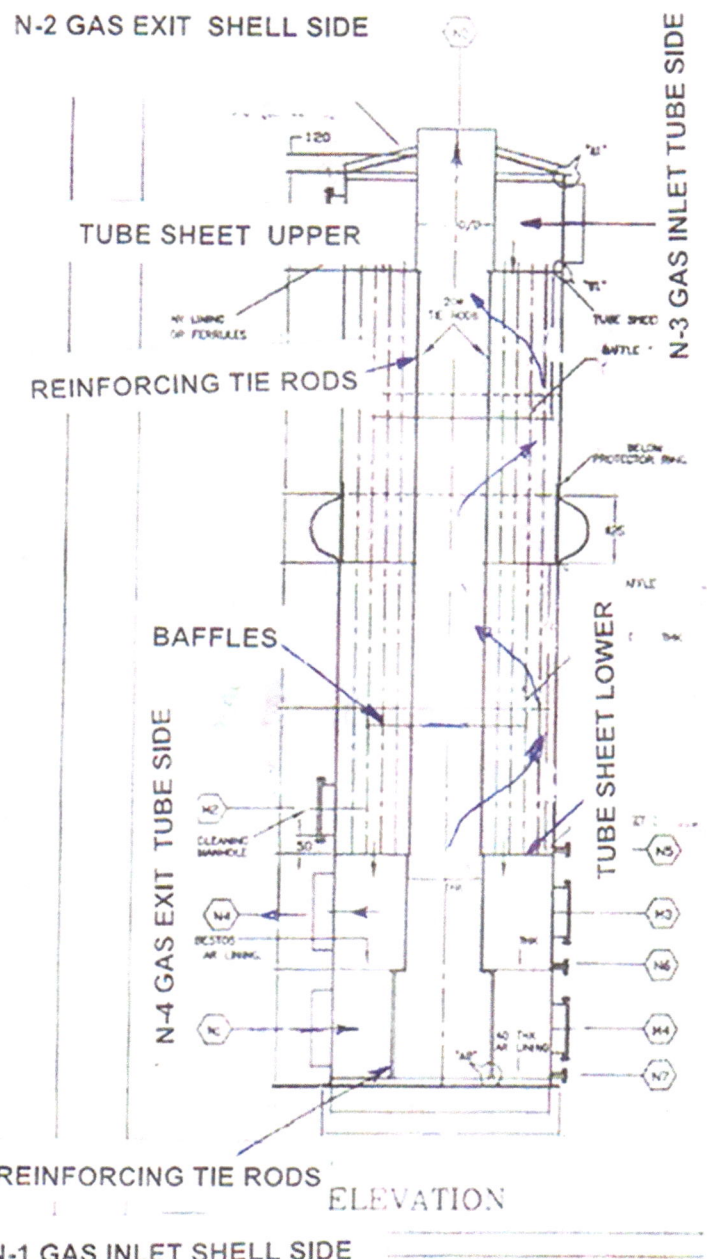

TUBE SHEET UPPER

REINFORCING TIE RODS

BAFFLES

REINFORCING TIE RODS ELEVATION

N-1 GAS INLET SHELL SIDE

Better Design of Heat Exchanger (Courtesy: V K Engineers, MIDC, Tarapur, India)

This is a Disc-and-Donut type heat exchanger. There is better distribution of gases on the shell side as there are no idle pockets (as in case of conventional heat exchangers having segmental baffles). This improves heat transfer and also reduces pressure drop on shell side. Orientations of gas inlet and exit nozzles can be adjusted without affecting performance of the unit. A similar unit was commissioned by the author.

10.7 Exploring More Sources for Heat Recovery *for Further Improvement*

- Study all exothermic reactions carried out in the plant.
- Examine (temperature, flow rate, composition, pressure, properties) of heat process fluids (leaving the process vessels) which are to be cooled and also examine these properties of cold process streams which are to be heated.
- Look into possibilities of heat exchange between these process streams.
- Suitable heat recovery equipments can be thought of and designed accordingly.
- This may not be possible always. Hence other methods of recovering the heat should be looked into.
- Explore other likely sources for heat recovery as below:

  - Hot process gases leaving process reactors, furnaces, discharge of compressors, incinerators, rotary kilns, rotary dryers, heat exchangers, process boilers etc.
  - Hot products from kilns, dryers, evaporators.
  - Hot condensates from jacketed reactors, process vessels.
  - Hot vapour gas inlets of condensers, absorption towers.
  - Hot liquids at exit of evaporators, boiler blow downs, absorbers, distillation units, process boilers.

Examples from Chemical Industries Are

Item	Name of unit/item	Hot stream can be used for
Sulphuric acid plant	Sulphur burning furnace	Generating steam by WHRB
Nitric acid plant	Reactors/Oxidisers	Generating steam by WHRB
Hazardous waste incinerators	Thermal oxidisers	Generating steam by WHRB
Sulphuric acid plant	Heat exchanger	As air-pre heater
Nitric acid plant	Heat exchanger	As air pre heater
Boilers	Blow down water	To pre heat incoming cold feed water

Item	Name of unit/item	Hot stream can be used for
Evaporators	Hot condensate	To pre heat incoming cold dilute feed
Sulphuric acid plant	Hot acid from absorption towers	To pre heat process water, to produce LP steam
Multiple effect evaporator	Vapours from evaporating section	For heating next section operating at lower pressure
Sulphuric acid plant	Hot gas leaving last pass of converter (entering absorption towers)	To pre heat boiler feed water by economiser

10.8 Selection Criteria for Energy Saving/Energy Recovery Equipment

10.8.1 Design of the System

Following matters shall be carefully looked in to:

- Process Operating conditions
- Calculate requirement of energy for operating the plant (from cold start, for sustaining the operations, for effluent treatment)
- Materials of Construction (wetted parts of all important process units and machinery)
- Energy Recovery (expected performance) potential (generally during steady state operation, and during shutting down for annual maintenance if possible)
- Existing and proposed Plant Layout with additional equipments
- Convenience during operation and maintenance

Details required for design
Process operating conditions

- Examine all operations and processes in the existing process plant
- Study all important parameters such as-(i) Flow rates, temperature, pressure, pH, presence of suspended solids, (ii) Composition at inlet and outlet of each unit
- Examine each individual process unit and machinery: (i) Residence time in the process unit. (ii) Batch time of the process (for batch process)---*heat recovery may not be at a steady rate in such cases*
- Physical and chemical properties of all hot and cold process streams being handled during operation of the plant
- Viscosity of fluids at different temperatures, specific heat, melting point, boiling point, pH, thermal conductivity, Flash point, toxicity, density of the materials handled, solubility of solids (and chances of precipitation from solution if the temperature falls below certain value)
- Properties and characteristics of each and every process stream particle/lump size of the components, composition, corrosive nature, moisture content

10.8.2 Define Required/Expected Performance from the Proposed Equipment

Estimate present losses of energy (which should be saved) from each process unit and machinery being used. One should try to collect data from other industrial units which are also manufacturing similar products using similar process. A comparison with own plant will indicate areas where improvement should be done for reducing energy consumption/increasing energy recovery.

Examine each process unit to assess the potential for energy recovery in the form of heat to be removed from system (from hot streams, from the heat evolved during reactions when the plant is running in a steady state)

10.8.3 Calculate Requirement of Energy

- When the process units/reactor themselves are to be brought up to operating condition from cold start **and**
- When raw material/process stream are to be admitted to the process units/reactor from the feeding point (storage yard or shed; storage tank) in order to initiate the reaction and thereafter for sustaining continuous operation of the reaction.
- Study the operating condition of process stream at exit end of a process unit and at entry of the next unit. The difference in **potential** i.e. pressure, temperature indicates the quantity of energy which can be recovered (and ease of recovery if the temperature/pressure differential is high enough)
- How it is proposed to recover and use the (heat) energy? The hot medium (fluid to be cooled) can be used to generate steam, pre-heat combustion air, pre-heat incoming cold fluid, or heat transfer oils. Select the best options after a detailed techno–economic study.
- When steam is used for process heating, the heat transfer area required in the cold side process unit is considerably less (as compared to using hot gases directly for heating) since condensing steam has a large coefficient of heat transfer.
- **However this may not be possible always because steam side operates under pressure and it will need higher steam pressure for achieving higher temperatures on the cold side. This will need heating jackets or coils to be suitably constructed for high pressure.**
- Heat transfer oils can recover heat from hot fluid and then can be used for process heating to reach higher temperature because these oils have high boiling points (around 300 °C) at atmospheric pressure. Hence these systems do not operate at high pressure.
- Pre-heated air is useful for better combustion efficiency (loss of un-burnt fuel is reduced/minimised) in oil fired and coal fired systems. It can be found useful for removal of moisture from wet solids in a drying system.

- Examine whether **it be possible to use all the recovered energy in the same plant** or in the same premises for some other useful operation. If not, can the excess energy recovered be exported (sold) to nearby industries?
- Generally such recovered energy is much cheaper than buying from external source of energy (as electrical power, fuel, steam, etc.) since the present consumption of fuel or electrical power can be reduced.
- However, the method and equipment to be used for energy recovery shall be technically and commercially viable. This needs to be carefully examined.

10.8.4   Materials of Construction

It should be corrosion resistant, able to stand the pressures and temperatures, pH, presence of erosive and corrosive particles when in contact with process streams (e.g. alloy steel tubes for boilers, Hastelloy C-276 for sulphuric acid service, high alumina bricks for recuperators)

10.8.5   Convenience During Operation and Maintenance

Following issues shall also be looked in to:-

- Will it be safe to operate such a system or elaborate safety procedures need to be observed during operation of such additional equipment like High Pressure boiler, Steam turbo- generator, Gas turbo- generator..
- Whether the new technology/details of machinery can be understood, operated and efficiently maintained by present employees (with their present level of skill) or new trained persons will have to be employed? Will there be a need for highly skilled personnel for operation? Or the present employees can be trained for the new machinery

Maintenance
- What is the expected cost of maintenance on yearly basis/ will it increase too much after 2 or 3 years of operation? Will any special spare parts will be required every now and then?
- Will it be easy to clean/replace the internal components (by suitable design and location of drain nozzles and cleaning manholes?
- Will the required spares be available regularly, quickly and at reasonable cost from Original Equipment Manufacturer?

10.8.6   Introducing New Units in the Plant

- Examine the General Arrangement (GA) drawing of the equipment, elevation and plan showing the planned position of the equipment in plant and the plant layout itself. Can it be accommodated in the available space or will it need dismantling of some existing units, connected piping, removal of structures? Cost of all such changes must be considered.
- Space required to install these (proposed) additional units and for connecting them to the existing units by gas ducts and/or liquid piping.
- Check the integration with existing plant—the process should continue to run smoothly without affecting the production rate, existing operations, safety of personnel and product quality after the additional units are commissioned.

10.8.7   Estimated Cost for the Energy/Heat Recovery Proposal

- Cost of additional/new equipment plus taxes and duties, licensing fees, transportation to site, civil foundation *if required*
- Cost for the following activities is to be added
- Dismantling of existing units, old thermal insulation, support structures, ducts, supports which are obstructing the erection of new machines
- Creating a parallel path to operate the new machine for a few weeks on trials
- New insulation and support structures required

10.8.8   Time Required for Implementation

- Estimate time required for all above and carries out as many activities while the plant is running. Typically these will include civil foundations (if drawings are available), laying electrical cables for connecting new motors, poles for additional lighting, arranging instrumentation etc.
- Creating a parallel path to operate the new machine for a few weeks on trials

10.8.9   General Considerations

- The rate of return on *funds invested for energy recovery* shall be such that the investment can be recovered generally within 3–4 years.
- Is it a proven design/technology?
- Is there any guarantee available for performance of the equipment?

- What is the expected time for implementation of the heat recovery scheme?
- Is equipment leasing available in case of high initial cost? Can some external party provide the additional equipment at their cost and recover it by having a share in the savings/additional profit generated?
- What is the expected life of energy recovery equipment?

If the new units do not work properly in the plant then:

- Will it affect the safety of personnel or existing units in the plant quality)?
- Will any additional effluent get generated?
- Will it be possible to attend them while the plant is working or by providing a bypass route?

10.8.10   Examples of Some Heat Recovery Systems

10.8.10.1   Hot Side.....Gas and Cold Side.....Gas

- Return cold gases from Inter-Pass Absorption Tower of a DCDA Sulphuric acid plant get heated by hot gases exiting from the second and third pass of converter (through the Cold and Hot heat exchangers)
- Heating of cold air by hot exhaust flue gases from an oil firing combustion system by an Air Pre-heater improves combustion efficiency

10.8.10.2   Hot Side.....Gas and Cold Side.....Liquid.

- Exit gases from sulphur burning furnace of a sulphuric acid plant are passed through a boiler to cool them as required at entry to first pass of converter. Steam is generated during this cooling of gases.
- Hot process gases are used to heat special thermic oils (known as heat transfer oils- which have a high boiling point) to 280–290 °C. This arrangement does not require operation of heating jacket under pressure.
- Hot process gases from exit of last pass of converter are used to boil 30% oleum in a Gas Heated Oleum Boiler to generate SO3 vapours. The cooled gases are now admitted to absorption tower.

10.8.10.3   Hot Side.....Gas and Cold Side.....Solid...

Rotary kiln for burning moist wastes in counter-current operation. Hot flue gases are sucked through the moist incoming waste by an induced draught fan. The moist waste gets dried by direct contact with the hot gases

10.8.10.4   Hot Side.....Solid and Cold Side.....Solid

Direct heat recovery may not be possible in some cases. The following example illustrates heat recovery through intermediate medium like ambient air.

Air at ambient temperature is blown across hot baked bricks loaded on trolley exiting from the Baking furnace. The air gets preheated while the bricks are getting cooled. The hot air is sucked by fans mounted on roof of the furnace and is blown across cold (moist green) bricks which are fed in from the other end of the drying furnace.. The dried bricks (which are pre-heated) are transferred to the baking furnace where further heating is done by oil firing. The exhaust air from drying furnace is sucked out and discharged through the chimney. (Please refer to Chap. 16, Figure for Heat recovery (as hot air) from hot bricks)

10.8.10.5   Hot Side.....Solid and Cold Side.....Gas

Combustion air can be pre-heated by blowing across hot product coming out from rotary dryer. This recovers heat from hot product while cooling it sufficiently making further handling easier. (Please refer to Chap. 16: Fig. 16.6 Heat Recovery from hot pellets)

10.8.10.6   Hot Side.....Liquid and Cold Side Also.....Liquid

- Preheating of 32% oleum before admitting in to Oleum Boiler is done by hot 22% oleum exiting from the boiler. (Please refer to Chap. 16: Fig. 16.5 Heat Recovery from hot oleum)
- Preheating of process water by hot sulphuric acid by means of Plate Heat Exchangers. The hot water was used elsewhere in the premises where steam was used earlier. This saved steam while cooling sulphuric acid.

10.9   Optimise Production Cycles

Reduce frequent changes in production rates and product mix.

Reduce frequently stopping and restarting the process unit.

Try to have longer production runs for a product before changing over to another product (if different *products* are to be manufactured in the same process units). This will minimise making changes in operating conditions every now and then.

10.9.1   Optimise Batch Size

It is generally necessary to completely stop the reactor for draining out, cleaning and restart after every batch of a product. Operation of the plant in small batches will require frequent stoppage. Short batches should therefore be avoided because energy and manpower are required for thorough cleaning and restart (especially when the reactor is used to manufacture different products as these may need some other raw materials).

Small amounts of products can also remain inside and get wasted during every cleaning of the reactor. *These small quantities may even contaminate the next batch.* Hence one may use either separate reactors for different products or use a reactor to produce a single product in more batches consecutively (to meet the demand in the market).

Typically this can be done for building up sufficient stock of the particular product as per demand by running the batches consecutively.

Other products can then be produced in next few days, and so on. This is to reduce the loss of materials, manpower, time and energy.

10.9.2   Avoid Frequently Starting and Stopping the Plant

Starting of a plant generally consumes fuel and power for bringing up the process units to the normal operating conditions (from a cold start). During this period very little product may be obtained. It may take many hours after *every start-up* to stabilise the plant and steadily get the product of required specification as per rated production capacity.

However more energy may be consumed (when calculated per tonne of product) if the plant is stopped and restarted frequently.

Process units (*which have to run at high temperatures during the process*) should not be frequently started and stopped as this can create thermal and / or mechanical stress in their refractory brick lining which this can reduce their life.

The loss in energy can be reduced by avoiding frequent stoppage and restart for product change over (since it needs readjusting the flow, temperature, control valves).

10.10   Saving Electrical Energy in Chemical Plants

- Minimise pressure drops by transfer of fluids through bigger diameter of pipes. Use long radius bends if possible in the piping lay out.
- Bigger diameter pipes will cost more and the initial investment will be correspondingly more due to cost of pipes, valves, supports etc. However the pressure

drop will be less due to lower velocity of the fluid. This will need less power for pumping. The saving in cost of power is to be weighed. against higher initial cost of piping. Readers may carry out the calculations at their end if the cost of power is more, and liquid is to be pumped for many hours in a day.

- Minimise loss of head (pressure) during entry in /exit from vessels by using flared nozzles (to reduce velocity head).
- select proper control valves—*minimise throttling of flow*
- Provide Hydraulic Systems for operation if viscous fluids are to be handled or high torque is required for start up of agitators in vessels or pushing pistons/ actuators for feeding system in to reactors. *Such operations may need motors of higher ratings, more accessories and consume more power.*
- Make provision of emergency power supply for (i) safety (ii) to prevent escape of pollutants (iii) for fire fighting and (iv) to prevent jamming of drives for process units or (v) choking of key equipments in case of main grid supply failure.

This will be helpful for smooth restart of the plant and thus save power ultimately

- Design flow of process material by gravity wherever possible
- Minimise unnecessary recycle of fluids back in to feed tanks/idle run of units.

However maximum amount of treated effluent is to be recycled

- Operation of process units with agitators

    (i) optimise designs of agitator blade shape, size, angle; and shape, numbers and size of baffles also
    (ii) carefully note the level of liquid inside, the speed of agitator and current drawn by the motor. Check the production rate and product quality when operating at different levels, when raw materials are being added, operating temperature etc. **Select the best condition and run the unit accordingly**. This can reduce power consumption in stirred tank reactors, mixers, dissolvers.

- Optimise running time of various units and stagger them if possible to minimise peak loads/maximum demand for heavy loads like crushing, grinding,
- whenever possible the material transfer or feeding in to the units is to be done in a staggered manner rather than running all units simultaneously
- provide metering pumps *in place of bigger* centrifugal pumps
- use screw compressors instead of reciprocating type
- install capacitor banks to improve power factors
- improve thermal insulation—internal lining and external cladding for process vessel heated by electrical heaters internally..
- provide sheds for preventing falling of direct sunlight on process vessels which are to be kept cool.

10.10.1 Variable Frequency Drives (VFD) for Speed Control

- Operate the equipments at lower speed as per the required output e.g. run a pump at lower speed to get less output–*instead of running at full speed and throttling the delivery valve.*
- A smooth speed variation with VFD can thus reduce power consumption.
- Optimise the additional investment in VFD by looking in to the potential for power saving.

10.10.2 Electrolysis Process Plants

- caustic soda & chlorine by electrolysis of brine
- copper refining by electrolysis of $CuSO_4$ solution
- electrolytic manganese dioxide by electrolysis of $MnSO_4$ solution
- aluminium metal- electro-reduction of Al_2O_3 with additives

10.10.2.1 Power Required at the Electrolysis Plant

- Consider Electrochemical equivalent of the desired products and Production rate per day (for calculating the theoretical power requirement in order to optimise the following):
- Production capacity per cell
- Number of Cells in series (voltage across each cell x no of cells)
- number of such arrangements in parallel (maximum current passing through each cell × number of parallel arrangements)
- Voltage drop in the cell: theoretical decomposition value, resistance of electrolyte, connections of bus-bar with electrodes etc.
- Rectifier capacity – Power supply to the cells, maximum current and voltage available at secondary side
- Minimise power loss through bus bars (mainly as heat) due to electrical resistance
- Electrical Switch Gear [Circuit breakers] and instruments for measuring current, voltage, power consumption etc.

10.10.2.2 Factors for Optimising Cell design and Their Accessories

MOC of electrodes:

- Cathode and Anode assemblies
- Distance between electrode assemblies

- Current carrying capacity of the electrodes and the Design Current density. (this should not exceed 60–70% of the maximum current carrying capacity for **reducing loss** during power transmission and for **better life** of electrodes)
- Corrosion resistance of electrodes to electrolyte
- protective internal lining of the cell
- Total electrical energy to be passed through
- Temperature of electrolyte for reducing viscosity
- Ionic concentration for improving electrical conductivity
- Additives to reduce resistance, temperature
- Incoming and exit arrangements (system) for electrolyte: feeding pumps, rotameters, flow control valves, composition, temperature and distribution of electrolyte
- Example: arrangement for collection and purification [H_2/Cl $_2$] compression of gases exiting from the electrolyte cell

10.10.2.3   Power Supply Required for Cells (*as per Production Rate Desired*)

- Material of Construction to be chosen for Bus bars and their properties.

- *Consider Copper, Aluminium flats/cables—electrical conductivity amp/cm² of cross section, corrosion resistance to chemical vapours and spillages which may be present near the electrolytic cells*
- Melting Point of bus bar material may also be considered if there are chances of overheating or operation near furnaces etc.
- Cross sectional area of bus bars shall be 1.5–2.0 times of required area as calculated from electrical conductivity amp/cm² of cross section
- Rectifier capacity: Voltage and current at incoming supply side (primary) and outgoing side (secondary)
- Series and parallel arrangement of Cells

Power required for other units and utilities
- Water treatment plant, Dissolvers, Steam/hot water generators
- Filters for electrolyte,
- Exhaust fans for cells and working areas
- Separation of products from exit streams (from outgoing electrolyte); and subsequent purification
- Electrical load balancing on the three phases in the plant is to be done by connecting the various drives for pumps, conveyors, blowers suitably. Individual power requirement and running time cycles are to be considered.
- **A small voltage drop can waste lot of power in such plants since the currents are of the magnitude of hundreds or thousands of amperes.** Hence the following should be looked in to:
- Controlling the flow rate of electrolyte to each cell
- Distribution of electrolyte around the electrodes.

- Prevent bypassing of the electrodes
- Electrolyte preparation section [Dissolver, dosers for additives and purifier/conditioner] for proper composition at inlet of the cell (proper ionisation); and optimisation of conductivity of the electrolyte
- *Additives to electrolyte-cryolite and flourspar for Aluminium production lowers the operating temperature*
- Temperature of electrolyte in the cell (to control viscosity of electrolyte)
- Distance between anode and cathode to control internal resistance of the electrolytic cell (by automatic adjustment)
- Removal of gases from vicinity of electrodes (gases are bad conductors) to minimise increase in internal resistance of cell.
- Use of bus bars with higher cross sections to minimise wastage of power
- Monitoring the performance of Rectifier regularly
- Regular cleaning of contacts in bus bars and electrodes when heavy currents are to be used for electrolysis.
- Geometry/shape of electrodes
- Use of adequate surface area for electrodes
- Remove deposits on electrodes by careful cleaning without damaging electrical connections
- Replace damaged electrodes by regular inspection

10.10.2.4   Some More Tips to Save Power in Chemical Process Plants

- Copper bus bars instead of aluminium bus bars (−since copper has a higher melting point and better conductivity than aluminium). *Optimise saving of power by Copper bus bar against higher cost of copper as compared to aluminium bus bar.*
- HT motors /energy efficient motors instead of LT motors,
- LED lights instead of incandescent bulbs
- Use of sunlight through transparent/translucent roof instead of electrical lights at the work place.
- Better thermal insulation to minimise load on air conditioning and refrigeration systems. Provide temperature controllers for AC Plants
- Electronic starters and chokes instead of conventional units
- Graphite electrodes instead of water cooled electrodes.
- Minimise use of motors of much higher rating than required.

10.10.2.5   Energy Saving Cooling Systems

- Conventional refrigeration plants – here low pressure ammonia vapour is compressed to high pressure vapour. It is condensed by cooling water to liquid NH3 which is then evaporated in chiller to produce low pressure ammonia vapour again. The latent heat for evaporation of ammonia is taken from incoming chilled water which gets cooled.

- Consider use of Chilled water (at +5 to +8 °C) instead of brine at (−5 to −2 °C) if the process plant units can be run satisfactorily. Refrigeration plants consume less power at higher temperatures of chilled water.

  Lithium Bromide based system is described in Refrigeration systems.

10.10.2.6   Tri-generation

In this set up the high pressure HP steam is passed through a steam turbine for power generation and the exiting low pressure LP steam from the turbine is used for process heating. The hot condensate from process heating is used in the LiBr based plant for producing cold water.

10.10.2.7   A Modified System

Uses concentrated NaOH solution instead of LiBr solution for absorbing water vapours. Other absorbents for water vapour can also be considered for producing chilled water.

Chapter 11
Examples of Modified Methods for Reducing Energy Consumption

Energy consumption study

Plant engineers shall look in to the possibility of Process Intensification (i) by carrying out some of the operations in lesser number of process units by modifying their design and construction; (ii) by reducing piping to save space in the plant and (iii) by using modified methods for certain cooling systems in order to reduce energy consumptions.

11.1 Examples of Process Intensification

They need less equipments, recover heat, reduce gas ducts and need less power and are safer to operate. This can increase the efficiency of the process plant.

- Inter-pass Absorption Tower IPAT system in a DCDA sulphuric acid plant: As per conventional design the acid mist in the gases leaving the IPAT is arrested by a demister provided in a separate vessel. This need gas ducts from IPAT to the vessel and from exit of the vessel to the downstream heat exchangers. When an arrangement is made inside the IPAT itself for arresting the acid mist, the additional vessel and connecting ducts are not required (Please see Figs. 11.1 and 11.2).
- Purification of crude CS_2 (removal of H_2S dissolved sulphur. The system illustrated in the figure shows separation of the three components (H_2S, CS_2 and Sulphur) in a safe manner in only two distillation columns. Column Number 1 also recovers heat from CS_2 vapours and uses it for pre-heating of incoming cold stream of CS_2 (Please see Figs. 11.3 and 11.4).

K. R. Golwalkar, *Integrated Maintenance and Energy Management in the Chemical Industries*, https://doi.org/10.1007/978-3-030-32526-8_11

Examples are also given for cooling methods used for process streams by conventional cooling towers, refrigeration plants and by unconventional cooling systems based on NaOH dilution or vapour absorption by LiBr solution.

Ref Figs. 11.1 (above) and 11.2 (below) The figure for modified IPAT shows FBME is internally installed and saves the separate vessel for demister pads, connected ducts (reduces power consumption and maintenance).

Note: (1) The separate demister vessel for housing demister pad is not required. (2) Internal Fiber Bed Mist Eliminator (FBME) (candle type) are provided in the IPAT. (3) This saves space and gas ducts in the plant.

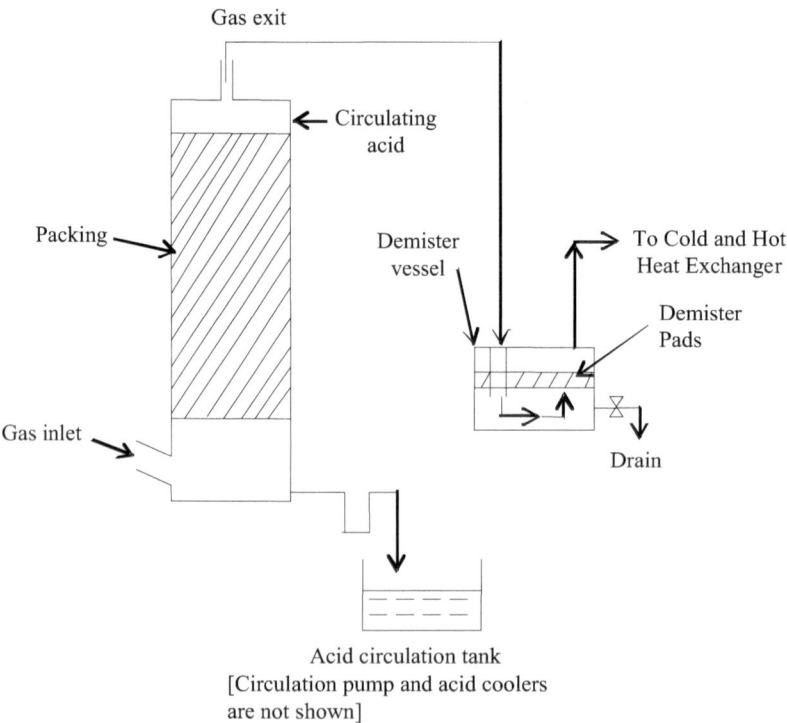

Fig. 11.1 Inter pass absorption tower (DCDA sulphuric acid plant)

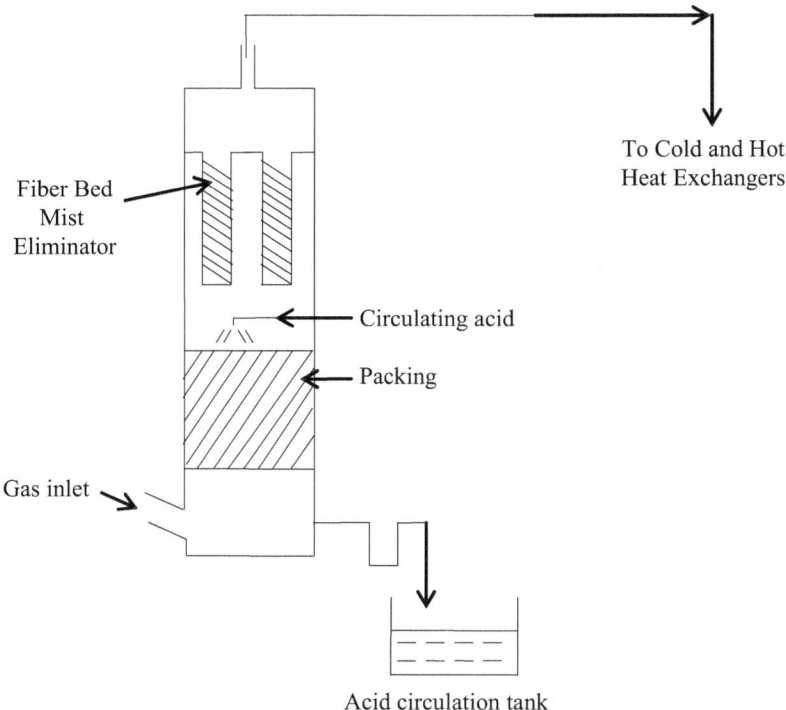

Fig. 11.2 Modified inter pass absorption tower (DCDA Sulphuric acid plant

Note for Figure 11.3: (1) Crude CS_2 contains dissolved H_2S and sulphur. It is fed into the central column where it meets rising hot vapours of CS_2. The H_2S gas is stripped. The exit vapours ($+H_2S$) are sent to reflux condenser where the CS_2 vapours condense and come back to the central column. (2) H_2S gas exiting from reflux condenser is sent to sulphur recovery unit (after passing through oil absorption to remove any residual CS_2 vapours). (3) The Distillation column No. 1 serves to remove H_2S. (4) It also preheats the incoming crude liquid CS_2 in the central column. (5) Hot rising vapours of CS_2 also impart some heat to the central column (which continues the pre-heating of the liquid CS_2 coming down to the boiling zone).

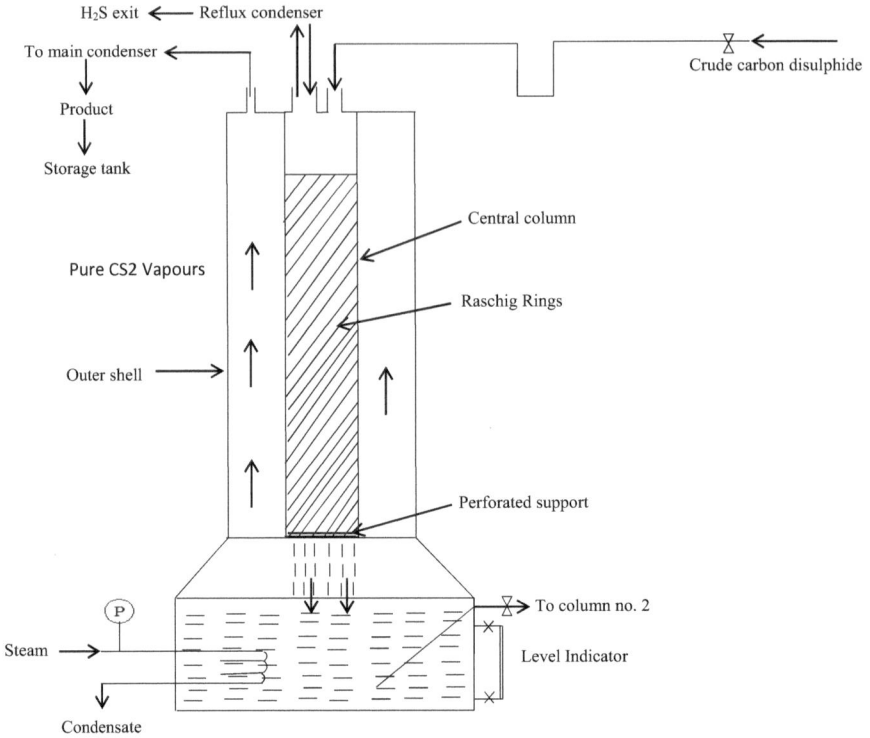

Fig. 11.3 Distillation column No. 1 (Process intensification)

Note for Figure 11.4: (1) In actual practice distillation column number 1 & 2, and sulphur separator are installed in close proximity. (2) The three components (H₂S, CS₂, and liquid sulphur) are separated by this arrangement. (3) The dilute solution of sulphur in liquid CS₂ is further concentrated in column number 2. (4) CS₂ vapours are separated during this concentration and condensed in product condenser. (5) Further separation of CS₂ and sulphur from the concentrated solution is done in sulphur separator. Sulphur is removed in liquid form, and CS₂ vapours are condensed in CS₂ recovery condenser (and recycled back to the column number 2).

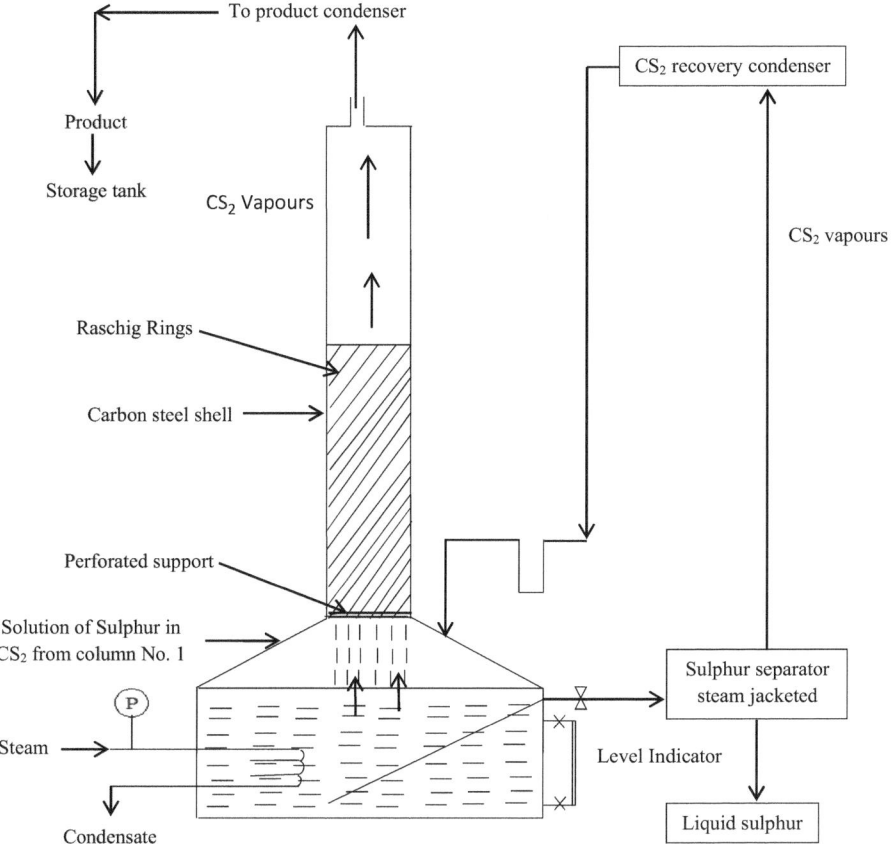

Fig. 11.4 Distillation column No. 2

11.2   Cooling Systems and Refrigeration

- Certain reactions are carried out at very low temperatures in some chemical industries.
- Many chemical industries have condensers to collect volatile products from reactors; and carry out distillation of crude products for purification to improve their quality.
- Condensers are used to recover valuable chemicals from exit gases also.
- Absorption towers also need cooling of circulating liquids for better absorption of components from gaseous streams and for pollution control.
- Storage tanks of volatile chemicals are cooled by spraying cold water.
- All such units need chilled water or brines at low temperatures.
- Spray ponds and Cooling towers are therefore provided in chemical plants. The cooling towers are generally used to cool the warm water at about 35–40 °C).
- Refrigeration systems are generally used for producing chilled water at (+5° to +10 °C) or brines at (−8° to −5 °C). The energy consumption is more when operating at lower temperatures while using the cooling media

11.3   Spray Ponds for Cooling Water

- Water is sprayed through nozzles for cooling and the cooled water is supplied to process units for cooling. This arrangement needs more space, and the water can get contaminated with dust or fugitive emissions.
- Maintenance: Regularly remove the settled sludge, algae etc. in the pond (this may need more effort as compared to ease of draining out sludge from cooling tower basins).But the plant operations can continue. Spray nozzles shall be cleaned regularly to maintain satisfactory cooling of water. Make water shall be provided by automatic control of level in the ponds.
- Advantage: they create a large storage of water in the plant for use in emergency

11.4   Cooling Towers

These are used for cooling the warm/hot water from exit of process condensers, from the exit of the steam condensers (which are installed in Co-generation plants for use of product steam generated by Waste Heat Recovery boilers); from condensers in refrigerating system. They are constructed of RCC basins with wooden structures on them; Fibre Glass Bodies with PVC fills etc. They are installed on ground floor or at a height (roof of process plant).

11.4.1   Observations during Operation

- temperature of cooled water at exit and warm water at inlet of cooling tower
- working of pressure gauges, temperature and level indicators..
- working of water pumps—discharge pressure, current drawn by motor
- all installed pumps should be operated one by one
- Water level and sludge level in basin.
- Analysis of incoming water which is used for make up to the Cooling Tower.
- Working of float operated control valve for makeup water [for maintaining water level]
- Analysis of water in the basin and the blow down water from cooling tower
- regular availability of makeup water and its analysis (ph, TDS, turbidity)
- operation of (float operated) valve in makeup water line which controls the level of water in the basin
- functioning of induced draft fan, drive motor coupling, gear box etc
- Running of ID fan, its speed and current drawn by motor.
- Observe the air discharge by ID fan [observe the fluttering of ribbon at top]
- Working of Vibration Monitor for ID fan, Gear box, Motor etc.
- any vibrations of the body of the cooling tower
- condition of fill and support grid for fill
- any corrosion or weakening of supports of cooling tower, and foundations of the water pumps if the unit is installed at a height

- Analysis of water bring circulated in system.
- pH, TDS, temp at exit and inlet of Cooling Tower.
- Pressure at discharge of Circulation Pump.
- Current drawn by Motor of Circulation Pump.
- State of readiness of standby pump [installed near Cooling Tower]
- Dry bulb and wet bulb temperatures at site.
- Working of high and low temperature alarms for Circulating Water.

11.4.2   Maintenance

Alternate arrangement needs to be made for supplying cooling water to process units, refrigeration system, cooling of tanks when the cooling tower is to be taken up for maintenance work...

Isolate the cooling tower completely OR the cells one by one if possible

Water pumps...
 (i) overhaul all pumps one by one
 (ii) check strainer in suction line, foundation bolts, coupling halves, any leak from glands, pipes etc.
(iii) provide fresh anti-corrosion coating on impeller and internals of the pump. This may reduce churning and turbulence inside the pump.

ID Fan
Check blade angle and optimise to reduce power consumption.

Fiber glass reinforced plastic can also be used instead of metallic blades to reduce consumption. Repair/replace damaged blades.

Provide fresh lubricant/grease to gear box, drive motor

Monitor the temperature of cooling water. Install arrangement for automatically switching off the ID fan when the temperature is low enough (in winter). This can save power.

Cooling tower basin   Do not allow build up of sludge or high concentration of solids in the circulating water which can the choke the tubes in process condensers.

Keep a record of water drained out as below:

Date	Volume drained out	Analysis of water drained out	Analysis of makeup water	Quantity of makeup water added [Water Meter readings	Remarks

Maintenance Work Done
- Repairs to Water Circulation pumps and motors.
- Repairs to Water pipe lines and Valves, pressure gauges on water lines.
- Repairs to ID fans, gears box, motor starter, cables etc.
- Repairs to basin & tower fill materials, temp indicators and alarms.

- Repairs to ladder, to work platform.
- Cleaning of Basin to remove accumulated sludge.
- Float operated (make up water) valve.
- *The makeup water shall not have more than 100 ppm TDS.*
- Cleaning of level indicator.
- Hot Water distributor (hot water coming from process/turbine condenser etc)
- Removal of any vegetative growth, algae, slimy material.
- Record the chemicals added for above treatment.

Tower basin...

  (i) bring down the level of water, and then drain it out.
 (ii) clean the basin thoroughly, and all accumulated muck is to be sent to drying bed.
(iii) wash the basin with minimum quantity of water; and drain it.
 (iv) try to conserve and reuse the drained out water elsewhere in the premises.
  (v) Repair the basin of fiber glass cooling towers (these are sometimes installed on roof of buildings housing process units)

Fill material

  (i) check whether all fill material is clean, or displaced from position, or some portions are missing.
 (ii) make up the loss by addition of fresh fill. It should be tested at operating conditions.
(iii) inspect support grids for tower fills. Repair/replace damaged portion of grid. Apply rust resistant coating after repairs. Alternatively, fiber glass reinforced grids may be used wherever feasible. Obtain sample piece of grid from vendor ant test in operating conditions.

11.5   Conventional Refrigeration Plants

11.5.1   Main Components and Machinery

- Compressor for the refrigerant (Ammonia/Freon/others).
- Condenser for the refrigerant (Ammonia/Freon/others).
- Cooling Tower, water pumps, valves and piping for cooling water –.
- Receiver for liquid refrigerant.
- Expansion – cum–Control Valve.
- Chiller [Evaporator]
- Chilled water tank/Brine Tank and Circulation Pump; Cooling Tower and Circulation pump
- Suitable insulation on liquid refrigerant and chilled water vessels and pipes.
- Instrumentation- Temp. indicators, alarms for high and low temperatures for chilled water, refrigerant, high pressure alarms, safety valves, safety interconnections and tripping system.
- H.P. and L.P. refrigerant piping
- Piping for Chilled Water/Brine.

- Instrumentation and Safety systems
- Electrical systems

11.5.2 Observations to Be Made for Safe and Smooth Working

- Compressor: Pressures of refrigerant at suction and discharge of compressor; compressor capacity in use; oil pressure, oil level in sight glass/in indicator
- Condenser: Pressure of refrigerant vapour at inlet, pressure and temperature of cooling water at inlet and exit of condenser. It is advisable to arrange a flow meter in the exit water line in such a manner that the flow of cooling water can be easily monitored.
- Gas leak from flanged joints, valves,
- Insulation–any damaged portion, ingress of moisture through leak.
- Chiller/evaporator: Flow of chilled water, pressure of chilled Water at inlet & outlet, pressure of refrigerant in the unit
- Alkalinity of circulating chilled water/brine; add anti- corrosive chemicals so that tubes, valves etc. are not corroded. Samples of chilled water should be checked every week for the corrosive nature and treated accordingly.
- Liquid receiver: Should have well protected level indicator with automatic closing arrangement for isolation valves in case of a leak.
- Control valve: For liquid flow to evaporator (for automatic adjustment of refrigerant flow as per cooling requirement and operating level in evaporator)
- **Pressure of refrigerant** shall be monitored at compressor exit, condenser inlet, evaporator inlet and exit
- **Load on electrical motors** should be monitored at (but not limited to these) compressor motor, cooling water pump, ID fan, chilled water pump.

11.5.3 Safety Valves

Should be available for following units:

- Compressor discharge
- Oil separator vessel
- Inlet of condenser
- Inlet of evaporator

Safety trips/interconnection for tripping of the compressor is to be provided for the following conditions:

- Compressor discharge pressure is high (cooling water flow may be less, temperature of cooling water may be high or heat transfer surfaces have become dirty/ scaling has taken place)
- lubricating oil pressure is low (problem with lube oil pump/frothing in oil due to dissolved gas)
- if cooling water pressure at condenser inlet is low (due to less supply from cooling water pump).

- If the flow of chilled water through evaporator is low
- If the temperature of cooling water at exit of condenser is high (due to low flow).

11.5.4 Some Precautions for Maintenance Work

- Evacuate the refrigerant completely and transfer to other unit or to cylinders.
- Condenser: The tubes are to be cleaned by brush and then washed by clean water. The shell side is to be cleaned by chemical cleaning and washed by clean water. After this both sides are to be tested at a pressure at least 1.5 times the maximum working pressure. Check the gaskets (at partition plate in end cover) in a multi pass system
- Now evacuate all air inside the system (preferably by vacuum dehydration) before filling up fresh refrigerant.
- Compressor: Check the cylinder heads, pistons etc. Check the screw in case of screw compressor. Clean and lubricate the bearings as per instructions from OEM.
- Take out oil sample from the crank case and check for any turbidity. Change the oil if advised by OEM. Also check the working of the oil lubrication system.
- Evaporator: Stop circulation of chilled water. Evacuate the refrigerant completely and transfer to other unit or to cylinders. Clean the tubes by brush and then wash by fresh water. Check the gaskets (at partition plate in end cover) in a multi pass system
- Receiver: Check the condition of level indicating gauge glass, connecting nozzles and the protective cover.
- Flow control valve for refrigerant: Check free movement and a pilot tapping from down stream side.
- Check thermal insulation on liquid receiver, evaporator and cold pipelines.
- All pressure parts and pipelines are to be tested at 1.5times of the maximum working pressure or as directed by statutory authorities.

11.5.5 Additional Safety Precautions

- Do not expose refrigerants, cylinders to direct sun light or temperatures more than 50 °C.
- Liquid ammonia can cause severe cold burns, hence take care not come in contact.
- Wash with copious amount of water in case of contact.
- Ammonia is also inflammable and can explode if concentration in air is more than 11%.
- No welding or gas cutting work should be done in presence of Flouro Carbon refrigerants as it can form toxic gases.
- Evacuate the system completely by removing all refrigerants before taking up any maintenance work. Bring the pressure to zero before opening any unit.
- Test the units, pipelines etc. by compressed nitrogen after the maintenance work is over.

- Gas detectors should be available to detect leak of ammonia. A wet cotton swab with HCl can indicate leaking spot of ammonia by generation of fumes around it. A piece of burning sulphur (SO_2) can also indicate source of leak by fumes.
- Gas masks, face shields and eye protection safety goggles, portable breathing apparatus and oxygen cylinders must be available for protection against such leaks

11.6 Unconventional Chilled Water Systems

11.6.1 Caustic Soda Based System

Water in an overhead water tank is boiled at very high vacuum and as a result it becomes very cold (8–10 °C). This cold water flows into the chilled water tank on ground floor by gravity. It is supplied to process for use.

Warm water (at about 15 °C) from process can be returned to this cooling system if required. Fresh make up water can also be added to maintain level.

The water vapours are absorbed in multi stage absorption column in a stream of strong caustic solution (which becomes dilute). The diluted solution is withdrawn from the system and used in process plant. An equivalent amount of strong caustic solution is continuously added. The water vapours are removed by absorption in caustic soda solution to maintain vacuum in the vessel. The system has following main components.

- Receiving tank for 45–48% caustic solution.
- Circulation pumps for strong caustic solution.
- Cooling system for strong caustic solution.
- Multi stage absorption column for water vapour
- Overhead water tank in which water is boiled
- Chilled water tank on ground floor
- Steam ejector for start up and to remove dissolved gases.
- Make up water system for overhead tank.
- Chilled water removal pump for supply to process
- Thermal insulation for tanks and pipe lines
- Instrumentation for monitoring operating conditions

There are no machines for compression of refrigerants in this system and power consumption is on lower side (for removing cooled water out, for circulating and cooling caustic solution).

11.6.2 Libr Based System

A similar system produces chilled water as described below.

- Water is boiled/evaporated under vacuum (at very low pressure) and gets cooled to a low temperature.
- This cold water is sprayed on heat transfer coils to remove heat from incoming cooling water from process which is circulating through the coils.

- The circulating cooling water is then sent to process plant.
- The water vapours produced are absorbed in strong LiBr solution to maintain the vacuum. The solution gets diluted as a result. The diluted LiBr solution is again concentrated by using waste steam/hot condensate and is reused
- Water vapours released during concentration of the dilute solution are condensed and reused as refrigerant.
- Low Pressure steam available in plant can be used here. In certain tailor made units, hot condensate available from process units (which are being heated by steam exiting from back pressure steam turbine)is used instead of LP steam.
- *LiBr is corrosive, and piping can get choked due to deposits. This method needs less power than conventional NH3 based plants.*

11.6.3 For Safety of Personnel

- Face shields and alkali/chemical resistant dress
- Full length hand gloves and gum boots
- Eye wash fountains and safety shower with dedicated water tank and gas detectors at appropriate places

11.7 Other Cooling Arrangements

- By controlled expansion of compressed propane gas through a throttling valve produces cooling due to Joule Thompson effect. This is for special application in a petro chemical plant.
- It can save electrical power if compressed propane is available.
- However special provisions are to be made for fire fighting and special precautions are to be taken (no open flame, no hot surfaces, use of non-sparking tools, proper earth connections with regular check of continuity) since propane is an inflammable gas. Propane tanks must be protected from high ambient temperatures, and should be designed and fabricated as HP vessels.
- Lightening arresters shall be provided.

11.8 Steam Jet Cooling

- Water is boiled at a very low pressure in a vessel. High pressure steam ejector is used to create vacuum and suck out the water vapour from the vessel. The heat required for boiling is taken from the water itself as a result of which it gets cooled (corresponding to the boiling point at the low pressure). The cooled water can be withdrawn from this vessel and make up water is to be added.
- This is a tailor made unit as it needs HP steam and an ejector. There should be a process unit in the premises for using the steam exiting from the vessel.

Chapter 12
Methods for Minimising Consumption of Energy

12.1 Project Planning

- **Selection of proper technology is a must** (discussed elsewhere in the book)
- **Production runs:** plan for longer continuous steady production runs (instead of running the plant at variable rates) can improve process operations, control and efficiency. (*please see Sect. 10.6 also*)
- **Site selection** harsh climate (severe summer/extremely cold winter/very heavy rains) can cause process disturbance for sensitive products.

- *Cool dry climate is helpful for better performance of cooling towers which in turn can result in better condensation of volatile materials. **The project site should be selected accordingly if possible.***

12.2 Reduce Consumption by Process Units by Better Design and Construction

- The design and construction of various process units (such reactor, agitated units, mixers, furnace, heat exchangers, distillation system, absorption tower etc.) including their internals should be done for meeting the required performance; shall be safe to operate, need less power to operate and the loss of heat shall be minimum. *This is to reduce wastage of energy and conservation of heat. There will be less corrosion of process units and hence lesser breakdowns when proper materials of construction for the process units are used.*

Minimise the energy input required
(Generally the main requirements are as power, fuel and steam)

A. **During Erection and commissioning of a new plant**

© Springer Nature Switzerland AG 2019
K. R. Golwalkar, *Integrated Maintenance and Energy Management in the Chemical Industries*, https://doi.org/10.1007/978-3-030-32526-8_12

- Minimise requirement of site fabrication
- Trial runs of various equipments
- Preheating of process units in plant
- Initial production runs and stabilisation of plant *till rated capacity is reached*

B. **During steady production run (after stabilisation)**

Consider the heat evolved by the reactions also as energy input and maximise its recovery.

12.3 Considerations for Reducing Energy Consumption

- New Plants: select better process designs and technology (operation at lower temperatures/pressures if possible)
- Existing process plant: use less flow rates/operate at lower pressures, minimise material handling by improved layout of the plant; explore innovative methods for operation
- Old (*idle*) plants: Modernise during revival or when planning for expansion after revival select a better process (examples-DCDA in place of SCSA for sulphuric acid, Membrane process instead of mercury cell process for caustic soda, chlorine)

The consumption norms/standards can be revised when better technology is adapted, when operations are improved and also when good suggestions are given by plant personnel.

- **Examples of better design in modern processes** (*please see Sect. 10.6 also*)

  (i) DCDA Process instead of SCSA Process for Manufacture of Sulphuric Acid.(reduced gas volume by higher SO_2% in process gas)
  (ii) Air blower installed **after** Drying Tower in a sulphuric acid plant to recover heat of compression (reduces cooling load of hot acid)
  (iii) Plate Heat Exchanger to recover heat from hot (75–85 °C) acid
  (iv) Caesium promoted instead of Potassium promoted catalyst
  (v) CS_2 Manufacture: modified brick lining and use of graphite electrodes in place of water cooled electrodes could save electrical energy. It is more safe also.
  (vi) Membrane Process instead of Mercury Cell for Manufacture of Caustic Soda

12.4   Energy Saving by Improved layout for Safe Convenient Movements

- Minimise handling of materials (raw and finished products)
- Maximise flow of liquids by use of gravity
- It becomes safer and convenient to operate the plant when there is no obstruction for keeping a watch on the process units from all sides, no obstruction in the escape routes, for movement of materials (whether manual or by fork lift, belt conveyors..), for cleaning of key equipments and repairs to their internals and clear passage is available for replacement of old units if possible.
- Elevated Storage Reservoir ESR, First Aid facilities, fire fighting arrangements and DG sets for emergency power supply shall be easily accessible to all sections of the plant. This improves morale of personnel and can make them more attentive.
- Process control laboratory shall be near as possible at (i) raw material receiving and feeding section (ii) near key processing units; and (iii) finished product storage. The results of analysis become known quickly and improves working efficiency
- Use incoming water (which can be at some pressure at entry point of supply to the plant) directly for supplying to cooling jacket of process units (if possible) *instead of pumps*; at inlet of filters in water treatment plant, for connecting to spray nozzles of absorbers of air pollution control units, for adding in to overhead tanks if possible. Try to design the layout of incoming water lines accordingly.
- At certain locations natural gas is available as fuel (under pressure directly from oil wells). It saves effort and cost of handling other fuels like furnace oil, coal.
- High pressure compressed air and steam lines should have minimum length

12.5   Controlled/Modified Operations for Reducing Energy Consumption

- Manufacture of Alum from Sulphuric acid and Alumina Hydrate AH—Controlled addition of water and Concentrated Sulphuric acid increases temperature to which AH is added slowly (can reduce steam consumption by minimising dilution of reaction mass)
- Boiler operation at reduced pressure will reduce exit flue gas temperature also since saturation temperature of steam corresponding to the lower pressure is also lower.
- Evaporator operation: control concentration of solution during evaporator operation to minimise deposit of salts on heat transfer surfaces. This can save steam.

12.6   Monitor Losses from Equipments and Ducts (Use Thermography Camera)

Some typical situations are given below:
- Hot surfaces of process vessels, furnaces with insufficient internal refractory and insulating lining (by bricks/ceramic paper)
- Ducts without proper internal insulation by castable refractory/ceramic paper and external insulation by mineral wool plus aluminium cladding
- Process pipes with insufficient external insulation or cladding (where ingress of moisture may occur). Excessive thickness of external insulation can result in more heat loss due to increased external diameter. Hence the thickness should be optimised.
- Insufficient capacity of equipments for heat recovery and hence there is heat loss
- Use of smaller diameter pipes for transfer of fluids to save initial cost (but power consumption is more due to increased pressure drop. A bigger diameter pipe will have lesser pressure drop and hence can save power. Optimise the higher cost of pipeline against lower power cost)
- Inefficient or improperly installed lighting makes working in plant difficult
- Heat recovery equipments like economisers, air pre-heaters, incoming cold feed water pre-heaters not provided; hence the heat is lost to atmosphere.

12.7   Check Losses During Plant Operation: (Some Typical Cases)

- Coal fired boiler: un-burnt fuel falling out through increased openings due to wearing out of grate for burning coal)
- Coal/Oil fired furnaces: supply of excess combustion air wasting fuel (needs adjustment of dampers in air lines for burners as per $O_2\%$ and $CO_2\%$ in flue gases).
- Loss of un-burnt oil due to incomplete combustion. *(needs reduction of the viscosity of fuel oil by adjusting temperature setting of oil heater for proper atomisation of oil).*
- Throttling of high pressure process streams (instead of using variable frequency drives for flow control) when less flow rate is required
- use of air injection for process cooling instead of installing sufficient equipments for heat recovery (air injection increases volume of gases to be handled which results in increase of power consumed by air blower for process plant) However, the benefit of heat recovery and reduced gas volume shall be weighed against investment required for the additional equipment.

- power losses in electrical circuits due to low power factor, high resistance in user units and power cables, bus-bars, electrolysis plants, continued use of inefficient electrical motors
- heat lost with hot outgoing gases (e.g. heat lost from hot flue gases) due to *poorly designed or maintained heat exchangers*; or excessive boiler blow downs required due to poor quality of feed water to the boilers.
- Using high flow rates of water for cooling instead of cleaning heat exchangers (more power is consumed by cooling water pump)
- energy loss through high velocity liquids or gases entering into process vessels (*use flared shaped nozzles at entry points to reduce velocity head loss*)
- Poorly maintained refrigeration and cooling towers
- Cooling water having high concentration of dissolved solids can cause inefficient heat transfer due to deposits of scales on heat exchanger surfaces.
- Implement LDAR (Leak Detection And Repair) programme to minimise losses.

12.8 Examine Potential for Energy Recovery: (*Please See Sect. 10.7 Also*)

- process streams at high temperatures (which are to be cooled) and hot condensate (to be collected from all steam consuming units in the plant)
- maximise heat recovery by Waste Heat Recovery Boilers
- install Steam Turbo-generators with accessories for Co-Generation (for providing electrical power and process heat to the plant), and export excess energy after meeting own requirement of power, steam, hot water etc.
- process streams at high pressure (install power recovery turbines while reducing pressure)
- possibility of providing air pre-heaters to reduce fuel consumption
- explore use of Gas Turbo generators for power generation
- undertake job work from external parties for drying, evaporation of their material by use of surplus steam/hot gases if excess energy cannot be exported.

12.9 Energy Audit

- This should be carried out every 3–6 months or whenever any energy or fuel consumption is more than standard norms (whichever is earlier)
- Study the major consumers of energy in the plant such as process reactors, refrigeration units, electrolysis cells, dryers for various items, high pressure pumps, compressors, crushers, grinders and investigate other energy consuming equip-

ments in the premises- material transfer systems, water treatment, and air pollution control
- An external agency may be employed (if own plant engineers do not have sufficient time to study in detail) to make an objective analysis of all the operations and processes being carried out in the plant.
- Certain matters (like unnecessary multiple handling operations, inefficient manual work instead of automation, short production cycles, inefficient use of steam, compressed air, hot water.) will get highlighted by such study.
- Energy audit of Electrical system to be done separately regularly and then discussed with plant operating engineers any suggestions for improvement in the matter.
- The external agency can give useful suggestions for using electrical power more efficiently and closely watching draw of power from external grid (without exceeding the maximum demand permitted) and advice for export of the excess energy available after co- generation.

12.10   Comparison with Ideal Conditions

- It will be useful to compare the data for energy consumption of one's own plant with the data for plants (*having comparable capacity and manufacturing similar products*) which consume less energy.
- **An important reason for the lower energy consumption could be availability of better inputs (which enable smooth and efficient plant operations) The management and senior engineers should explore whether some of these inputs as given below can be arranged for their own plant.**

   1. Modern tools and materials for quick timely maintenance (repairs) of process units and machinery which *enables restoring efficient operations earlier.*
   2. Availability of better raw materials (see Table 12.1)
   3. Raw water of better quality (having low TDS, suspended matter, low hardness).These waters will need less treatment before use in chemical industries for cooling of process streams and for making solutions for manufacturing the products.
   4. Steady power supply results in uninterrupted run of plant (a number of restart operations are required when power supply is interrupted and this needs frequent adjustment of process controllers. e.g. adjustment of process parameters like gas flow, cooling water, heating by steam, raw material feeding etc. Such frequent adjustments may not be required when plant is running steadily.)
   5. Well trained and disciplined labour...commit less mistakes, are more attentive

Table 12.1 Selection and procurement of raw material. It is mainly done for reducing the pre-treatment cost/energy consumption

Sr. No.	Procure raw materials as indicated below	Advantage
1.	Molten, pre-filtered Sulphur instead of solid Sulphur	It saves time, energy (used by steam for melting), and efforts for purification by operating filteration units
2.	Steam coal with low moisture (<5%) and low ash content (<10%) instead of cheap coal available	Saves energy loss due to wet coal, efforts required to crush/pulverise, and reduce ash deposition on boiler tubes.
3.	Rock phosphate with very low CaF_2 (<0.5%) content for superphosphate manufacturing	Reduces emission of fluorine compounds in exit gas. Hence, less energy is spent on scrubbing system.
4.	NaCl of good quality (with low Ca/Mg compounds) for NaOH production	Needs less efforts to purify brine in NaOH plant and it reduces the choking of membranes in electrolysis cells
5.	Hard wood charcoal with low moisture (<4%) and bigger pieces (>18–20 mm) instead of cheap charcoal from low grade wood	Improve working of electric furnaces for production of CS_2. It will also generate less gaseous effluent with H_2S
6.	Small bauxite pieces (moisture <5%) with low Fe content (<2%)	Saves the equipments and energy to crush, grind and control the dust when big lumps of raw material are procured. Needs considerable steam for evaporating dilute solution
7.	Alumina hydrate powder $Al(OH)_3$ instead of bauxite lumps	Reduce energy for crushing, grinding, dust control, clarification of reaction mass and filteration; needs much less steam (there is no dilute solution). Also needs less equipments and gives better product. This can offset higher cost of $Al(OH)_3$.
8.	Crude petroleum with low wax instead of heavy crude oil	Reduce energy for heating, cracking and distillation in refineries
9.	Clean white cellulose sheet instead of powdered cellulose with dust and other impurities	It improves production of Alk-cell in Rayon plant. It needs less energy for filtration of cellulose xanthate solution. It gives more smooth working of spinning machine and reduce consumption of NaOH
10.	Raw water with low TDS (<50 ppm) instead of poor quality water having high TDS (>200 ppm)	Less energy and efforts required to produce demineralised boiler feed water which also reduces the blow down from boiler
11.	Standard pre-weighted packets of stabilisers/anti foaming agent	Saves effort to take out from transport vehicle, fill in small packets, and weigh before adding to product.
12.	Obtaining alum as a ready to use solution at the point of use	Saves energy to crush and dissolve the solid lumps
13.	Obtaining additives in small carbouys of known weight (5–10–20 kgs)	Saves manpower, time and energy to transfer from tankers (5000–10000 ltr) into small carbouys

12.11   Examine the Loss of Energy During Maintenance Work

- When any major maintenance job is to be done the entire plant may have to be shut down and concerned unit is to be isolated, cooled and internal materials are to be flushed or gases are to be removed by exhaust fans. Put blinds in inlet and exit ducts of other hot process units (if no work is to be done on them) **to retain heat inside them** while maintenance work is being done on the unit taken up for repairs.
- Minimise energy required for carrying out repairs, taking trial runs after the maintenance work is over and to restart the plant after *a short plant stoppage*. This can be done by carefully examining the practices followed so far; and improving them if possible.
- Minimise the energy required to restart the plant *after annual maintenance* activities are over (since all units in the plant are to be completely shut down and cooled to ambient temperature). *This restart may need different procedures (than the procedures followed after a short plant stoppage, or during **initial commissioning** of the plant).*

Certain energy saving steps could be possible. These shall be explored in consultation with Plant Designers, senior experienced engineers and manufacturer of machinery.

- ETP units may have to handle extra loads during the maintenance jobs since more effluent can get generated due to washing/flushing of plant units; and production of off-spec products (some of which cannot be recycled or reused) *till the operations are stabilised on restarting the plant.* The energy consumption in ETP units can be more in this period than during steady plant operation. *Hence plant engineers shall try to minimise generation of the effluent during this period.*

12.12   Minimise Energy Consumption by Procuring Better Raw Materials

- Considerable energy is required for crushing and grinding if big hard lumps are present. These operations also need dust control and air pollution control when lot of crushing and grinding of the raw materials is required. It requires energy to operate dust control and air pollution control units.
- Raw material with excessive dust: this can deposit on surface of catalyst making it inactive and thus wasting energy for operating the plant.
- They also cause disturbance in operation when there are more deposits of dust on surface of heat exchanger tubes, sludge formation in process tanks.

- Unsatisfactory quality of raw materials (having more dust, harmful impurities) forces operation of pollution control facilities on higher load.
- Purchase department should search for obtaining **better raw materials** (which have less moisture, less impurities, smaller lump size, less ash content etc.) since such materials need less heat for drying, less efforts and energy for melting; and less power is required for crushing and grinding.
- If such raw materials are not available then pre-treatment of raw material before feeding to reactors becomes necessary. It may include filteration of molten raw materials (or their solutions); crushing of big lumps for ease of melting of solids, drying of wet materials to remove excess moisture.

12.13   Situations to be Addressed

Some typical reasons for higher consumption of energy are given below. These should be looked into and addressed by the Management and Senior engineers

- short production runs-(batch run instead of continuous operation)
- incorrect product mix planning (frequent changeover to different products, not arranging the required raw material in time, product mix not planned as per demand in market or difficult climatic condition,)
- delay in procurement of efficient equipment
- delay in cleaning and maintenance of equipments (specially heat transfer units)
- effluent treatment facilities not ready before trial runs: (this can force the plant to be operated at lower production rates). This can waste energy in operation of certain heavy equipments since they cannot be operated efficiently below a certain load.
- certain necessary start up inputs not available in sufficient quantity e.g. fuels, scrubber chemicals, raw materials, circulation liquors for absorbers – thus forcing plant to be operated at lower production rate.
- some statutory clearances not available before commencing heating (commissioning of the plant may have to be aborted even after heating up process units. This causes wastage of fuel spent earlier)
- heat exchangers design issues (improper design causes inefficient operation,)
- heat transfer surfaces not clean (dirty surfaces reduces heat transfer/recovery)
- incorrect piping design with small diameter pipes, sharp bends can result in excessive pressure drops with consequent high power consumption
- high velocity entry and exit in process units (higher loss of velocity head)
- choked filter leaves, bag filters (high power consumption)
- higher feed rate to process followed by excessive recycle (high power consumption)
- lower temperature driving force for cooling being compensated by higher flow rates of cooling water (leads to high power consumption)

12.14   Examples of Saving Energy and Heat Recovery

Conventional method of cooling hot circulating acid by cold water supply from cooling tower.

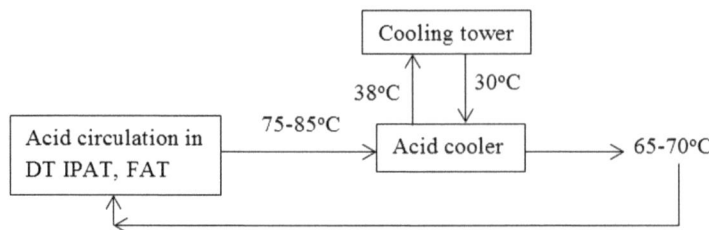

**Figure for Conventional cooling method
of circulating acid in a sulphuric acid plant**

The heat energy of hot acid is completely lost, because the warm water at about 38 °C cannot be used for any useful purpose in the plant.

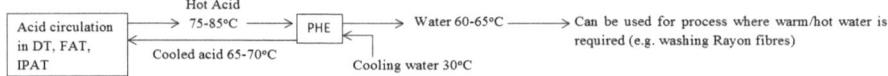

Figure for Improved method for heat recovery from hot acid

A plate heat exchanger is used to recover the heat from hot acid. Process water at about 30 °C is heated up to 60–65 °C and used in the premises elsewhere.

Note: It is advisable not to pre-heat boiler feed water by the PHE as above, because in case of any leak the feed water will become acidic and can damage the boiler.

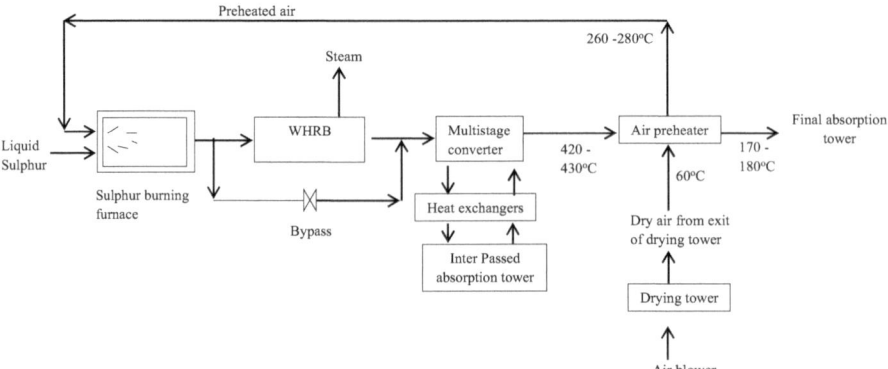

Figure for Heat recovery as Preheated Air in a Sulphuric Acid Plant

Notes:
There is gas flow on shell side as well as tube side of air pre-heater [APH] which causes pressure drop on both shell and tube side (hence more power is consumed by Air blower).

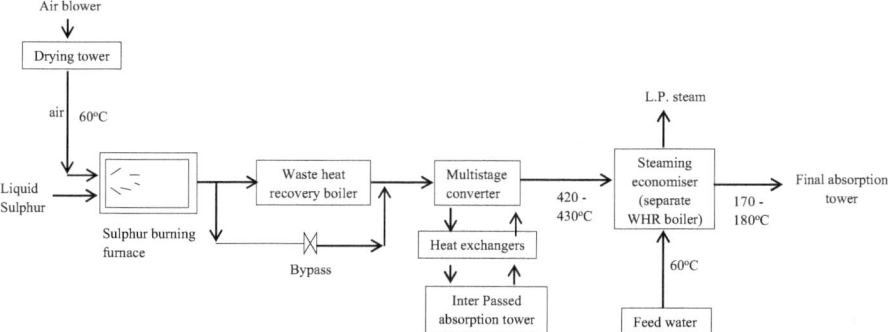

Figure for Heat recovery as Low Pressure steam in a Sulphuric Acid Plant

Notes:

1. LP steam is produced directly.
2. Gas side pressure drop is less since Air preheater is replaced by economiser (separate WHR boiler). Therefore, air blower consumes less power.

12.15   Process Boilers

These units are not used for generation of steam. These are used for generation of vapours from a solution (SO_3 from strong oleum), separation of volatile components dissolved in a liquid (H_2S from crude CS_2), extraction of products from mixture of liquids etc.

They are used for saving steam consumption (by recovery of heat from hot process gases).

- The design and construction should be as if these are pressure vessels (follow standard codes ASME-Sec-VIII.)
- All safety devices like safety valves, rupture disc, high temperature and pressure alarms, emergency vents shall be provided. The gases from these must be absorbed in suitable absorbent and should not be released to the atmosphere as such.
- Quality Assurance Plan must be as required by technical teams of purchasers. It shall be confirmed by vendor/fabricators and implemented.
- Direct firing by fuels is not usually done.
- Heating is generally done by low pressure steam or by hot process gases.
- Material of construction can be stainless steel 316/special corrosion resistant alloys since some of the materials can be corrosive chemical solutions

Shell and tube type Process Boiler:
Example: when steam is passed through the tubes for boiling oleum there is possibility of sudden rise of pressure in case of a tube leakage, due to reaction of oleum with steam (heating medium) which can be dangerous.

When process gas is used for heating, contamination of the process gas can occur or the process in downstream units can get disturbed (due to ingress of oleum from a leaking tube to the gas side).

In case of a leak from shell: the process fluid will come out and can spread toxic, corrosive, inflammable material in working areas resulting in accidents or fires.

- *Proper louvers shall be provided around (which will provide protection from such spillage but will not affect ventilation)*
- Provision of demister/entrainment separator is necessary to arrest liquid droplets and allow only pure vapours to go out.
- The control of feed composition and temperature at exit (as per corrosive nature of the hot liquid) is important.
- Provision of sight glass and light glass is desirable to observe internals. However this will depend on the actual operating conditions.
- Provision of pressure control and safety valves on steam supply lines
- Provision of valves in inlet and exit hot gas connections to the process boiler.
- Provision of a bypass line to carry the hot gases should be available to control the rate of heating.
- It is advisable to have another arrangement (air pre-heater/economiser or a boiler etc.) to cool the hot process gases in case the process boiler is not available due to any reason and the process plant has to be run.
- External thermal insulation and cladding
- Adequate support structure, working platforms, ladders, shed for protection from rain, snow, direct sunlight and proper lighting must be provided.
- Eye wash fountains and safety showers with supply of water from dedicated overhead tank must be always available.
- Fire fighting arrangement should be always available.
- It is advisable to have good ventilation around these units and collection basin below them for preventing spread of liquids in case of a leak.

Instrumentation for Process Boiler:
- Temperature indicators for incoming and outgoing liquid
- Temperature indicators for incoming and outgoing hot process gas
- Flow rate of incoming and outgoing liquid
- Level indicator and high level alarm
- Pressure of incoming and outgoing hot gases
- Pressure and temperature outgoing vapours

12.16   Optimising Power Consumption for Agitated Vessel

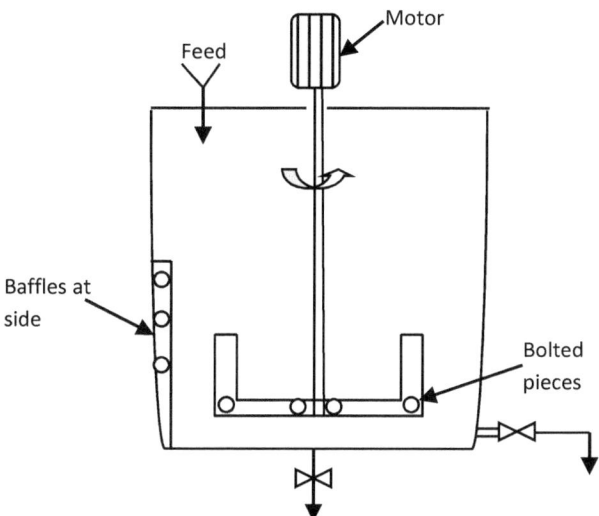

Figure for Optimising Power Consumption in Agitated Vessel

Optimising Power Consumption in Agitated Vessel. Can be done by changing the bolted arms of agitator, shape and number of baffles at side, speed of agitator, operation at different liquid levels etc. The current drawn by the motor is to be observed for every change made (while maintaining the production rate) and the best condition is to be chosen. Rate of feeding material shall also be adjusted.

12.17   Dual Drive System

The Dual drive system consists of Steam turbine, driven machine, and driving motor. High pressure steam (whose supply can be unsteady since it is generated by a WHRB) is fed to the steam turbine which is used for driving the machine (which is also driven by an electric motor as shown in figure). The exit steam from turbine can be used elsewhere in the plant

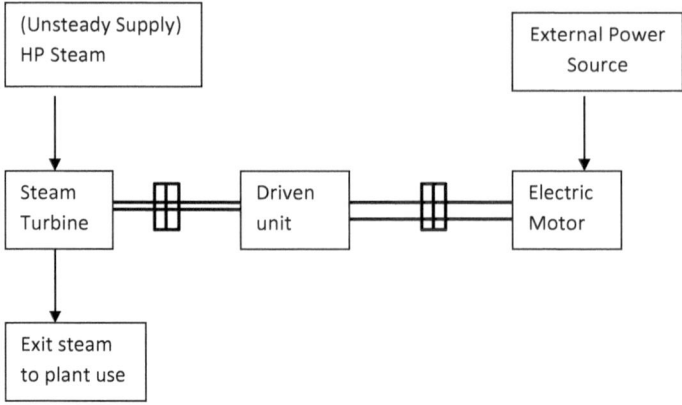

Figure for Dual Drive System

Notes
- Motor is always kept on to maintain speed of the driven unit.
- Part of the load on the electric motor driven unit is shared by the steam turbine.
- Electrical generator is not provided. This reduces maintenance requirement also.
- This arrangement reduces power drawn by the motor from the external source.
- *During operation of the chemical plant the WHRB may be bypassed sometimes to further raise the temperature of downstream process unit as per operational requirement. This may reduce the steam generation and consequently the supply to the steam turbine. However it will not affect the speed of driven unit because the motor is running at constant speed.*
- This arrangement uses whatever steam can be produced by the WHRB.
- A sudden tripping of the driven unit is avoided in this system because it is very rare that the steam turbine and electric motor would stop simultaneously.

Chapter 13
Retaining Thermal Energy While Simultaneously Protecting Equipment

13.1 Refractory Materials and Their Properties

Purpose
- For protection of process units operating at high temperature or corrosive conditions like sulphur burning furnace, converter, heat exchanger bottom plates of sulphuric acid plants, Electric furnaces for CS2 manufacture
- To prevent contamination of products due to contact with metallic surface.
- To conserve heat in the equipments.

Considerations for providing refractory lining
- Maximum Operating temperature inside the equipment.
- Material being handled and their chemical compositions; corrosive and erosive nature *(refractory lining of special composition is applied to reduce erosion while processing such materials. Consult the manufacturer of refractory before use.)*
- Melting point of materials being handled and tendency to form hard deposits on solidification.
- If reaction is highly endothermic internal heating by oil or gas firing/electrical heating may have to be done. In this case (and when conducting slag is formed) it is necessary to provide impervious and non-conducting lining. Mica sheets/Ceramic paper/asbestos sheets soaked in Potassium silicate may be placed between the shell and the lining.
- Brick-lining of equipments is generally done at *site since the lining can get disturbed during transportation (even over a short distance)*

Selection criteria
- Alumina percentage in the material (higher $Al_2O_3\%$ can withstand higher temperature.)
- Porosity (Not more than 23% for Fire Bricks, 65% for insulating bricks).

© Springer Nature Switzerland AG 2019
K. R. Golwalkar, *Integrated Maintenance and Energy Management in the Chemical Industries*, https://doi.org/10.1007/978-3-030-32526-8_13

- Density (dense bricks are less porous and hence less by penetration of corrosive gases)
- Dimensions (should be as per specifications given by purchaser.).
- Thermal conductivity KCal/m°C.hr,
- Refractoriness under load (RUL) in Kg/cm^2 at ....high service temperature.
- Permanent linear change shall not be more than 1.5% at 1250 °C
- Cold Crushing Strength in Kg/cm^2
- Pyrometric Cone Equivalent (PCE) value to be 32 onwards
- Iron content (higher iron % can lower (RUL).
- Resistance to Acid/Alkali (% loss in weight in 24 hours when exposed to operating conditions. This shall not exceed 1.0–1.5%)

Insulating Bricks
- Used between Fire (high alumina) bricks and equipment shell
- IS:2042-1972 Type-2 bricks or equivalent international standards
- Pyrometric Cone Equivalent: 27 (ASTM)
- Bulk density (max): 0.9 gm/cm^3
- The length, breadth and width/radius of curvature shall not differ by more than 1.5–2.0 mm from each other amongst the lot of bricks.
- Apparent porosity (min): 60%
- Cold Crushing Strength shall not be less than 15 kg/cm^2
- The brick shall be able to withstand 1250 °C on continuous duty.
- Each brick shall be packed carefully and individually since they are delicate.

High Alumina Bricks
- IS:8-1983 Type-2 or equivalent international standards
- Shall contain not less than 45% Al_2O_3 and the Fe_2O_3 content shall not exceed 1.8%
- Pyrometric Cone Equivalent (ASTM):32
- Refractoriness Under Load (RUL): 1400 °C
- Apparent Porosity not more than: 20–22%
- The length, breadth and width/radius of curvature shall not differ by more than 1.5–2.0 mm from each other. Examine about 5–7% numbers from the lot of bricks.
- Permanent linear change shall not more than 1% at 1250 °C – (*to minimise cracks during operation*)

Calderys Refractories India Ltd, Mount road, Sadar, Nagpur (India) are the leading manufacturer of such bricks.

Typical compositions and properties of their products are given below which are being used by many industries.

Product AC–45
Maximum service temperature....1400 °C
    Al2O3.....42–45%
    Apparent porosity...17–22%
    Cold Crushing strength...350–550 kgs/cm^2

Product AC–80
Maximum service temperature...1550 °C
  Al2O3.....77–80%
  Apparent porosity...16–19%
  Cold Crushing strength...750–950 kgs/cm^2

AC-INS-110
- Maximum service temperature..1100 °C
- Al2O3.....20–25%
- Apparent porosity..60–65%
- Cold Crushing strength 25–40 kgs/cm^2

Product AC–INS-135
- Maximum service temperature...1350 °C
- Al2O3.....27–31%
- Apparent porosity...60–66%
- Cold Crushing strength...30–45 kgs/cm^2

In addition to above products, Calderys have many more products which are best suitable for Chemical Industry. Details of these can be available from Calderys demand. Apart from this, Calderys has few excellent monolithic refractory materials which can contribute in conserving heat to great extent.

SKG Refractories Ltd, Butibori (near Nagpur, India) are also leading manufacturer of refractory bricks. The high alumina bricks produced by them have the following typical analysis and properties.

SKG-45
- Al_2O_3 content...45%, Density 2.2 g/cc
- Cold crushing strength...300 kgs/cm^2, PCE 34, RUL 1450 °C

SKG-70B
- Al_2O_3 content...70%, Density 2.6 g/cc
- Cold crushing strength...400 kgs/cm^2, PCE 34, RUL 1470 °C

Castable Refractory
- Used for difficult shapes, in corners of lining, for repairs, at sight glasses, thermowell nozzles
- Cast on appropriate anchors of stainless steel
- The grain size shall generally not exceed 3–4 mms
- Ability to withstand service temperatures (1200 °C at least.)
- Twisted fibres of Stainless Steel 316 (of 2 mm diameter and 25–50 mm length) can be mixed with the castable refractory to the extent of 2–3% for better strength of the lining.

Special Refractory
- Carbon blocks
- Chrome–Magnesite
- Phosphate bonded–

These are used as per requirements of high strength, for resistance to severe corrosion/erosion when aggressive materials are being handled. Consult manufacture before procurement.

- Curved bricks for manhole arches
- Special shapes for furnace roof internal

Ceramic Paper
- These are flexible sheets of alumino-silicate high purity washed fibres having uniform structure and hence can be easily handled
- Non flammable
- Low thermal conductivity
- Good dielectric strength (check before use with conducting slag)
- Excellent corrosion resistance (it is resistant to most chemicals except hydrofluoric, phosphoric acids and concentrated alkalis).
- Density: 160 kg/m^3
- Chemical Composition: Al_2O_3: 46–48%; Total Al_2O_3 and SiO_2 > 97%; Fe_2O_3: < 0.5%
- Sizes in which it is generally available: Thickness: 0.8 mm, 1.6, 3, 5 mm 6 mm etc.
- Width: 600 mm and 1200 mm
- Tensile Strength: 1.5 kg/cm^2
- Working temperature: 900 °C
- Maximum exposure temperature shall not exceed: 1200 C
- Used in petroleum industry, furnaces, converters operating at high temperature,
- As a thermal insulation layer, as gasket material for cover plates on nozzles
- As expansion joints in refractory brick lining can reduce the stresses created when the bricks expand as the temperature in the vessel is increased
- As a separator gasket on both sides of a blind inserted in high temperature ducts
- For fire and spark protection
- Insulation in ducts carrying high temperature gases: it can reduce required thickness of castable refractory layer and hence the weight of ducts

Procurement of Refractory
- Obtain test certificates for composition, CCS, RUL, Thermal conductivity, porosity corrosion resistance. Inform the manufacture details of operating conditions. Sampling plan for drawing samples of bricks and cement from the lots shall be discussed with supplier. The items can be tested in an approved laboratory acceptable to both parties.
- Consult manufacturer for cement to be used and precautions for storage as well as procedure for installation. Generally 8–10% extra numbers of bricks are procured.

• **Cost reduction**—use curved bricks for small radius vessels, at manholes.. Standard rectangular bricks can be used for large radius vessels

13.2   Installation of Refractory

• The surfaces to be lined with refractory are to be cleaned by wire brush, acid wash, smooth grinding of the welding seams. Any internal dust or oxides of iron are to be completely removed– *without which the protective lining may not adhere to the surface properly.*
• The bricks should be stored carefully in a shed well protected from rain, dust, ash
• Use only clean potable water for making cement paste.
• Avoid too thick joints (shall not be more than 3 mm) between bricks.
• Very smooth internal surface (to reduce erosion) can be created by using screened high alumina cement
• Provide Retainer Plates at intermediate levels in tall furnaces to prevent sliding down or the lower layers getting crushed under own weight
• Stainless Steel 304/310 anchors to be welded on vessel before castable lining. Design of anchors should provide suitable grip at the service temperature.

13.3   Typical Curing Procedure

• It is necessary to slowly remove the moisture from the refractory layers before commissioning the vessel. The layer nearest to the shell is to be slowly heated by oil firing. The flame should not directly impinge on the lining. Sufficient air volume is to be used for proper contact of hot air in all parts of the vessel. Cure each layer separately after initial set. Do not remove temporary supports provided for the layers till then. The higher volume of hot air ensures slow uniform heating of the lining.
• Provide Vents: for escape of moisture during curing/initial heating.
• The temperature rise in first 24 hours shall not exceed 150 °C, and then 250 °C in the next 24 hours, 450 °C in the next 24 hours.
• Soaking period–maintain at 500–600 °C for 24 hours
• Cooling period—allow cooling at 150 °C/24 hours till ambient temperature is reached. This will remove moisture without thermal shock. Please consult Manufacturer for the procedure. Preserve the records of heating and cooling temperatures for future reference.

Some precautions
• Clean the inside of furnace/process unit thoroughly before commissioning them.

- Always increase operating temperatures slowly to avoid thermal shocks. Never allow direct impingement of burner flame on brick lining. Regularly observe through peep holes
- Regularly monitor external shell temperature for any hot spots. Thermo-graphic Cameras may be used for monitoring from a safe distance.

13.4   Observations during Running of the Process Units

The brick lining and its curing is generally not done by fabricator of the equipment These activities are to be carried out by other specialists *who may not be engaged by the fabricator*. The purchaser may have to select suitable party for this.

Fire brick lined furnaces/process units   The lining is provided to protect the shell and to prevent loss of heat. In case of more than one layer of lining, the moisture is to be removed slowly from the layers **one by one**. This is to prevent cracks in the lining as the moisture from the earlier layer may escape by developing cracks at the joints. It is advisable to provide suitable vents through which the water vapour can escape. These vents can be plugged after all moisture has been removed.

After removal of initial moisture   Slowly heat to a certain higher temperature (known as soaking temperature) and maintain it as per instructions from OEM/ refractory manufacturer – *This is to ensure that all layers of bricks have been sufficiently and uniformly heated up prior to plant start up. The temperature of brick lined units should not be increased at a very fast rate during curing and commissioning.*

Monitor the temperature of external surface of the process units.

Monitor the temperature of external surface of the inlet and exit ducts of process units where combustion of fuels is taking place.

A hand held temperature measurement instrument or an infra red detector can be used. Dial thermometers can be permanently provided on the external shells at important locations. The pockets for the dial thermometers shall have some high boiling oil inside for correct indication.

One may sprinkle a few drops of water on the shell. If they boil off immediately the temperature may be more than 100 °C. Now mark a line with solid sulphur at the same spot. If it does not melt then the temperature is less than 120 °C.

Regularly monitor the external shell for any abnormally hot spots. A piece of sulphur when touched on them will ignite if the temperature is more than 250 °C. It can indicate internal damage to refractory.

Examine flame from oil burner (firing will not be proper if spray nozzle is damaged or atomizing air pressure is not sufficient), any localized feeding of raw material, excessive power input to the unit etc. in such cases.

13.5 Maintenance of Refractory Lining

- Suck out (exhaust) all hot gases through suitable process units or scrubbing system. Remove all clinkers, solid deposits inside the unit by appropriate mechanism, safe tools while using proper personal safety devices, protective dress; and ventilation.
- Slowly cool the process unit to ambient temperature.
- Examine presence of any toxic or dangerous gases inside. Flush out by air/nitrogen etc. Enter carefully while carrying portable breathing apparatus. Provide rope ladder for quick escape from the vessel if required.
- Use low voltage flame proof lighting inside.
- Let some persons stand outside to continuously watch and provide immediate help to rescue the persons who have gone inside.
- Check damage to refractory lining in the vicinity of oil firing burners, heating electrodes, areas corresponding to hot spots detected externally.
- Remove only the damaged/fused refractory material starting from upper side. Remove damaged anchors and retainer plate portions.
- Replace by new anchors and new portions of retainer plates.
- Provide new bricks wherever possible.
- Use castable refractory in places where the damage is in an irregular shape.
- Cure by slow gentle warming initially and thereafter by slow heating.
- Make sure the moisture has been removed.
- Restart the process after thoroughly cleaning from inside.

Acid Resistant Bricks (for protection of the shell from hot acid)
- As per IS:4860–1968 Class-I or equivalent international standards
- Chemical analysis shall be done of brick samples.
- The length, breadth and width/radius of curvature shall not differ by more than 1.5—2.0 mm from each other amongst the lot of bricks.
- Surfaces shall be smooth (without scratches) and edges shall not be broken or damaged. *Salt glazed surfaces are preferred.*
- Water absorption max 2.0%
- Compressive strength 700 kg/cm^2 (min)
- Resistance to acid (sulphuric 98%) i.e. loss in weight not to exceed 1.5% in 24 hours when in contact with 98% sulphuric acid at 100 °C
- The brick lining shall be cured as per advice from OEM before taking in to use.

13.6 External Thermal Insulations

Furnaces and process vessels operating at high temperatures can have a surface temperature of 60–65 °C by provision of brick lining inside. This is to reduce heat loss from the shell.

However in certain cases thickness of the refractory layer is designed so that the shell temperature remains above dew point (more than 80 °C) of the corrosive gases inside. This is to minimise the condensation of acidic material if the unit is run intermittently.

At certain locations a shed is erected above the vessel to prevent rain water directly falling on the vessel and cool the shell.

13.6.1 Ducts Carrying High Temperature Gases

A high surface temperature can cause burns to working personnel around the ducts. Hence a layer of special glass fiber or magnesia lining (about 50–80 mm thick) is provided on the external surface. It can be covered by aluminium sheets of SWG 18–20 to prevent further heat loss. Special care is to be taken so that no gaps are left in the cladding through which ingress of rain water, snow can take place, etc.

Good steel anchors are welded on the external surface for fixing the insulation material. Ready pads of insulation material can be kept ready for quickly providing thermal insulation if required.

It is advised to monitor the temperature of the aluminium clad surface regularly.

Chapter 14
Equipments for Energy Recovery

14.1 Selection of Waste Heat Recovery Boilers (WHRB)

Production engineers shall work out inquiry document for the boiler and econo-miser as follows. They may add more points to the inquiry if required.

- Gas volume, composition, dust content, maximum temperature at inlet and required temperature at exit.
- Type of boiler required: Water tube type or smoke tube (fire tube) type
- Normal working pressure at which steam is to be generated.
- Maximum pressure drop allowed on gas side.
- Pressure at which hot gases will be available at WHB inlet.
- Temperature at which feed water will be available. (Boiler Manufacturer shall give specification for the feed water) if no economiser is installed.
- Ask Boiler Manufacturer to provide all standard accessories, water level control-lers and safety devices.
- **General Arrangement GA drawing** and orientations of gas inlet box and outlet box (to be provided by the Boiler manufacturer).
- Gas side bypass arrangement to be made

   Information of (i) overall length × breadth × height, (ii) empty and full weight of the boiler (iii) space required for operation, cleaning and maintenance to be given by Boiler Manufacturer.

- All Statutory requirements in the country of installation must be complied with.
- Statutory Approval for Design and Drawings, welding electrodes, Fabrication (IBR and ASME Sec-VIII Codes) Radiography of all welded parts must be obtained

© Springer Nature Switzerland AG 2019
K. R. Golwalkar, *Integrated Maintenance and Energy Management in the Chemical Industries*, https://doi.org/10.1007/978-3-030-32526-8_14

- Stage Inspections and Final inspections should be offered.
- Quality Assurance Plan–MOC of tubes, Tube Sheets, Boiler shell with test certificates; approved Design and Drawings for all pressure parts, Test certificates for all bought out components (feed pipes, valves, flanges, level indicators…etc.)

14.1.1 Scope of Supply (Shall Be as per Battery Limits Agreed Between Vendor and Purchaser)

- Design, manufacture, testing, loading and transport to site, unloading at site, Erection and Commissioning.
- Civil foundation drawings and Foundation bolts
- Structural like saddle supports, work platforms
- Feed water pumps with motors, base plates as preassembled units
- All feed water piping with valves, NRV, conductivity meter for feed water
- Water level controllers and safety alarms and trip devices
- Refractory lining in gas inlet and exit boxes
- External insulation and cladding. External insulation—(after hydraulic pressure test)—*may be done by purchaser also*
- Spares for 2 years smooth operation – various valves, feed water pump parts, level gauges and controller spares, CA/ID fan spares, motors etc.
- **Other items and accessories** (if some other fuel will also be used as a support fuel).
- Fuel feeding and firing system, air pre-heater, combustion air fans,; FD Fans [Primary/Secondary], ID Fans
- Pollution Control system – Ash collection and disposal [Bottom Ash], ESP [Fly ash collection]/WESP
- Chimney with accessories [Lightening arrester, aviation warning lamps, internal bricking, etc.]
- Flue gas desulphurisation arrangement should be provided by the Boiler Manufacturer
- Desuperheater
- Instrumentation – Pressure gauge, level controller, conductivity meter, steam flow recorder]
- Thermal insulation

14.1.2 Performance Guarantees (as agreed mutually)

14.1.2.1 Commercial Terms

Including Bar chart for Implementation, transport to site, insurance, erection and commissioning assistance, delivery period etc.

14.2 Selection of Economiser

14.2.1 Considerations and Scope of Supply

Shall be as per battery limits agreed between vendor and purchaser

- Location in plant
- Gas volume, composition, maximum inlet temperature and required at exit.
- Allowable pressure drop through the unit on the gas side.
- Normal working pressure and the maximum pressure at which the boiler will be operated (**the economiser shall be designed accordingly**)
- Temperature at which feed water will be available (from the feed water tank).
- General arrangement of the unit, while showing gas inlet and outlet nozzles and internal fire/acid resistant linings
- Provision of cast iron gills or stainless steel fins on the tubes (for protection from corrosion).
- The bends for connecting adjacent tubes and welded joints shall be easily accessible for maintenance; hence preferably shall be located outside the gas enclosure.
- Overall Length × Breadth × Height
- Total weight when empty and when full.
- Standard fittings and mountings.
- Drain nozzles on the gas inlet & outlet boxes for draining out any (acidic) condensate.
- Design, manufacture, testing, loading and transport to site, unloading at site, Erection and Commissioning shall be in Vendor's scope.
- Civil foundation drawings to be supplied by with empty and full weight of the boiler
- Space required for operation, cleaning and maintenance
- External insulation and cladding offered by vendor.
- Structural like saddle supports, work platforms should be included in scope of supply
- External insulation—(after hydraulic pressure test)—*may be done by purchaser also*
- Spares for 2 years smooth operation– various valves, feed water pump parts, level gauges and controller spares, CA/ID fan spares, motors etc.

14.2.2 Performance Guarantees (as agreed mutually)

14.2.2.1 Commercial Terms

including Bar chart for Implementation, transport to site, insurance, erection and commissioning assistance, delivery period etc.

14.3   Problems for WHR Systems

- Deposits of dust, hard clinkers on heat transfer surfaces
- Attack from Corrosive gases
- Unsteady flow rates of hot streams
- Unsteady temperatures/variable Calorific values
- Bypassing WHR system due to process requirement (internal/external passage)
- Mismatch between Steam generation, process needs and efficiency desired

14.4   Precautions for Operation of Boiler

- <u>Water Treatment Plant</u> – Shall be always ready to supply make up water of speci-fied quality to BFW Tank. Water softeners, Cation/Anion resin columns and Mixed bed shall be regenerated well in time. Capacity of this plant should be confirmed by taking trials by feeding it with the raw water as available at site. It should be at least 40–50% more than the requirement of the boiler to ensure that sufficient treated water of specified quality will be always available in the BFW Tank. Water softeners, Cation/Anion resin columns and Mixed bed shall be loaded with sufficient resins so that frequent regeneration will not be required
- Calibration of Conductivity meter, TDS meter must be checked every week.
- Level in Boiler Feed water tank shall be checked every hour. It is advisable to maintain a stock of treated water to last for 8–12 hours always

  - Check the working of the Water Treatment Plant at the beginning and end of every shift.
  - **The Boiler Feed Water should not contain chlorides since they are corrosive**.
  - The stock of raw water in the main tank should be enough for 2 days at least.

- Run the installed feed water pumps (generally two are installed) alternately to make sure both are in working order. Attend immediately if anything is wrong.
- Monitor Water level in WHR Boiler regularly. Gauge glasses shall be regularly flushed clean. Check working of water level controller regularly.
- Feed Water quality must be analysed every 8 hours and must meet specifications for TDS, pH, conductivity etc. No free acidity shall be present. The pH shall not be allowed to fall below 8.0 at any time.
- Maximise condensate recovery from various points in the plant—specifically from steam turbine condenser and process units (evaporators, tray dryers.)
- Analyse the water inside the Boiler shall be regularly. Adjust the blow down quantity, phosphate dosing, alkalinity etc. accordingly.
- Check low level alarms and very low level tripping device at least once a week. When the water level goes very low, the tripping device shall trip the firing sys-

tem/stop hot gases input to the WHRB. *(should trip the air blower and sulphur feeding pump in a sulphuric acid plant)*
- Operation of all valves for level gauges, HP safety valves, NRV and feed check valves to be checked regularly.
- Settings of Safety valves shall never be tampered with. The initial setting shall be done by authorized Boiler Inspector only.
- Regularly monitor steam pressure and temperature at boiler exit
- Regularly monitor inlet and outlet steam pressure, temperature & flow of steam and water through Super heaters and De-super heaters This is to ensure proper steam quality at the inlet of Turbine/user point.
- Regularly monitor gas inlet and outlet temp and presence of SO2, HCl, moisture from WHRB.
- *Gas inlet temperature shall not exceed 1100 °C/the higher limit given by Boiler Manufacturer, whichever is lower.*
- Gas outlet temperature shall not fall below dew point of the gas mixture at any time to prevent corrosion of the downstream equipments.
- Analyse hot gases for SO_2, HCl, moisture, acid droplets regularly.

14.5   Co-Generation

Hot process gases are to be cooled in many chemical industries for further processing.

Considerable heat is recovered when these gases are passed through Waste Heat Recovery Boilers WHRB which generate saturated steam. It is further heated in super heaters to a higher temperature. When the steam is superheated to higher temperature more heat is available for power generation.

The HP superheated steam can be used to generate power by passing through turbines which drive an electrical generator. Exhaust steam from the turbines is condensed to recover it as condensate which is recycled to the WHRB feed tank.

14.5.1   Condensing Type Turbine

In this system the entire steam is used for power generation by passing it through the turbine. It is then condensed and the condensate is recycled. Latent heat of steam is lost to the cooling water which is circulated through the steam condenser. Since the latent heat of condensation is quite large, the system is not very efficient. Only the difference in enthalpy of steam at inlet and exit of turbine is available for generating power.

14.5.2 Back Pressure Turbine

Steam is taken out at a certain pressure from the exit of turbine and used for process heating. The pressure depends on the temperature required for heating. The steam is first passed through the turbine to generate power.

The latent heat of steam is (used) recovered during process heating. Less power is generated by this method since difference in enthalpies of steam at inlet and exit of turbine is lower. However, the overall system efficiency is better than above system having condensing type turbine.

14.5.3 Extraction Type

In certain plants, part of the steam is taken out at an intermediate stage from the turbine for heating and the rest is allowed to pass through for power generation (it is condensed). The steam taken out at the intermediate stage can be further heated (**Reheat type**) to generate more power.

14.5.4 Estimation of Power and Steam Requirement

The plant engineers shall work out the power required for operating plant equipments and steam requirement (for process heating) for the present and future product mix.

This will enable selection of proper type of Steam Turbine Generator for generating power and steam for process heating when both requirements are taken in to consideration.

Various examples for Cogeneration are given in the following book.

14.5.5 Numerical Example

Condensing turbine, steam superheated to 400 °C
- Steam conditions at inlet of turbine 40Bar and 400 °C, enthalpy = 3213.6 kJ/kg;
- Steam condition at exit of turbine 0.1 Bar and 45.8 °C, enthalpy = 2584.7 kJ/kg;
- Energy available for power generation = 3213.6 kJ/.kg – 2584.7 kJ/kg = 628.9 kJ. kg;
- Energy availability efficiency = 628.9 kJ/kg/3213.6 kJ/kg = **19.57%**.

Back-pressure turbine, steam superheated to 400 °C
- Steam condition at inlet of turbine 40 Bar and 400 °C, enthalpy = 3213.6 kJ/kg;
- Steam condition at exit of turbine 6.0 Bar and 158.9 °C, enthalpy = 2756.8 kJ/kg;
- Energy available for power generation = 3213.6 kJ/kg – 2756.8 kJ/kg = 456.8 kJ/kg;
- Enthalpy of condensate at 6 bar pressure = 670.6 KJ/kg.
- Therefore, enthalpy available for process heating = 2756.80–670.6 = 2086.2 KJ/kg.
- Total energy available = 456.8 kJ/kg + 2086.2 kJ/kg = 2543 KJ/kg;
- Energy availability efficiency = 2543 kJ/kg/3213.6 kJ/kg = **79.13%**.

Reference: (for more information and examples)
Production Management of Chemical Industries by this author.
Published by Springer International Publishing, Switzerland.

14.5.6 Some Important Considerations for Selecting Steam Turbines

- Steam conditions at inlet and exit of turbine (pressure & temperature at the inlet and outlet of the turbine). Superheated steam can generate more power.
- Steam Consumption for generating power (kgs/KWH)
- Normal operating speed
- Accessories like gear box, generator, base plates,
- speed controllers
- hand operated valve,
- standard accessories
- Steam condenser, ejector and connecting pipes, valves
- **Evaluate carefully the Instrumentation and safety devices/safety trips offered**:
- Erection and Commissioning services shall be in scope of the vendor.
- Operation and maintenance instruction manuals
- Spare parts for 2 years smooth operation list.
- water pumps, cooling tower, condensate recycle system generally by vendor
- Steam supply line moisture separator and regulator for the turbine (generally by purchaser);
- Battery limits of supply and exclusions from scope of supply by vendor.

Another option for consideration is Reheat Turbines:-HP steam turbine can exhaust medium pressure saturated steam which can be reheated by hot process gases if available and more power can be generated.

14.5.7 Steam Turbine Driven Generators

These can be considered in following typical situations
- When power requirement in the plant is steady with less chances of increased demand
- When power requirement in plant slowly picks up from cold start *of the plant*. When requirement for process heating by steam can be more (continuous heating of reactors, operation of evaporators for large quantity of dilute solutions.)
- When hot gases generated in the plant (during reactions etc.) are having considerable dust particles, or corrosive components and cannot be used in gas turbines.

Steam Turbo Generator: System components
- Hot gases available from reactions in chemical plant unit
- HP Steam Generator (could be Water Tube type with protection of CI Gills on the tubes OR Smoke Tube type with special alloy steel tubes)
- Boiler Water Feeding arrangement with HP pumps, and automatic control of level as required
- Incoming, exiting and bypass arrangement for Hot Process Gases as per Plant Design and Operational requirement (if temperature of downstream process units is to be maintained at a desired value)
- Corrosive hot process gases can be passed through the HP boiler but they shall not be allowed to cool up to dew point.
- Steam Turbine with accessories
- Electrical Generator
- *Pressure Reducing valve if it is required to use HP steam in the plant for process heating also.*
- *However this a*rrangement may not required for using the steam from Back pressure turbine exit for process heating
- Appropriate instrumentation, thermal insulation, and electrical switch gear to use the power.

14.5.8 Check List Before Commissioning of Steam Turbine

- Civil foundation...Length × Breadth × Height should be as per drawings
- Pockets for foundation bolts and their depths should match with base plate for the Steam Turbine Generator assembly
- Clearance from Civil engineer that foundation is fully cured
- The foundation bolts should be supplied by the OEM. Confirm that they are firmly grouted and tightened;
- Proper lock nuts/cotter pins should be provided
- Check GA drawing for the assembly of Steam Turbine and Generator, steam condenser;

- Alignment of Steam Turbine Generator assembly
- Strainer, steam traps, pressure and flow control valves (water, steam..oil),
- Check speed governor (mechanical/electronic)
- Inspect the complete piping for steam and cooling water,
- pressure and temperature gauges for steam, water, oil etc. should be calibrated and tested
- Working of Lube oil pump, Oil circulation and cooling system
- Battery back up (should be fully charged),
- Oil quantity and quality (to match Original Equipment Manufacturer specs), vapour release vent for oil
- grease in bearings
- Nozzle orientations for Steam Turbine–steam inlet and exhaust to Condenser or for Process heating purpose,
- Condensate exit piping, hot well pump, return piping to boiler feed water tank
- Ejector for creating and maintaining vacuum in condenser
- Pressure test for condenser and all piping
- P & ID...drawing
- Location of equipments on ground floor, first floor, second floor
- Thermal insulations on steam and **condensate recovery** lines
- Trial runs of oil circulation pumps, cooling water pumps,
- Distribution of return water (from condenser) in to Cooling water
- Trial runs of ID fans in Cooling Tower
- Bolts on casings of all units should be tight
- Check provision and setting/calibration of all pressure release devices, high speed trips, emergency trips, manual trips low water supply trips, high oil temp/ low oil pressure to gear box etc. These should be done as per OEM recommendations.

14.6 Commissioning

- Slowly rotate STG assembly by hand or mechanical means to confirm it is free to rotate
- Slowly open steam supply valve to warm up steam lines. Check working of all steam traps
- Blow off all dust, welding debris.
- Keep ST exhaust valve fully open.
- Slowly increase steam supply pressure to STG
- Observe speed of ST-gear box-generator
- Observe Voltage, current, KW etc. at generator terminals
- Connect to power supply MCC for the plant.
- Gradually add load by supplying power to small drives first; then bigger ones
- Synchronise with grid supply OR
- Use an Induction generator to reduce the draw of power from with grid.

- Before taking the steam in the turbine, the inlet lines have to be blown thoroughly by steam. A temporary arrangement with a silencer for blowing should be made.
- To check if the lines are clear, target plates made of SS or Brass are installed in the lines during the blowing.
- They are removed after about 2–3 hours of blowing and the surfaces are checked by the roughness and the number of dents.
- There should not be more than 2–3 dents of small size per square inch.
- Lube oil circulation needs to be done by using fine mesh to prevent any dirt going to the bearings. On line filtration can be applied so that the oil is filtered.
- Governor calibration is required.
- After turbine is ready for admitting steam, Overspeed test is to be carried out in decoupled condition. Steam is taken in and turbine started. Speed is increased through Governor till the allowed over speed limit. The turbine should trip at this speed.
- The Over speed test (OST) has to be done three times to ensure repeatability. Older turbines and smaller turbines are provided with Mechanical OST device.
- Now-a-days, turbines are provided with Electronic governors and trips.
- If the turbine does not trip at the pre-defined speed, manual adjustments will have to be done.

14.6.1   Maintenance of Steam Turbine

- Isolate from incoming and outgoing streams, electrical side, steam supply.
- Allow to cool and slowly depressurise.
- Take samples of lubricants from casing, bearings etc. and analyse for presence of metallic, carbon particles, ash. Preserve the samples for future reference.
- Open the casing and inspect internal parts; and their clearances from body any damage to wearing rings, turbine blades [check if any ash/particulate matter has deposited], piston rings, cylinder liners, shaft,,.
- Check axial and radial movements and refer to original permitted values as given by OEM. In case of excessive play/movement, examine all parts for any internal wearing out. *Consult OEM and request for sending an expert at site (if required).*
- . Any cracks in welded parts shall be attended by properly supporting the unit; pre-heating the damaged position and welding by electrodes of recommended composition only. The welder must be an experienced person (ASME Section IX). Post Weld Heat Treatment to be carried out as per advice from OEM. Use proper slings for lifting to avoid warping/distortion of shafts.
- Assembly of rotating parts may be completed after testing.
- Check by slowly rotating (by hand) if possible.

- Confirm direction of rotation [as per orientation of impeller vanes, turbine blades etc.] and run for a few minutes only.

    – The steam turbine needs pure clean superheated steam –(i.e. without any moisture or silica particles) for long life of the rotor blades and nozzles.
    – The super heater working shall be monitored very closely during operation and it should be cleaned during shutdown.
    – The spray of demineralised water in the attemperator [de-super heater] needs careful control. Hence the control valve for DM water spray must be checked whether it is getting actuated correctly as per steam conditions (pressure, temperature, flow) at inlet and exit of the attemperator.

Demister [Strainer] in steam lines
- This must completely remove (an efficiency of 99.99% is desirable) all mist particles from the steam.
- As a matter of abundant precaution, a minimum superheat temperature about 100 °C in excess of saturation temperature at the working pressure is desirable.

Excessive Superheat (more than 550 °C) can warp/damage the steam turbine blades due loss of mechanical strength. This must be avoided.

- Check and reset the controller accordingly. A bypass line may be provided to reduce heat input to the attemperator (where by hot process gases will not enter the unit). This arrangement is provided in chemical plants where waste heat boilers are installed to recover excess heat available from hot process gases is used for Co-generation.
- Examine gear box and speed governor.

14.7   Options to Be Looked in to for the Chemical Plant

Plant Engineers shall look in to the following options for Waste Heat Recovery along with power generation when external grid supply is also available.

- Estimate Power and Steam requirement in the process plant for the present and in future
- Consider process conditions, hot process gases composition, temperature, dust content, corrosive nature.
- Variations in gas flow due to changes in process-(production rate, temp, product mix, need to raise temp of downstream units)
- Heat available from hot process gases
- Gas Turbo Generators (GTG) can be considered when clean fuel (natural gas, Propane, LPG..) is available at site; **or** when the hot process gases are clean, do not have dust and are not corrosive.

- A Low Pressure steam generator can be provided after the (GTG) for further heat recovery
- The diagram shows option for energy recovery as electrical power by means of GTG and STG also when the process gases are clean. If the gases are corrosive, contain dust then a heat exchanger should be used to heat up the compressed air (heat exchanger is not shown).
- In case of corrosive process gases the energy/heat can be recovered by HP boiler with alloy steel tubes and appropriate Steam Turbine for generating power and process heating.
- Back Pressure Turbine drives certain selected process machines (air blower, boiler feed water pump). The steam from exit of the turbine can be used for process heating for increasing overall efficiency.
- Some considerations for choice of Electrical Generators are: (i)There is more requirement of power, but less need of steam for heating.......select synchronous generator. (ii) There is less need for power but more requirement of steam for heating.......select induction generator. (iii) *availability of steam at a steady rate may not be possible if the Waste Heat Recovery Boiler is partially bypassed occasionally for directly heating downstream process units.*
- A Dual Drive arrangement can be considered in such a case. The steam turbine shares the load on the motor as per availability of steam; and thus reduces draw of power from external source. (Please see the figure for Dual Drive arrangement at Chapter 12.17)
- Boiler capacity and type planned–steam kgs/hr., pressure, super heated/ saturated
- Process heating requirement– required, rate of heating, steady or intermittent

14.8 Gas Turbine Generators

14.8.1 *These Can Be Considered in Following Typical Situations*

- When normal requirement for power is less in the plant; with only occasional increased requirement–Gas Turbo Generator can supply the increased demand quickly.
- When power is to be quickly generated from cold start *of the plant*
- When requirement for process heating by steam is limited (drying of products with less moisture, melting of raw materials in small amount)
- When clean fuels like Natural Gas, Propane, LPG are easily available in sufficient quantities at reasonable cost.

14.8.2   Gas Turbo Generator: System Components

- Combustion Chamber (refractory lined) for generating hot gases by burning of clean liquid or gaseous fuels.
- Temperature should be around 900–1000 °C at exit of combustion chamber
- Feeding and firing arrangement for **gaseous** fuel (Natural Gas, Propane, LPG) with gas pressure, metering system and flow regulator as per temperature required at inlet of Gas Turbine.
- Feeding and firing arrangement for c**lean liquid fuels** like High Speed Diesel, Aviation Turbine Fuel (receiving and day oil tank, pumps, burner, and feed rate control) as per temperature required at inlet of Gas Turbine
- Filter at air inlet (suction) side of the Air Compressor
- A heat exchanger if **corrosive hot process gases** from chemical plant units are to be used to increase temperature of compressed air. This is not shown in the figure given below.
- **The hot process gases can be mixed with compressed air and used directly in the Gas Turbine only if permitted by the manufacturer.**
- Gas turbine-generator
- The gases shall not be allowed to cool up to dew point at exit of gas turbine
- Heat Recovery Steam Generator HRSG (at Gas Turbine exit gases) with accessories
- Boiler Feed Water Pumps
- PRV for steam and arrangement for using this steam for Process heating or for export
- Appropriate Instrumentation
- Thermal Insulation
- Electrical switch gear to use the power.
- Final exit gases from HRSG can be let out through a Chimney as such

OR
- a*n Air Pre-heater for* Final exit gases *can be used for getting hot air*

14.8.3   Combustion System of Gas Turbine

Check alignments of compressor- generator – turbine- gear box before start

- Air Filter at inlet of compressor – clean thoroughly; replace any damage filter media.
- Fuel supply controller to burner/Fuel pump/Gas supply regulator for heating the compressed air to optimum temperature at inlet of the gas turbine. It should never exceed the upper limit advised by OEM.

14.8.3.1 Observations During Operation

- Air pressures and temperature at exit of compressor, combustion chamber, turbine inlet and exit;
- Rate of fuel consumption.
- Shell (external surface) temperature of Combustion Chamber.
- Speed (RPM) of all rotating components.
- Vibrations or any abnormal noise.
- Check safety vent valve on Combustion Chamber.
- HRSG (Heat Recovery Steam Generation) after the turbine – (Operating conditions).

14.8.4 Turbine

Check the following during inspection for maintenance:

- Nozzle apertures (whether opening are clean or choked; or increased in hole size)
- Fitting of blades on the rotors (whether any blade has become loose on the rotor, has warped, whether angle of fitting has changed or particulate matter has deposited on the blades).
- *In such conditions the flow of steam (or gas) may not take place in a desired manner and the rotor will not rotate with correct speed.*
- Smooth operation of hand wheels
- "Over speed Trip" mechanism.

Lubrication system Oil pump, quality of lube oil, oil cooler, availability of emergency power through battery back up.

14.9 Type of Generators

- A Synchronous Generator is self excited, and stand alone type. It will keep the plant running even when the external grid supply trips. However, it should have the capacity to supply power as per need and may not be used if the supply of hot inlet gases/steam is not very steady.
- An Induction Generator is used to reduce draw of power from external grid. It keeps working even in a situation when supply of hot inlet gases/steam is not very steady.

- If Induction generator is used, it will draw reactive power from the grid, and the power factor will be low. Capacitors may have to be installed to compensate this.

Final decision should be taken after the electrical engineer checks the site conditions thoroughly.

Operational convenience, cost and maintenance
This will depend on site conditions. Plant engineers should examine the various options shown in the diagram below as per need for power and steam in the plant; and whether excess of these can be exported to external parties/external grid.

(i) **HP boiler and Steam turbine**
(ii) **Gas Turbo Generator system with Heat Recovery Steam Generator (HRSG) boiler at GTG exit**

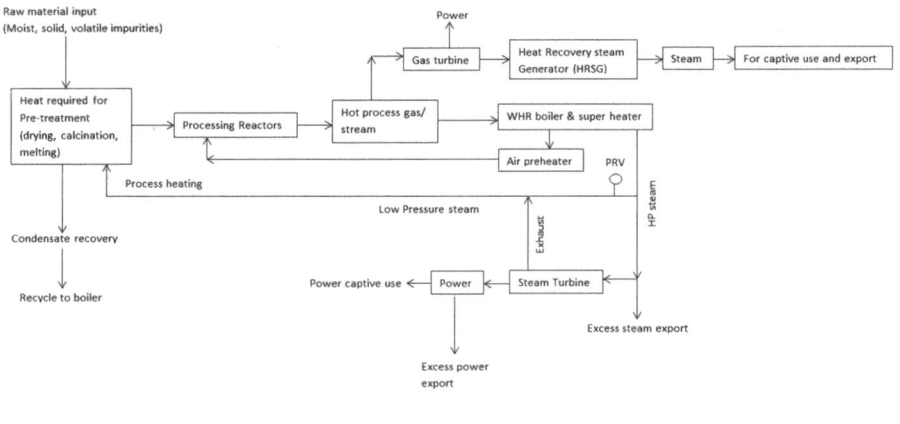

Options for co-generation

14.10   Cleaning of Boilers

- Process gases may carry considerable dust which can get deposited in the inside of boiler tubes (in case of smoke tube boilers) or on their external surfaces (in case of water tube boilers). Scaling can also take place due to corrosive process gases.
- The deposit of dust/scaling can cause serious operating problems. The dust and scales have low thermal conductivity and can cause poor heat transfer to water in the boiler. This can result in high wall temperature of the tubes and may even puncture them.
- **WHRB in chemical process plants**: the efficiency of heat transfer goes down and hot process gases are not cooled as per requirement of downstream units if the tubes are having deposits/scaling. Hence cleaning of boiler tubes becomes necessary to restore the efficiency for heat recovery and to meet operational requirement of process in downstream units.

- Generally the scales are of iron oxides, calcium carbonates and oily deposit or organic compounds (more during initial commissioning)
- Additional chemicals may be required if the scales consist of other compounds different than calcium carbonate. **Boiler tubes are generally cleaned by brush while the shell side is cleaned by chemical method.**

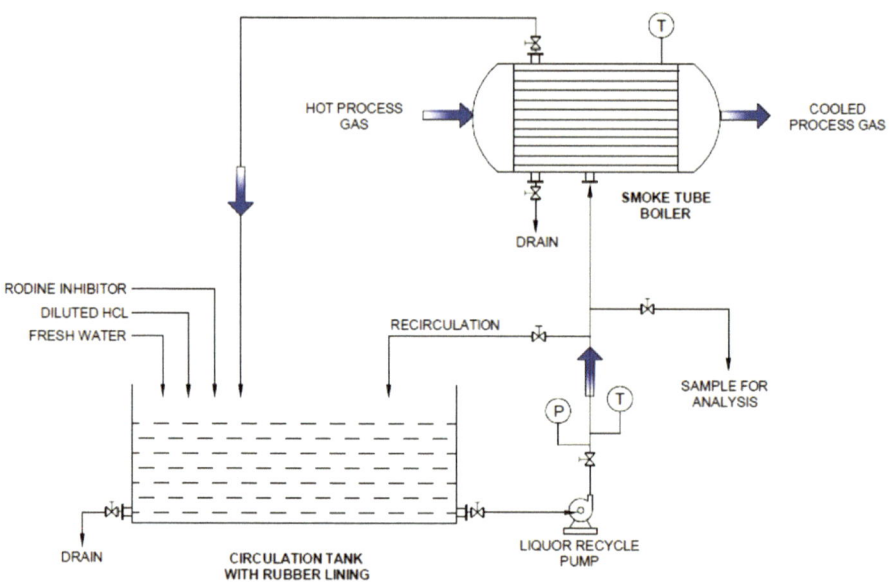

CHEMICAL CLEANING OF SMOKE TUBE BOILER

General procedure for cleaning:

The process plant shall be stopped before the last working date permitted by statutory inspector for boiler. The boiler should not be operated beyond this date unless permitted in writing by statutory authority in exceptional circumstances.

Slowly depressurise the boiler to ambient pressure and then isolate it from hot gases. Purge out any remaining gases from inside. The gas inlet and exit sides should be opened and good ventilation to be provided before entering inside. Check by gas detectors that all traces of process gases present inside have been removed. Use gas masks and breathing apparatus

Procedure for Smoke tube boiler

Fill up the boiler with cold feed water and allow it to cool to ambient temperature. Flush out the water from the boiler.

Arrange a circulation system with a **(rubber lined)** tank, pump, piping and valves. Connect the pump discharge side to water inlet of boiler and return line from the vent at top of boiler.

Fill the circulation tank with clean filtered water to about 75% of its capacity and add HCl and Rodine inhibitor slowly with stirring induced by recirculation from the recycle pump. Adjust concentration of HCl to about 3.5–4.0% and 0.4–0.5% of the Rodine **inhibitor**. The pH shall not fall below 3.0–3.5

A sample of rusted piece or tube with deposit of scales can be suspended for about 12–16 hours in the inhibited HCl solution to check its effectiveness and corrosive properties.

The concentration of HCl and Rodine can be adjusted as per need.

Transfer this batch of descaling solution to the boiler. Prepare more such batches and fill up the boiler till the liquor comes out from the vent in to the circulation tank. Check the concentration of each batch and also that of liquor coming out from the vent.

Continue circulation of the descaling solution in the boiler for about 12–16 hours.

Check the cleaned surfaces of tubes, as well as turbidity of the boiler water sample. This will indicate satisfactory descaling of the tube surfaces.

After satisfactory descaling of the tubes the acidic water in the boiler is to be neutralised by dilute alkali (3–5% NaOH) and then drained out. The boiler is to be rinsed by demineralised water now.

In case the scales are not removed properly, the procedure may be repeated taking care that no boiler tube gets punctured.

Provide good ventilation while handling the hydrochloric acid since it can produce irritating fumes during cleaning the boiler.

Tube side of the Smoke Tube boiler can be cleaned by wire brush and flushing by water. The cleaned tube surfaces can be observed by a torch, which also show if any tube is sagged.

Water Tube Boiler

The accumulated acidic condensate on shell side should be drained out. The refractory lining on inside walls shall be inspected carefully and repaired wherever necessary by providing new bricks or by castable refractory. The shell side can be cleaned by opening the cleaning manholes and removing all deposits and clinkers inside.

The tubes can be cleaned by circulation of inhibited HCl/descalant solution in a similar manner like Smoke tube boiler.

HCl acid and inhibitor

Suitable grades of Rodine are added to the dilute HCl acid solutions. The Rodine can inhibit base metal attack. It can offer protection from excessive-pickling by the acid and reduces fuming. These inhibited HCl solutions can now be used for descaling of boiler tubes.

Sulfamic acid is also used as an acidic cleaning agent either as such or mixed with other chemicals as a constituent of a proprietary mixture offered by vendors.

It can remove rust and calcium carbonate deposits more conveniently as compared to the more volatile HCl which gives off irritating vapours. *HCl can be however cheaper.*

It also finds applications in the cleaning of some industrial equipment. Although it is considered less corrosive than HCl certain corrosion inhibitors are often added while using it as a matter of safety.

Instrumentation

All pressure gauges, temperature indicators, conductivity meters, level indicators and controllers shall be inspected and calibrated. The probes shall be cleaned if any deposit is noticed on them.

14.11   Boiler Maintenance

- Carry out hydraulic test of pressure parts after cleaning. Any sagging or erosion of tubes should be closely examined. Examine thoroughly—replace sagged tubes, damaged parts
- flush out gas side by air, arrange good ventilation before entering the vestibule at inlet and gas exit box
- Check the condition of refractory lining in gas inlet and outlet boxes; and attend.
- Examine and attend the following fittings and mountings: Safety valves, Blow down valves, Non-Return valves, feed check valve, vent valve, main steam stop valve, level indicators etc. must be checked thoroughly.
- Blind all nozzles and carry out pressure test by using certified pressure gauge only in presence of Boiler Inspector
- Safety valves are reset by statutory boiler inspecting personnel after the boiler is cleaned. Hydraulic pressure test of the boiler is to be carried out after fitting them.
- Soot blowers for removal of deposits of unburnt carbon on heat transfer surfaces.
- Steam lancers with adjustment of angle for removal of deposits.
- Working of combustion air blowers, dampers in air lines.
- Working of oil pumps for oil firing.
- Oil burner and its assembly.
- Overhaul the boiler feed water pumps, feed check valves, conductivity meter and non return valves.
- Check shims for BFW pumps.
- Confirm proper working of water level controller, low level alarm and tripping system at very low water level.
- Air pollution control with ESP, SO_X, NO_X removal systems.
- **Comply with instructions from statutory inspecting officer for replacement of tubes and any other jobs to be done.**
- Repeat the hydraulic pressure test in presence of the statutory inspecting authority.
- Test Certificates shall be taken for tubes, welding rods and all spares. Welding Job to be done by Certified Welder only.
- Resetting of safety valves is normally done by Boiler Inspector
- Clean steam drum –if separate
- Remove Ferrules on tube sheet and clean surface thoroughly before inspection by Boiler Inspector.

- Inspect all fittings, level indicator and controller
- Clean Feed Water tank, working of feed water pumps (including stand by units) and feed water lines
- Repair refractory lining and replace ferrules
- Replace external insulation after permission from Boiler Inspector.
- Fill up with water, confirm satisfactory working of pumps, feed water valves, level controllers, blow down valves etc.
- Refit all mountings and valves and repeat pressure test before taking the unit in to use again.

14.12   Cleaning of Economiser

- Open the external bends (which connect the adjacent tubes). Clean tubes by brush, descalant solution and wash by water, dilute alkali (5%); followed by washing by demineralised water.
- The accumulated acidic condensate on shell side should be drained out.
- Inspect the gas inlet and exit nozzles, refractory lining on inside walls and bottom side. Repair by providing new bricks/castable refractory.
- Clean the Cast Iron gills on the tubes by pressure spray of de-scalant solution to remove all deposited scales on them, and then by dilute alkali. Finally wash by clean water. Replace damaged C I gills
- .Clean all pressure taps and thermo-wells on gas inlet and exit lines.
- The shell side can be cleaned by opening the cleaning manholes and removing all deposits and clinkers inside.
- Clean all pressure parts and carry out hydraulic test.
- Clean and change *if necessary* the cast iron gills on tubes

14.13   Economiser Maintenance

- replace damaged tubes by certified tubes.
- Replace external insulation after permission from Boiler Inspector.
- Fill up with water, confirm satisfactory working of pumps, feed water valves, level controllers, blow down valves etc.
- Refit all mountings and valves and repeat pressure test before taking the unit in to use again.
- Inspect and service the safety valves, blow down valve, vent valve, water inlet valve and outlet valve. Carry out hydraulic pressure test for all these valves
- Overhaul and service the drain valve.
- Attend external thermal insulation and cladding

Chapter 15
Safety Precautions During Maintenance and Energy Recovery

15.1 Safety Precautions During Maintenance: General

Structural supports

The process units and machinery should be erected on strong supporting structures. The design, fabrication and construction of these structures should be carried out by qualified, licensed and experienced structural engineers and technicians. They shall be protected by corrosion resistant coatings/cements and regularly checked.

Any damage noticed to the protective coatings or cements should be immediately attended to ensure safety of the personnel and plant units.

Carefully designed, fabricated units with the right materials of constructions and operated within safe limits are very important for reducing breakdowns.

Storage of dangerous, toxic or inflammable materials should be at locations at a safe distance from hot units like furnaces, reactors and high tension electrical equipments. It should be possible to quickly isolate such dangerous materials from each other.

Avoid excessive storage of hazardous material in the premises.

• Dyke walls shall be provided around storage tanks to prevent spreading of dangerous liquids in case of a leak. The enclosure must have sufficient capacity to collect all the liquids which may leak out.

These precautions reduce the risk and therefore can increase the life of the process plant. It can become safer to operate and maintain the plant.

© Springer Nature Switzerland AG 2019   313
K. R. Golwalkar, *Integrated Maintenance and Energy Management in the Chemical Industries*, https://doi.org/10.1007/978-3-030-32526-8_15

15.2   Typical Guidelines for Safety Precautions

These precautions are of a general type. They shall be modified to suit the individual process plant after considering the properties of the chemicals being handled; the operating conditions and design of the equipment. Special precautions are necessary when entering a closed vessel and carrying out maintenance work.

A well designed **Work Permit system** shall be always followed.

- It is necessary to make sure that all necessary precautions have been taken before starting maintenance work such as entering a process vessel, welding a leaking duct, dismantling an equipment for major repairs or replacement.
- *Standard set of Formats **(Work Permits)** should be prepared for individual units. This shall be done by senior engineers and safety officers. **Please see 15.8 for more details**.*
- Isolate from all input and output connections to other equipments. (cooling water or nitrogen flushing may not be disconnected)
- Only closing of the valves in connecting ducts or pipelines is not enough because some valve may be leaking. Hence blinds shall be inserted in the ducts at appropriate places; and the position shall be recorded. The blind plate should have a protruding tail piece
- Provide temporary support to connected ducts on both sides when a pipe length is opened. Cover all flanged joints by appropriate covers
- Completely flush out all liquid chemicals/exhaust the vapours from inside. Get confirmation from production engineer in the work permit. Keep ready drums for collecting the waste material to be drained from the vessel

15.3   Checks by Safety Officer

The safety officer should check the following even after getting work permits from production engineer, electrical engineer etc. (in case of work inside a process vessel)

- Remove fuses from main supply to prevent accidental starting of motors (for agitators inside the vessel) or pumps which may pump in liquid in the vessel. Get confirmation from electrical engineer in the work permit.
- Keep dry chemical powder/CO_2 cylinder/hose pipes, sand and ash buckets ready.
- Personal protective appliances: face shield, gas mask, gum boots, full length hand gloves shall worn by all personnel working on such jobs.
- Check expiry dates and record of use of gas mask canister. Use only appropriate canister for the process gas. Do not use a gas mask canister (*even before the expiry date*) if it has been already used for maximum permissible time.
- Provide fume extraction hoods and exhaust fans to remove toxic, inflammable gases and vapours from working area. Check presence of dangerous gases in the working area by gas detector.

- Do not use the inlet, outlet, sampling nozzles or cleaning manholes for tying by link chains before lifting. Use only lifting lugs provided for this purpose.
- Use chain pulley blocks of sufficient load lifting capacity
- Only tested link chains and slings are to be used. Do not suspend from trusses of roofs or gas ducts or liquid pipe lines.
- Use rust remover to open nuts. Do not use chisels or gas cutters if inflammable vapours are present in the vicinity of the vessel.
- Fix slings around the body of component which is to be taken out of the equipment
- Cordon off area on ground floor when some weight/heavy equipment is being lifted.
- Check limit switches on hoist/crane to prevent over lifting.
- Check automatic braking system (to prevent failure of the hoist fails midway while lifting the equipment)
- Do not use steel hammer/chisels in presence of inflammable vapours. Use wooden hammer and non sparking tools.
- Check sufficient lighting, ladders, railing around the working platform.
- Confirm availability of standby power from DG set if required for continuous maintenance activities as per need.
- Follow OEM manual while dismantling complicated machines.
- Cordon off area when carrying out cleaning by using long rods or chemical cleaning by circulating acid.
- Arrange safety interconnections in case certain plant conditions become dangerous......(high/low levels in tanks; high/low temperatures of process streams; excessive pollutant concentration in exit gases, high pressure in units;) which need to be immediately controlled

    – Check the earth connections for equipment being welded/repaired. Check continuity of cables and resistance of earth connection
    – High Pressure Unit (under repair) shall be de-pressurised *slowly* at a rate as per instruction of OEM
    – Fire fighting arrangements (portable fire extinguishers appropriate for the type of fire) shall be ready with trained persons on the spot.
    – Plant engineers should add more precautions as appropriate for their own plants. These can be based on experience and analysis of –causes of accidents; and methods for prevention
    – Use special tools when required for work inside absorption towers, furnaces, process reactors.
    – Collect the waste washing liquid generated during shutdown and send to ETP.

15.4   Maintenance of Equipment Handling Dangerous Items/ Work Inside a Closed Vessel

- All sources of heating the vessel shall be stopped except where the material to be removed may solidify or becomes viscous on cooling.
- Keep an exhaust fan running to remove gases from inside.
- Keep an air blower running to supply fresh air into the vessel.
- Cooling water spray and smoke detectors should be available for all such vessels
- Keep a portable rope ladder suspended in the vessel from top to enable a person who is working inside to come out immediately in case of emergency.
- Check expiry dates and record of use of gas mask canister. Use only appropriate canister for the process gas. Do not use a gas mask canister (*even before the expiry date*) if it has been already used for maximum permissible time. *This precaution is very important and hence repeated here.*
- Portable breathing apparatus should be available for the person working inside.
- At least one person should be standing outside the vessel all the time and keeping a watch on the person who is working inside
- Provide fume extraction hoods and exhaust fans to remove toxic, inflammable gases and vapours from working area and from internals of the vessels under repair.
- Check presence of dangerous gases inside the vessel by gas detector.
- Electrical motor used for driving the agitator, feeding raw material, running a circulation pump etc. shall be disconnected by removing the fuses from supply. This will prevent accidental starting of the motor even if one switches on by mistake.
- Spillage of inflammable materials or vapours from such liquids should not come in contact with any open flames in the area by any chance. These spillages must be collected in containers and stored safely. Do not use gas cutters with open flame in such cases.

15.5   Personal Protective Equipments

These should be in perfect working condition always as they will be required during maintenance work to be done in presence of toxic/inflammable vapours or corrosive substances; or during work in closed vessels.

- Safety Showers which should be able to fully drench the person by copious amounts of water. Many such showers should be located in areas where strong acids/alkalies, corrosive substances are being handled
- Eye wash fountains.

- Face shield, safety goggles, safety shoes, full length acid/alkali resistant hand gloves
- Shoes with rubber soles (to minimise chances of sparking)
- Portable breathing apparatus
- Safety lamps operated by 24 volt supply
- Rope ladder
- Waist belt with a long rope
- High temperature resistant dress
- Emergency bell
- Portable fire extinguishers
- Buckets filled with ash, sand etc.
- **Dedicated Overhead Water tank –** *for supplying water to safety showers, eye wash fountains*

This should have Float-operated water inlet valve. It should always be kept full. Remote indication of water level should be always available. There should be water supply connection from at least two independent sources (separate pumps/tanks).

15.6   Alarms, Safety Devices and Interconnections (for Electrical Tripping)

Some typical examples are given below for safety and pollution control. These are also very useful for minimising breakdowns of process units and machinery. Plant engineers and experienced operators can think and work out more such warning devices for their plants.

Warning lamps, hooters (audio visual alarms) should be installed at important process units and mechanical equipments for following conditions:

- High level in over head tanks for acids, fuel oils
- Low level in feed tanks for process units
- High Press in steam supply line to heating jackets
- Low discharge pressure of pump for cooling water supply.
- High temperature in process units which may cause fires
- Low temperature in units which may reduce or stop flow of liquids due to freezing
- Effluent treatment plant: High concentration of pollutants in incoming streams, high pressure at inlet of pressure sand filter, active carbon filter, low pH (acidic condition), of effluent.
- Air pollution control: High SO_2, HCl, content in exit gas, low pH of scrubbing liquor, low level of alkali in supply tank, low discharge pressure of circulating pump, low delivery rate of circulating pump, low current drawn by the circulating pump motor.

- High temperature of the shell (furnace, converter, reactor etc.) to be monitored by thermocouples installed at suitable locations (to prevent damage to shell).
- Very hot spots getting developed on furnace shell, to be monitored by close circuit TV (CCTV) or thermographic camera.
- High pressure at inlet of filter press, reverse osmosis membrane inlet.
- Agitated vessel: Very high current drawn by motor (agitator shaft is jammed).
- Agitated vessel: Very low current drawn by motor (either agitator shaft is broken).
- High temperature of gases at inlet to gas turbine (overheating in combustion chamber).
- High temperature of steam at inlet to steam turbine (overheating by superheater or malfunctioning of de-superheater)
- High pressure of steam at inlet to heating jacket of reactors (pressure control valve not working).
- Low air pressure at inlet to oil firing unit (chance of unburnt oil carried over or explosion).
- Release of uncondensed vapour from safety vents (insufficient cooling by condenser)
- Special Safety precautions during opening, cleaning, repairing of the units are to be taken as per advice from OEM. Typically these relate to the order in which the parts are opened, cleaned, inspected, repaired/overhauled and then assembled again.
- Instructions from OEM for testing the parts after repairs and thereafter the entire assembly are to be followed. Follow the procedure for restarting individual units and for the entire plant also.

15.7   Safe Commissioning of Energy Recovery Units

Guidelines for safe energy recovery units
- Various types of energy recovery equipments are used in the chemical industries such as air pre-heaters, Waste heat recovery boilers and economisers, steaming type economisers, solution pre-heaters, gas generating process boilers (SO_3 from boiling oleum) etc. These shall be designed to operate safely in case of extreme weather conditions also.
- It is essential to study the physical and chemical properties of all materials involved for their design, fabrication and operation. The process boilers may be treated as pressure vessels since they generate vapours under pressure when heat recovery is done through hot process gases. *Hence they should be designed, fabricated and tested as per Codes applicable to pressure vessels.*
- The layout should be designed with proper ventilation of buildings and proper emergency exit facilities as well as enough space for maintenance or handling emergency situations.

15.7.1 HAZOP Study

Make a detailed flow sheet for the individual heat recovery sections showing all inputs, outputs and *expected* process conditions at the heat recovery units. Identify all likely hazards due to generation of high pressure, high temperature, excess flow of process gases, excess or no-flow of fluid being heated, tripping of feed water pump etc.

These shall be resolved by making changes in the P & ID after detailed discussions among designer, plant engineers and experienced technicians.

The units may be provided external reinforcements for additional strengthening.

Structural Design
The heat recovery units should have strong mechanical supports designed for adequate structural stability (considering the maximum weight of the equipment as if the vessel is full of material and with weight of fittings, connected pipes and weight of persons working there).

Capacity and stability of the supporting structures should be validated through licensed structural engineers for adequate safety before installation of the heat recovery units and connecting ducts/piping. They shall be protected by corrosion resistant coatings/cements and regularly checked.

15.7.2 Checks Before Erection

• Study the process and operating conditions considered for design
• Study the specification given to vendor while ordering equipment
• Confirm that all drawings (fabrication, layout, connecting ducts and pipes) have been approved by statutory authorities.
• Confirm that the fabrication of energy recovery equipment has been done as per approved codes and approved Materials of Construction have been used for fabrication
• **Confirm the following are ready**: (i) civil foundation and all supporting structures at the finalised location (ii) ladders, work platform, railings for ease of operation and maintenance. Get the design and construction of support structure checked by a licensed structural engineer. (iii) Proper lighting, escape routes at working areas

• *At least two independent escape routes must be available from the working area. These shall not have any obstructions.* (iv) Foundation bolts have been firmly grouted, pockets filled up with cement concrete and cured. (v) It is advisable to have a bypass route (duct/pipeline) available so that the heat recovery unit can be bypassed in case of a fault. Valves may be provided in bypass routes for uninterrupted smooth operation.

15.7.3   Checks Before Commissioning

- Post Weld Heat Treatment should have been done and records shall be available. *No cutting or welding is to be done on any pressure part after PWHT.*
- Pressure tests, radiography etc. should be carried out prior to commissioning @ 1.5 times maximum working pressure or as directed by local inspecting authorities–whichever is higher
- All personnel who will operate the units should have valid license
- Check easy visibility and accessibility to instruments and operating control valves
- Connecting ducts, and pipes should be pressure tested
- All parts/vessels (which may get pressurised during operation) and safety valves, rupture discs provided on them shall be tested as per statutory regulations.
- Economiser must have the connecting bends for tubes outside the shell (for ease of removal). Cast Iron gills shall be provided on outside of the tubes for protection if corrosive fluids will be present on shell side.
- Thermal insulation on hot surfaces shall be in place to prevent burn injury to operating personnel
- Confirm satisfactory flow of hot and cold fluids before commissioning
- Availability of alternative arrangement (e.g. standby water pump) for flow of cold fluid for heat recovery to be confirmed.
- Confirm setting and working of audio visual safety alarms (for high/low pressure, temperature, levels in connected process units/overhead tanks/feed tanks) to be checked and confirmed.
- Check safety vents after rupture discs (which will release gases away from working area in to a safety pot with 10% alkali solution)
- Confirm the calibration of all instruments for pressure, temperature, flow rate
- Confirm working of safety interconnections (which are provided for tripping certain units/motors in case of high temperatures, high pressures to take care of dangerous situations). These could be as per recommendation of OEM or as decided after HAZID/HAZOP studies.
- **Process controllers**—examine and attend operation of corrective/actuating links (such as electrical, pneumatic, mechanical) so that excessive pressure/temperature does not develop in heat recovery or connected process units.
- Smooth working of all valves in hot gas ducts and liquid pipes, working of bypass valves, Non Return Valve and main stop valve in steam, feed water lines should be confirmed. All of them should have external indication for inner flap, plug, stem etc.
- Electrical earth connections to be provided for all vessels and pipes handling dangerous fluids. These earth connections for safety shall be checked for continuity i.e. they should not be broken anywhere.

Shell and tube type heat recovery units
- Provide ferrules (ceramic, stainless steel, alloy steel) on tubes
- acid and heat resistant lining in inlet and exit gas chambers

- Sensing points (for pressure, temperature, composition measurement) before and after energy recovery units should be clean for getting truly representative sample

15.7.4   Observation During Operation

- Temperatures of all process streams
- Temperature of equipment shell by special temperature sensing chalks/hand held instrument
- *These shall be compared with design values/expected conditions*
- Carefully observe the changes which occur in operating conditions (pressures, temperatures, flow rates) after the heat recovery unit is taken in to use and compare with earlier values.
- Is any occurrence of excessive build up of pressure, abnormal temperatures, or vibrations taking place?
- Is any frequent blowing off safety valves, repeated tripping of motors, surging of blowers, overflow from any process tank taking place?
- monitor sudden change/low levels in process tanks and flow rates of process streams, current drawn by motors, steam pressure etc.
- Run standby units like process/cooling water pumps on alternate days/weeks if possible). At least run them on trials and keep in working conditions
- Check foundation bolts, coupling halves, terminal boxes of motors, earth connections, trip systems every week
- Visual checks of external shells of units operating at high temperature or pressure.
- **Important**: Monitor performance of downstream units after commissioning the heat recovery unit and compare with expected performance (calculations made earlier)

15.8   Safety Organisation

Safety organisation should have the following members

- Safety Officer (and Designer, Consultant to be available)
- Deputy safety officer
- Liaison Officer for external agencies and nearby organisations for additional help
- Legal Officer
- Representative (senior experienced technicians) from operational team, mechanical maintenance and electrical maintenance teams, fire fighting expert.
- Experienced representatives from workers unions should also be members of the safety organisation

- Representative from instrumentation, civil, raw material procurement, stores department and marketing department shall be available for discussion if required.
- The safety officer should confirm availability of all personal protective appliances

15.8.1 Safe Expansion of Capacity

- A study is to be carried out to assess the maximum **safe permissible** operating capacity of existing units and machines in the plant. This shall be compared with the required capacities for increasing the production rates. The short comings can be made up by modifying/upgrading the existing units or by incorporating new machines. Management shall train the production and maintenance personnel for the new process unit/machine being procured to ensure that they will be able to operate and maintain the new items without any mishap.
- **Plant expansion** shall be carried out with proper attention to safety matters since some of the existing systems may get overloaded like refrigeration plants, electrical main bus bars, cables, incoming transformer, pressure vessels, heat exchangers, condensers, effluent treatment plants even when operated carefully—because they may be already operating near their maximum capacity – *and this can create dangerous situations.*

15.8.2 Important Inputs from Safety Officer

Safety officer should assist the Production engineers, Maintenance engineers and Management for the following activities:

- Compliance with all Statutory Rules
- Condition Monitoring and Preventive Maintenance of key equipments
- Standardize operating procedures—revise if necessary
- Regular testing of all safety devices (safety valves for releasing excess pressure, rupture discs, tripping systems etc.)
- Safety officer should **again** check necessary safety devices, safety interconnections and proper handling of all dangerous materials during maintenance activities.
- Ensure all workmen are given personal safety devices
- Display important safety instructions prominently
- **Design of Work Permit** system for Maintenance jobs

Standard set of important precautions to be taken should be prepared for individual units. These should include (but not limited to) isolation and thorough flushing of the unit, removal of residual chemicals and gases, electrical disconnection; and

further precautions like provision of fire fighting, personal protection devices, checking by gas detector
Small amounts of inflammable, toxic or corrosive materials could be present in the crevices or idle pockets of process units or ducts even after flushing them. It is therefore necessary to confirm that all necessary precautions have been taken before starting maintenance work in closed vessels, in areas where toxic, inflammable, corrosive materials could be present. The job may involve internal repairs of a process vessel, welding a leaking duct, dismantling equipment for major repairs or replacement.

The work permits shall be issued by concerned section head/senior engineer before carrying out maintenance work. It should mention the time of expiry.
Hence it will cease to be valid on next day.
A fresh work permit should be again obtained before starting maintenance work on next day

Investigations of mishaps and Corrective actions
• Record all mishaps and incidents (even without mishaps)
• Categorize as per seriousness of the incidents
• Objectively investigate all mishaps and incidents
• Establish reasons—human error, design fault, equipment failure
• Take appropriate corrective actions and confirm by trial runs, safety drills.
• Revise the above if necessary

15.8.3 Assistance by Safety Organisation

• Review design of equipments, Process Conditions (pressure, temperatures etc.)
• Review Operation methods, Plant lighting, location of valves, sensors for instruments and working of Field instruments etc.
• Modify/change wherever necessary and implement them after approval by Factory Inspectors, Electrical Inspectors, etc.
• Maintain carefully and repair faulty unit immediately.
• Experience gained during working of process and while carrying out the maintenance work must be discussed during a meeting after the work is over. Suggestions for better preventive measures and working methods may be obtained.
• Review of such suggestions shall be done in the subsequent meetings of the safety organisation for continuously improving safety in the light of experiences gained during implementation of earlier methods.
• *Management shall provide adequate funds for safety related matters.*

15.8.4   Emergency Response Training to Personnel

- Safety officer should train all plant personnel in consultation with production and maintenance engineers for handling emergency situations by carrying out safety drills at regular intervals, and in First Aid
- Safety officer should make all plant personnel aware of their duties (what to do, in case of a mishap/emergency situation) and familiarise them about location of escape routes, fire fighting procedures.

15.8.5   Safety Organisation to Arrange the Following

- Sign boards with clear safety instructions.
- Sign board for assembly point in case of emergency
- Names of contact persons with their cell phone numbers
- Details of outside help team with address and phone numbers.

15.8.6   Coordination with External Parties

The safety officer should also have close coordination with other nearby industries, fire fighting stations, police, nearby hospitals and other agencies who can provide more help and vehicles in case of emergency.

15.9   Guidelines for Storage of Petroleum Products and Fuels

The following arrangement should be provided at the facility for storage tanks of petroleum products and fuels.

- Provision of lightening arrester for each storage tank.
- Preferably two earth connection for each storage tank shall be provided. These earth connections shall be tested every 15 days or earlier (as per statutory directions) for continuity.
- Arrangement for cooling water spray with fire extinguisher foam on outside.
- Regular testing of wall thickness of storage tanks by ultrasonic tester.
- There shall be dyke walled enclosure for all the tanks and the capacity should be sufficient to hold the contents of the biggest tank.
- All such fuel storage tanks should be minimum 50 meters away from the area where process plants, furnaces or electrical installation are located. There should be a boundary wall 5 m all around the tanks with barbed wire fence.

- All connecting pipelines should be designed, fabricated and tested at least 8 kgs/cm^2 pressure before commissioning and tested at regular intervals.
- All valves shall be tested for at least 8 kgs/cm^2 pressure regularly.
- Keep the upper 15–20% of the storage tanks empty for vapours.
- Alarm for high level should be provided for each storage tank.
- No permission shall be given to start or operate a vehicle with Internal Combustion engine (petrol/diesel) in vicinity of storage tanks. All such vehicles moving in the other parts of the premises should have flame arresters on the exhaust pipes.
- Fire fighting facility shall be available with fire hydrants, foam extinguishers and Dry Chemical Powder units.
- All electrical fittings, lighting etc. should be flame proof.
- All gas cylinders must be pressure tested and expiry dates should be painted on them.
- Colour codes must be strictly followed for storing gases in them.
- They should never be exposed to high temperature areas, electrical installations, rain or direct sunlight.
- These are some guidelines suggested for safe workings; however the plant management should comply with the rules and regulations as well as instructions given by local statutory authorities for storage of petroleum products, fuels, and Gas Cylinder Rules in force in the country.

Chapter 16
Examples of Waste Heat Recovery in Chemical Industries

16.1 Sulphuric Acid

- Exit gases from sulphur burning furnace at about 950–1050 °C and containing SO_2 are cooled in WHRB #1 to 400–420 °C (before entering converter stage I) to generate steam.
- Exit gases from converter stage I at 590–600 °C (and containing SO_2 and SO_3) are cooled in WHRB # 2 to generate more steam. Gases from last pass of converter at about 410–430 °C are cooled in economiser to generate hot water for feeding to the boilers. The HP steam is used for generating power or for driving a back pressure turbine to operate the air blower, to melt raw sulphur and for process heating. A super-heater (not shown in the figure below) is also incorporated for further heating the steam. Excess steam, if any, can be exported to nearby industries (Fig. 16.1).
- In certain plant designs atmospheric air can be preheated by the hot gases from exit of converter last pass and used for process drying for some other products in the premises. Separate heat exchanger is provided in the plant for this purpose.
- Earlier designs preheated the dry air from the drying tower before admitting to the sulphur burning furnace (Fig. 16.2).

Notes: (1) There is gas flow on shell side as well as tube side of air pre-heater [APH] due to which there is pressure drop on both shell and tube side. (2) When economiser is used there is no pressure drop on tube side. The volume of air from exit of drying tower does not increase till it reaches the furnace

© Springer Nature Switzerland AG 2019   327
K. R. Golwalkar, *Integrated Maintenance and Energy Management in the Chemical Industries*, https://doi.org/10.1007/978-3-030-32526-8_16

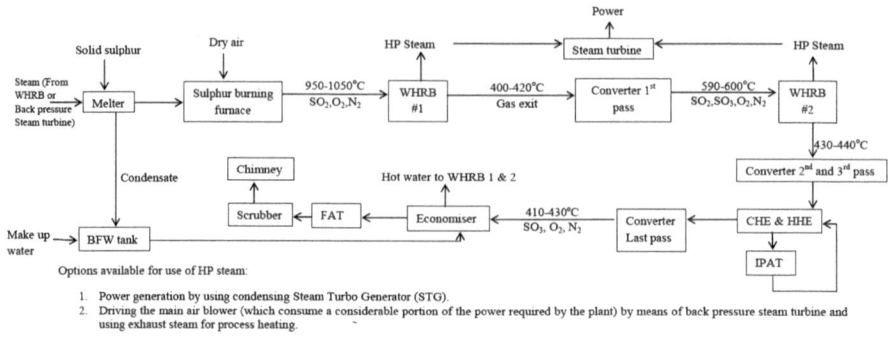

Fig. 16.1 Waste heat recovery in sulphuric acid plant

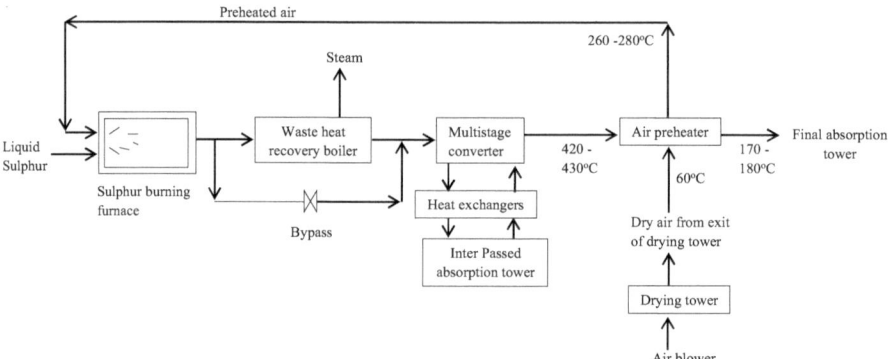

Fig. 16.2 Heat recovery as hot air in a sulphuric acid plant

16.2 Nitric Acid

- Exit gases from ammonia oxidiser at about 900 °C containing nitrogen oxide are cooled to generate steam in WHRB. The steam is used to generate power and for evaporation of liquid ammonia before feeding to the oxidiser.
- Exit gases from the WHRB are used to preheat incoming air to the oxidiser (Fig. 16.3).

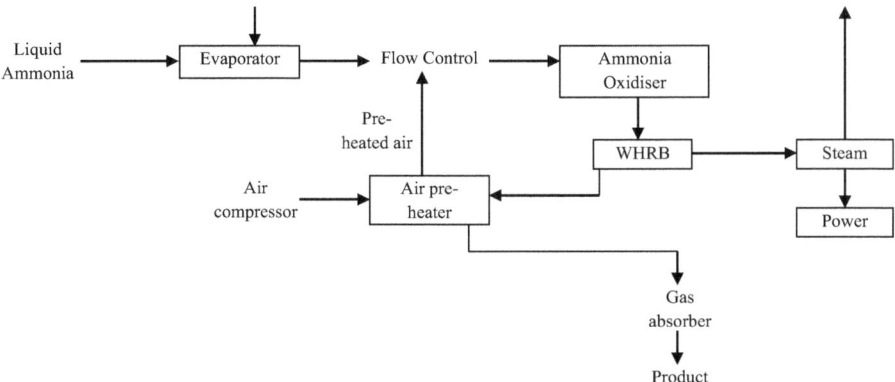

Fig. 16.3 Heat recovery in nitric acid plant

16.3   Hazardous Waste Destruction Plant

- Exit gases from thermal oxidiser are cooled by WHRB (to generate steam) before admitting to scrubbing units. These gases are at about 1100 °C and contain corrosive components like SO_2, HCl, dust, moisture etc. The cooled gases can now be scrubbed more efficiently. The steam is used to generate power.

16.4   Distillery Spent Wash Incineration (Fig. 16.4)

- Exit gases from incinerator are cooled by WHRB to generate steam for producing power and for concentrating dilute spent wash before incineration. *The supply of steam to the spent wash concentrator can be from exit of the back pressure type steam turbine or from the boiler through a pressure reducing station. (not shown)*
- Supplementary fuel in the form of coal is also fired if necessary
- Readers may contact Thermax Ltd., Pune (India) for more information since this technology is developed by them.

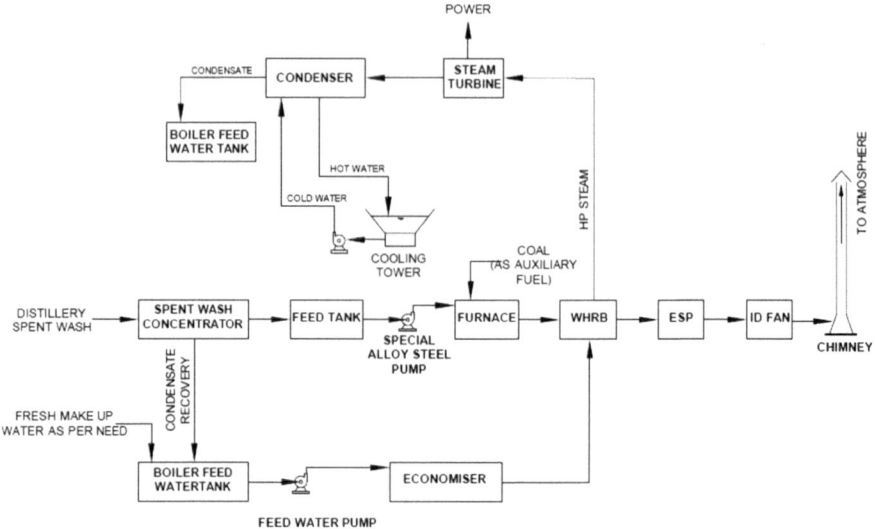

Fig. 16.4 Heat recovery from spent wash incinerator

16.5   Sugar Industry

• Spent bagasse (after extraction of juice) is burnt to generate steam by boilers. The steam is used to generate power and to concentrate dilute juice in evaporators.

16.6   Production of SO_3 by Boiling Oleum

• Hot returning depleted oleum from the oleum boiler at about 130 °C is cooled by incoming stronger oleum at about 48 °C. The stronger oleum gets preheated before entering the oleum boiler (which is heated by hot process gases to generate SO_3 vapours). Some more heat becomes available from process gases as a result and it is used to generate more steam/hot water (Fig. 16.5).

16.7   Rotary Dryer Unit of Rice Husk Ash Pelletisation Plant

The moist pellets are dried in a counter current operation. The product comes out at a high temperature from which heat recovery becomes necessary. Part of the drying air current is used for heat recovery (Fig. 16.6).

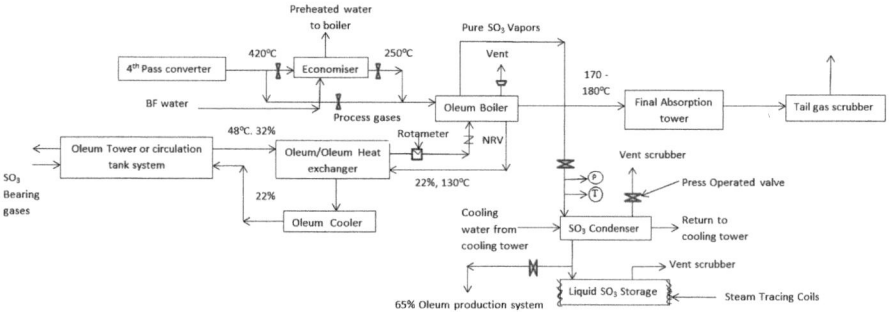

Gas Heated Oleum Boiler (GHOB)

Fig. 16.5 Heat recovery from hot oleum

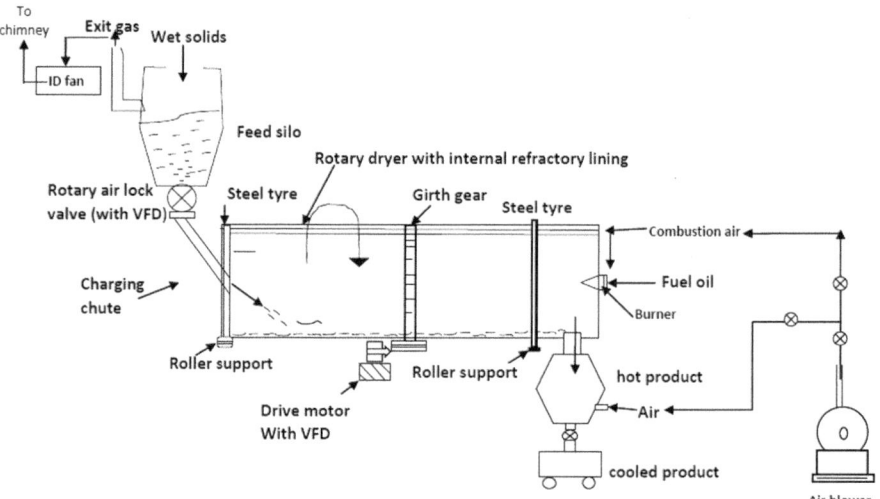

(Counter current dryer with heat recovery from hot flue gases and hot product)

Fig. 16.6 Heat recovery from hot pellets

Waste Heat Recovery in Refractory Brick Manufacturing Plants

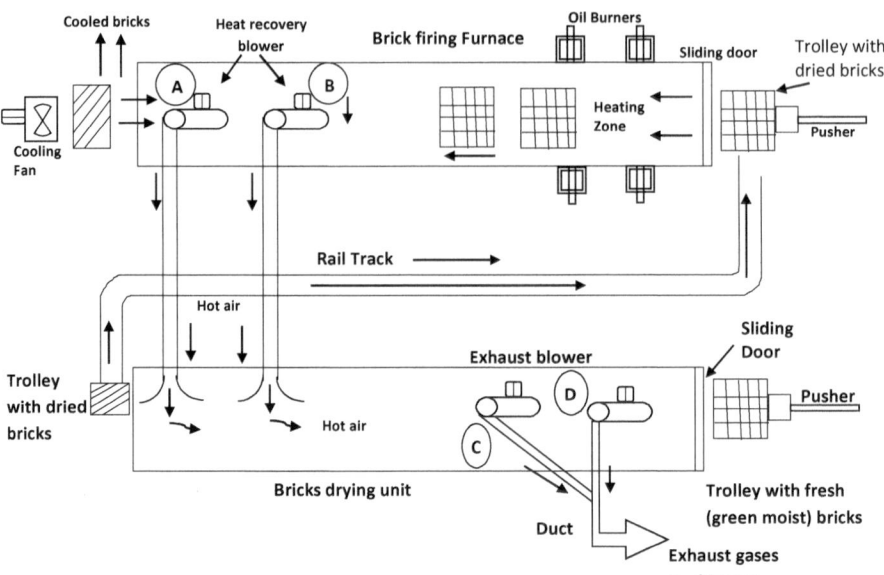

16.8   Heat Recovery (as Hot Air) From Hot Bricks

Note:
- Trolleys loaded with fresh bricks (moist) are pushed in one by one
- After drying the bricks the trolleys are taken to baking furnace equipped with oil burners
- A & B are Heat Recovery Blowers installed on roof of the baking furnace, fresh air blown in by cooling fan gets heated when it passes over hot bricks. The hot air is sucked by A and B and sent to drying unit for fresh (moist) bricks.
- C & D are exhaust blowers to remove the air with moisture.
- Combustion air fan not shown.

Appendix I: Lubrication

Lubrication of machinery is necessary for smooth running of the process units (agitated vessels, stirred tank reactors, absorption towers, pumps, turbo-machinery, effluent control units, ID fans of cooling tower, Air pollution control systems)

The main aims of lubrication are: (i) to ensure smooth movement of machinery surfaces in rotating contact (ii) to minimise erosion and corrosion of the surfaces – since wearing of the surfaces can reduce the contact area partially; while rust formation on machine parts can be harmful as it creates additional friction between the surfaces (iii) to reduce power consumption (iv) to reduce noise level (v) to increase life of the machine parts in contact with relative motion (vi) to reduce temperature of machine parts in rotating contact

Methods for lubrication
- keeping the moving parts sufficiently submerged/in good contact with a thin film of the lubricant by splashing lubricants on the moving parts.
- Maintaining satisfactory circulation of lubricants by means of pumps (positive displacement type are preferable). These pumps can be directly driven by a separate motor or through a separate drive mechanism from the main drive shaft of the machine itself. The lubricant pump motor can have an emergency power supply from a battery also.

Desirable properties of lubricants
- Optimum density and viscosity at all operating temperatures
- Should have higher boiling point (which can reduce loss of lubricant as vapour)
- Should not cause corrosion of the machine parts
- It should be non-toxic [should be safe to handle]
- Non-explosive and non-inflammable (should have high flash point)
- Shall not be too thick (highly viscous) which are not suitable for high speed machines
- Shall not be very thin (low viscosity) which are not suitable for heavy duty, slow speed machines.
- Moisture content should be below 200 ppm.

© Springer Nature Switzerland AG 2019
K. R. Golwalkar, *Integrated Maintenance and Energy Management in the Chemical Industries*, https://doi.org/10.1007/978-3-030-32526-8

Additives used to improve the performance of lubricating oil
They minimise the (i) decomposition/oxidation (carbonising) of oil into particles
(ii) reduce leaks from shaft seals, and (iii) maintain lubrication at all operating temperatures (iv) control viscosity of oil

Such additives permit lubricants to perform better under severe conditions, such
as extreme pressures and temperatures and due to contamination; and thus help in
mechanical performance of the machines.

These are necessary properties which can reduce power consumption and minimise their wearing *due to friction of machine parts* (hence can reduce breakdowns).
These are very important requirements for gear boxes, bearings for high speed
machines and for protection of reciprocating pistons etc.

The breakdowns of machinery must be minimised in a continuous process plant
since they can cause loss of production, affect product quality, waste materials and
energy; and cause environmental pollution.

Pour Point Depressants are helpful for maintaining proper flow of oils at low
temperatures [in severe winter conditions or use of machines at high altitudes].
These additives do not allow oils to become very viscous at low ambient temperatures and hence the surfaces which are in motion (while in contact) continue to get
oil films.

Antifoaming agents are added to reduce air bubbles in the oil. Such bubbles can
cause loss of lubrication due to loss of contact of oil with metallic surface. Pitting
or corrosion of surfaces can occur due to entrained air bubbles (foam).

Sealing property modifiers can cause swelling of gaskets/seals at moving shafts
which can reduce ingress of air in the machine. They also reduce loss of oil through
leaks at seals.

Additives for reducing Wear These additives reduce wear of moving parts.

Necessary advice may be taken from manufacturers of lubricants after informing
them the operating conditions.

Corrosion Inhibitors help to minimise corrosion of the moving parts.

Antioxidants The lubricating oils can get oxidised during use and may get carbonised (producing particulate matter and sludge). This can make them viscous and
affect lubricating properties. Typical additives are organic chemicals which can
retard oxidation.

Carefully select the anti-oxidants *that are compatible with process fluid and also
with machine components* for applications where oils can get mixed with process
fluids. Consult manufacturer of machinery for the right type of anti-oxidants
(retardants)

This is to be done for process machinery where the licensors specify "food
grade oils".

Controlling chemical breakdown
- additives with detergent properties are used to clean and neutralize impurities in oil which can deposits on vital machine parts.
- Rust can increase friction, hence rust inhibitor additives are used.
- Additives for passivating metallic surfaces to prevent the metal from oxidising oil.

For viscosity
- **Viscosity Modifiers** can increase the viscosity of the oil to a higher value at elevated temperatures. This can reduce the tendency of the oil to become very thin at high temperature-(which can reduce the thickness of oil film to a very low level and lubrication may suffer. However such viscosity modifiers when added to oils operating at low temperature do not allow them to become very viscous (which can reduce lubrication)

Cleanliness of lube oil is required to be maintained to improve machine life.

The oils can get contaminated sometimes due to leakages within the system/wear of machine components, etc. Filtration of oil by using 10 microns filter in the oil system can be done and external filtration system can be used if required. The degree of filtration is dependent on the application. National Aerospace Standard (NAS) 1638 is used for measuring cleanliness of oil.

Reclamation of spent lubricants This can be done by filteration of oil (with controlled heating) to remove particulate matter, sludge and moisture, if any. Measured quantities of the components in oil (which have got depleted during operation) are added to restore the properties of the oil. Advice from manufacturer of lubricating oil may be taken.

The spent oils which cannot be reused are to be sent to authorised agencies for recycling; or sent to authorised parties for destruction as per statutory guidelines.

Appendix II: Vibration

Vibration is the unintended high frequency small amplitude movement of equipment and machine components such as shafts, springs, covers; shells of vessels, connected piping, gas ducts etc. The vibratory movement can occur even in structural supports when machines are operated at high speed, at high pressure, under heavy load etc.

Likely reasons

Following likely reasons for vibration in the system or individual components should be carefully looked in to and appropriate actions are to be taken to minimise any damage.

1. **weak civil foundation or loose foundation bolts**
2. **base plates are not rigid** or thick enough
3. **alignment between drive motor and driven machines** (pump/compressor) is not properly done
4. **improper operation**-suddenly opening or closing valves in high pressure pipe
5. **anti vibration pad not provided at all or existing one** is damaged
6. **improper assembly** (with some degree of unbalance) of machine parts which may have to run at very high speed
7. **weakening of structural** support or some reinforcement member has become loose
8. **spring in support for pipes** is damaged
9. **sudden charging of large amounts of material** in reactor having a rotary agitator
10. **accidental loss (draining out) of lubricant** from drain plug of gear box.
11. **excessive axial or radial play in shafts** of pump/compressor
12. **loose coupling halves or shaft keys**
13. **operation of reciprocating pump against high discharge pressure** if dampening pot is not properly working/not provided at all
14. **pipelines are supported/clamped without anti-vibration pads**

© Springer Nature Switzerland AG 2019
K. R. Golwalkar, *Integrated Maintenance and Energy Management in the Chemical Industries*, https://doi.org/10.1007/978-3-030-32526-8

15. **Non Return Valves** not provided for individual pump while more than one pump is discharging in the same pipeline in parallel. There could be back pressure on the pump which is not able to develop sufficiently high pressure

Dynamic balancing

Unbalance is measured as gm-mm. (which means weight in gramme at a distance in millimeters from center of the rotor.).

The unbalance is identified by rotating the assembly on a balancing machine. The desired grade of balancing, weight of rotor, speed at which it is to be checked (*at a slower speed than operating speed for rigid rotors*), dimensions of the rotor, planes of balancing are to be fed in the software designed for Dynamic Balancing. When the rotor is run, the software displays the location and weight of the unbalance on the screen. Either equivalent weights are added to a side or removed from the opposite side in presence of Original Equipment Manufacturer's engineer

As good engineering practice the unbalance of all rotors should be checked before assembly/fitting in to the volutes (body) and corrected.

The pulley or the flywheel of reciprocating machines/compressors are to be balanced separately before replacement.

Resonance

System components have a natural frequency of vibration. Nearby (connected) items may have a similar frequency of vibration and at a certain point of operation they suddenly start vibrating with much more amplitude than individual values. This can be dangerous if the system as a whole vibrates too much; hence keep a strict watch while increasing speed slowly.

How vibrations affect working of chemical process plants

Raw material feeding to process units

- Crushers, pulverisers, centrifuges generally tend to vibrate more *causing internal wearing out or may even damage some components if not attended in time*
- Settings of metering pumps, control valve for oil burners or valves in combustion air lines can get disturbed.
- Day tanks/Receiving silos may overflow because of malfunctioning of level controllers due to vibrations and supply from main storage will continue in spite of high level.

Conveyors speed of screw conveyors, bucket elevators will need resetting if spillage of material is taking place in some spots due to vibrations

Pneumatic Transfer arrangement

The position of level sensors in discharge end silos may get disturbed which may cause spillage due to overflow;

Rotary Air Lock valve for feeding may need resetting (if disturbed from original setting).

Agitated vessels

- Excessive vibrations can damage inner glass lining; sight glass & light glass assembly, blades of agitator shaft may become loose (which can cause improper mixing), sensors for temperature & pressure may get damaged or displaced from their position (thus indicate incorrect reading).

- Misalignment of agitator shaft with driving gear box can cause excessive vibrations resulting in bearing failures, gland packing getting worn out rapidly.

Process reactors and accessories
- sensors for temperature & pressure may get damaged or displaced from their position (and then indicate incorrect reading)
- incoming feed of raw material may not get uniformly distributed if the distributor plate/nozzle for such distribution gets tilted or even detached from the support
- protective lining of glass, refractory or acid resistant bricks may develop cracks (the equipment life can get reduced due to seepage of corrosive chemicals through these cracks)
- pressure relief valves (pop-up type) will blow off frequently even when the pressure is less than the setting. This can cause waste of reactants and products; and environmental pollution. Pressure release safety valves are very important and hence care is taken at all installations to fix the setting properly on these valves.

Fixed bed Converters
- settling of catalyst beds may occur due to vibration (which can cause more compact beds which reduces porosity and hence the resistance to gas flow increases). This can increase pressure drop in the bed
- disturbance of the sensors (located typically at the interface of catalyst with support bed of pebbles) for temperature indicating instruments can give erroneous readings for temperature changes occurring in the bed. This can lead to incorrect inference about the degree of conversion taking place;

Fluidised bed Converters disturbance of sensors for process conditions may occur and working of process controllers will be affected (process control may become difficult)

Electrical systems-motors, bus-bars, interconnections
Vibrations can result in loose contacts at junction boxes of motors, control circuits, lighting, and important safety interconnecting circuits etc. These can create serious problems like sparking, damage to equipment, short circuits and fires in rare cases.

Instrumentation Sensors/instrument probes may get slightly shifted from present location (can result in incorrect indication of temperature, flow rate, level in tank, pH), setting of controllers can get disturbed and process control can get affected;

Contacts with connecting cables (for indicating instruments in control room) can become loose and indicate erroneous reading in control room.

Compressed air system: moisture trap (for automatic draining) will blow off frequently

If the high pressure trip switch for compressor provided on air receiver is disturbed, the vessel may get over pressurised/will not have enough compressed air (under pressurised)

Absorption towers
Settling of tower internal packings, tilting of distribution trays can occur.

Spray nozzles for incoming absorbent liquid can tilt or come off from incoming pipes (which can result in improper distribution of absorbent liquid over the cross section of absorption tower–hence inefficient absorption/bypassing can take place)

Acidic mist from absorption tower exit can escape in case of displacement of demister pads due to vibrations (caused by high gas flow rate). This can corrode downstream equipments (e.g. in a DCDA Sulphuric acid plant); result in environmental pollution, corrosion and harmful effect to surroundings,

Boiler water level control level controller setting can get disturbed resulting in high level of water (which may even enter steam line) or low level in boiler which may reduce its life/cause overheating of the tubes since there is no water available to cover them (e.g. Smoke tube boiler).

Pressure release safety valves are very important and hence care is taken at all installations to fix the setting (should not get disturbed due to vibrations) on these valves.

Oil firing systems
Setting of temperature control for oil heaters (for heating fuel oils during winter),discharge pressure of oil supply pumps, dampers in air lines for primary air (for atomising air) secondary and tertiary air for complete combustion; pressure relief valve (generally spring loaded) on the furnace may get disturbed.

There can be incomplete combustion of oil, carryover of soot to downstream units, flame may hit the inside walls of combustion chamber, and chances of explosion may arise due to presence of excess oil vapour.

Gas ducts, process piping
Glass pipes, Cast Iron pipes can develop cracks and process fluid can leak out. Inner protective lining of glass lined pipes can get detached and pipes will leak/their life will get reduced.

Vibrations can loosen clamps or pads provided for glass piping/cast iron pipes and they may get damaged (if vibrations are excessive) after some time

Damage to gaskets, supports to pipelines or Plate Heat Exchangers can result in leak from flanged joints: corrosive liquids or dangerous gases (toxic, inflammable, corrosive) can come out and spread in the premises and cause accidents.

Systems operating under vacuum will have air ingress and these units may not work efficiently.

Bypass valves for process vessels are adjusted as per requirement. Vibrations can close or open them further which makes it very difficult to maintain process conditions

Steam piping
leak can occur from flanged joints in pipes, or damage to supports will result in heavy leak of steam (due to excessive vibrations)

Steam turbines, Compressors

Vibrations can disturb settings of controllers for steam flow, pressure, high speed alarms and trip settings.

The alarms for low pressure of lubricating oil and tripping of turbine in case of very low oil pressure may also get disturbed resulting in nuisance tripping of the turbine and thereafter sudden loss of output from driven unit/disturbed working of important process units.

Air Pollution Control APC and Effluent Treatment Plants ETP

- Sensors for analysis of exiting gas/exiting treated effluent may indicate incorrect readings if they are displaced from installed locations.
- Setting of pH controllers for scrubbing liquor in APC, and for primary treatment of incoming effluent can get disturbed.
- Level indication of alkali solution tanks may become erroneous and they may get emptied during operation or overflow during filling.
- VFD controllers for aeration blowers can also get disturbed

Gas supply systems–N_2, O_2, Fuel gases

- controller settings may get disturbed and process vessels may get over pressurised,
- N2 blanketing may fall short of requirement
- Excessive supply of fuel gases can create explosive conditions in combustion chambers

Intentional Vibrations are provided for

- Feeding of solid powder from silos
- Removal of ash/dust deposits collected by the bags in Bag Filter, deposits on Electrostatic precipitator, boiler tubes
- Removal of air bubbles from cast cement layers to make strong foundations, removal of air bubbles from castable refractory layers to make them capable of protecting metallic surfaces, anchors, external shells of furnaces.
- Digging out hard deposits from process vessels, or Sulphur melter pits during annual plant overhauls for cleaning

Preventive measures

1. **Strengthen** the foundation of machinery/agitated vessels. Use well designed foundation bolts as per OEM advice. Use rigid base plates
2. **Check accurately** the alignment between drive motor and driven machines
3. **Avoid improper** operation **Do not** suddenly open or close valves in high pressure pipe
4. **Provide anti vibration pads** as per advice from OEM
5. Check complete assembly of machine parts which may have to run at very high speed
6. Provide reinforcement member at weak spots of structures. Get the design and fabrication tested at least three times the maximum expected weight during operation.

7. Replace damaged springs which are provided for support of pipelines. Cover them by grease or corrosion resistant paint
8. Avoid sudden charging of large amounts of material in reactor having a rotary agitator.
9. Display Standard Operating Practices prominently on boards in process plant
10. Prevent accidental draining out of lubricant from drain plug of gear box by providing a locking pin or lock nut for the plug.
11. **Never allow** excessive axial or radial play in shafts of pump/compressor
12. **Check any loose** coupling halves or shaft keys and attend them immediately
13. **Do not operate** reciprocating pump against high discharge pressure if dampening pot is not working properly.
14. **Glass, Cast Iron pipelines** must be supported/clamped with proper anti-vibration pads on support
15. **Non Return Valves must be** provided for individual pump if more than one pump is discharging in the same pipeline in parallel. There could be back pressure on the pump which is not able to develop sufficiently high pressure

Appendix III: Some Typical Non-destructive Tests

General

- **These tests shall be done in presence of Purchaser's representative**
- Visual checks of all dimensions, nozzle orientations, fittings etc.
- Ovality checks (refers to maximum diameter minus minimum diameter.)
- The pressure vessel may be de-rated if the ovality is beyond acceptance limits given by designer.
- Only certified (calibrated through standard test agency) pressure gauge should be used for the test. Test records for calibration of the pressure gauges (used during NDT) shall be available to purchaser as well as Statutory Authorities.

Dye Penetration

This is used for detecting surface flaws. It is a quick low cost test, but shall not be used to detect faults in welded joints.

Thoroughly clean the surface to be tested and remove all oil, grease, dust etc. Clean solvents, dilute alkali solution may be used for this purpos. It is then wiped off and dried.

Now the coloured penetrant liquid is applied on the surface and allowed to soak for 15–30 minutes. Excess penetrant shall not be applied. Now a clean solvent is sprayed to remove the penetrant fluid from the surface. This is followed by application of a developer which is a white material. It takes out (draws out) the coloured penetrant from the flaw and indicates any cracks etc.

Ultrasonic Thickness Test

This can be done by an instrument to check the thickness of wall of a vessel and warn the operator of a leak that may occur. In case of a vessel becoming too thin, it should be taken out of service and the weak portion shall be attended by patch welding if feasible. However in case of a pressure vessel or one which is required to handle dangerous fluids it should not be used again for the same purpose even after repairs to prevent mishap.

© Springer Nature Switzerland AG 2019   343
K. R. Golwalkar, *Integrated Maintenance and Energy Management in the
Chemical Industries*, https://doi.org/10.1007/978-3-030-32526-8

Hydraulic Test
- All valves, level indicator connections etc. are to be removed and the openings are to be closed by blinds. The inside air is removed out from the vent at topmost point.
- The vessel to be tested shall now be filled up completely by demineralised water.
- It is to be slowly pressurized to 1.5 times the maximum working pressure, or as instructed by the inspecting statutory authority. It is to be kept pressurized for a time as instructed. The pressure is to be carefully monitored every 30 minutes. The vessel is to be inspected from all sides by soap solution to detect any small/minute leak by small bubbles. The leaking spot is to be marked and welded if permitted. The test is to be repeated after this.
- In case of certain leaks permission may not be granted to use it as a pressure vessel
- In case of a minor leak, the inspecting authority may derate the vessel and allow it to be used at a lower pressure or for handling fluids which are not dangerous.
- If there is no fall in the pressure the vessel can be slowly depressurized to ambient pressure as per instruction.
- All water inside is to be drained out. Final washing is to be done by *demineralised water.* Test water sample by Silver Nitrate solution for presence of any chlorides in the water (as they are corrosive.)
- The vessel is now to be completely dried by warm air or nitrogen to remove all traces of water from crevices inside since it can react with acidic gases during operation of the plant to form corrosive products inside.

Spark Test
It is carried for testing the integrity of glass, rubber, fiber glass reinforced lining of a vessel. One electrode of the test kit is connected to the outer shell of the vessel and the other electrode is connected to a smooth conducting wire brush.

A voltage of 3000–7000 volts (as instructed by the designer and as agreed by the vendor appropriate for the thickness of lining provided) is applied. The brush is slowly moved all over to detect any sparks. If no spark is detected the lining is ok.

Very high voltage can damage the lining, hence one should be careful.

Radiography
This test is used for detecting flaws in welded vessels. The surface of the vessel is cleaned. A source of radiation is used to send a beam of radiation to the surface. A sensitive film is placed at the other side to detect the intensity of radiation passing through the welded part. More rays can pass through flaws in the welded part while less rays will pass through properly welded parts.

Protective devices must be used and the test carried out under supervision of an expert.

The exposed film shall be preserved for future reference.

Appendix IV: Disposal of Hazardous Waste

Many chemical industries generate effluents which are harmful to the environment. These should be processed and recycled in to the same plant or are to be rendered totally harmless before discharge.

However some of the effluents have constituents in them which are very dangerous to the environment due to characteristics such as:

- Being corrosive to surroundings
- Chemically reactive when they come in contact with surrounding materials or rain water and can release toxic compounds or gases
- Can self ignite in ambient conditions due to their inflammable nature (low flash point)
- Explosive nature
- Toxic to living beings and vegetation.

Most of the countries have identified and listed industries which can generate hazardous waste. The dangerous constituents and their concentration limits are defined by the pollution control authorities in these countries and very stringent regulations for storage, handling, recycling, packing, transportation, and disposal of such hazardous wastes are prescribed.

If the generator of waste does not have suitable arrangement for the proper disposal as per prescribed rules then they may be permitted to hand over the wastes to **authorized** Common Hazardous Waste Transporter for subsequent storage at their specially designed facility for Secured Land Fill or Thermal Destruction. The authorized party will analyse the wastes for the physical and chemical properties, calorific value and suitable pathway for the treatment will be worked out accordingly.

The waste can then be sent to the secured land fill (with leak proof lining) after suitable chemical treatment or sent to thermal destruction.

Generally two stage procedures for disposal are prescribed as follows:

(i) incineration at 800°C in suitably designed furnaces lined with high temperature resistant refractory and equipped with a proper charging system

© Springer Nature Switzerland AG 2019

K. R. Golwalkar, *Integrated Maintenance and Energy Management in the Chemical Industries*, https://doi.org/10.1007/978-3-030-32526-8

(ii) the incinerator will have to be followed by another furnace operating at 1100°C, with a residence time of at least 2 seconds for the complete combustion and destruction.

(iii) The exit gases from the second furnace are to be passed through properly designed air pollution control units.

The ash generated after incineration is to be further treated (since it can be toxic) and finally disposed off in secured landfills.

Plasma Gasification Vitrification process
This is the modern Plasma Gasification & Vitrification process which is used for destruction of those wastes which are difficult to treat by conventional physical, chemical methods or incineration in furnaces.

This process also generates very negligible amount of ash.

The wastes are mixed with suitable fluxes and charged at a controlled rate in to a reactor where only a limited oxygen is available through strictly controlled air stream which is subjected to intense heat generated by electrically operated plasma torches. The ingress of air is minimized during operation.

The waste is destroyed due to very high temperature air and generates an ionized combustible gaseous exit stream called Syn gas. This gas stream is then passed through another unit operating at 1100°C + for completing the destruction of pollutants.

It is followed by Waste Heat Recovery Boiler to generate steam which can run a steam turbo-generator to minimize the draw of power from external grid. The plant can even become self sufficient in power when the calorific value and the composition of the waste is appropriate.

The exit gases are passed through venturi scrubber, packed tower, electrostatic precipitator and polishing tower for complete removal of the pollutants from the gaseous stream. An induced draft fan operates the plant under negative pressure and thus prevents any escape of pollutants.

The treated gases are continuously monitored and then finally discharged through a tall chimney whose height is as per rules prescribed by local pollution control authorities.

The fluxes added to the reactor (*during charging of waste*) combine with other constituents of the waste and a molten mass is taken out from the reactor. It is then allowed to cool.

The cooled mass has a glassy impervious surface and hence no toxic/heavy metals can come out unlike the ash generated by conventional incineration.

This vitrified mass can be used for lining of drains, pavements etc.

The main features of the process are:

• Electrically generated high temperatures instead of heating by firing of fuels.
• Complete destruction of hazardous waste to produce combustible syn gas.
• Minimum generation of finally disposable molten mass
• Heat recovery from exit gases- (to generate steam for further use)
• Air pollution control by multiple process units

- Safety of personnel and machinery is ensured by (i) multiple layers of refractory lining suitable for high temperature (ii) strict control on power input to the torches (iii) complete chemical and physical analysis of waste before charging ensures correct feeding programme of waste (iv) electrical interconnections, warning alarms, safety valves at appropriate locations (v) multi stage air pollution control

Maintenance

Careful inspection and maintenance is required for the following units in this plant since it handles corrosive wastes at high temperatures.

- Waste feeding system
- Plasma Gasification Reactor (special grade of refractory)
- Plasma torches and their electrical controls
- Thermal Oxidiser furnace (special grade of refractory)
- Waste Heat Recovery units (steam generation at high pressure)
- Air Pollution Control units (venturi scrubber, packed tower, electro static precipitator, polishing scrubber.)
- Special instrumentation and safety devices
- Plant auxiliary units like compressor, water treatment etc.

This technology has been successfully developed and implemented by **SMS Envocare Ltd (a Nagpur based company in India)** at their plant at MIDC, Ranjangaon, District Pune (Maharashtra State), India.

The author has no intellectual and any other rights on this technology.

Readers may contact this company on following Head Office address for further details.

SMS Envocare Ltd, 20, I T Park Road,

Nagpur 440022 (Maharashtra State), India.

Appendix V: Hydraulic Motors and Systems

Very useful for providing more torque for agitators in process reactors, mixers, dissolvers if the liquid is viscous or the blades are stuck up due to any reason. It may otherwise require a motor of higher rating only for start up even when the running load is much less. This needs more investment in case of need for flame proof motors and accessories; and may consume more power.

Also used for providing more force for pushing/pulling operating rods to dampers or for providing rotary (both clockwise and anti clockwise) movement for operation of valves or charging doors of furnaces, reactors.

Typical Components

Hydraulic Oil Tank, Oil filter, High Pressure Circulation Pumps, Control Valves [for forward and backward movements of pistons/rods; clockwise and anti-clockwise movement of shafts.], manual over-ride arrangement for operating pistons/rods; hydraulic motors attached to agitator shafts/loads which may need large torque at start

Cooling system of oil (external or internal cooling coils for the oil tank, cooling tower, water circulation pumps), high pressure flexible piping (hose pipes with stainless steel braiding), Pressure release valves, rupture discs, electrical motors (generally flame proof), plant lighting, standby power supply arrangement for emergency, structural steel support members, alarms for high temperature and pressure of hydraulic oil and other required instrumentation etc.

Observations to be made during running of the system

These observations can indicate whether the system is working properly or not.

- Oil pressure and temperature at discharge of oil pumps, inlet and exit of hydraulic cylinders.
- Current drawn by oil pump motor.
- Cooling system for oil [temperature of oil at pump discharge and in the main tank].
- Fluctuations, if any, in oil pressure in various lines.

© Springer Nature Switzerland AG 2019
K. R. Golwalkar, *Integrated Maintenance and Energy Management in the Chemical Industries*, https://doi.org/10.1007/978-3-030-32526-8

- Jerky motion of pistons/equipment shafts.
- Working of control valves in oil lines, limit switches for piston movement
- Setting of stroke length for operating pistons (which move to and fro).
- Any jamming of shafts (to be rotated/to be pushed/pulled)

Maintenance
- Inspect Oil Tank, Oil pipe lines, gaskets for any leak
- Mechanical shaft seals for oil pumps.
- Confirm smooth operation of all valves; including control valves.
- Clean Oil cooler, water circulation pumps, oil filter
- Working of pressure release valves, high pressure and temperature alarms
- Limit switches on movement of piston
- Some parts (bearing, volute, discharge nozzle) getting hot.
- Lubricant/grease coming out from bearing.
- Shaft Seal getting hot. Change if necessary
- Drive belts/gear box becoming hot.
- Add make up quantity of correct grade of hydraulic oil from reputed manufacturer

Restart of system
Take trial of the oil circulation arrangement at no-load by running both running and standby pumps one by one.

Check
- if there is any oil leaks or valves in oil lines are not operating properly.
- the running of hydraulic motors in both clockwise and anti-clockwise directions
- setting and operation of safety valve to release oil pressure back to circulation tank. current drawn by motor.
- cooling arrangement for oil before taking in use.

Appendix VI: Protective Lining for Process Units

Protective linings are provided for certain process units like reactors, furnaces, absorbers, circulation tanks etc. They are also provided for units in air pollution control systems and water treatment plants.

The main purpose is to provide protection for the metallic shells against:

- chemical corrosion and extreme heat.
- prevent contamination of products due to contact with metallic surface.
- To conserve heat in the equipments.
- To reduce plant stoppages due to leaks from process units.

Materials for linings

- Refractory lining for thermal protections. These are fire bricks, castable refractory, insulating bricks and castable refractory with appropriate anchors of stainless steel They are used for units like furnaces operating at high temperatures (500–1200°C). Generally more than one layer is provided to meet the requirement.
- Rubber, Fibreglass Reinforced Plastics, Glass for protection against chemical corrosion. These are used in comparatively thin layers (2–5 mm thick). These lining can get damaged due to sharp particles in the material being handled. They are generally used at lower temperatures and provide protection against corrosive chemicals like alum solution, acidic liquors, dilute acids.
- Stainless steels, special alloy steels lining for thermal and chemical protection. These lining are used in thin layers (as alloy/as cladding material) and have good mechanical strength. Can be used in presence of corrosive gases, particulate matter, solids at somewhat higher temperatures (400–500°C).
- Lead bonding with base metal
- Glass lining

PTFE (Poly Tetra Flouro Ethylene) lining
This is not attacked by most of the chemicals. Hence it is suitable for highly corrosive chemicals, for preventing product contamination due to equipment walls/shafts/agitators.

© Springer Nature Switzerland AG 2019
K. R. Golwalkar, *Integrated Maintenance and Energy Management in the Chemical Industries*, https://doi.org/10.1007/978-3-030-32526-8

Generally a thickness of about 2.0–3.0 mm is used.

However, it is not suitable for use above 250°C, and in presence of sharp particles in process liquids. Heat transfer through PTFE lining is also poor.

General: A small drain hole may be provided in the unit at bottom to detect any failure/cracks in the lining if any continuous flow of liquid is seen.

It is preferable to remove the entire lining if a damage is seen. New lining shall be installed.

Selection criteria for Lining Materials

- Operating conditions inside the equipment.
- Material being handled and their compositions; corrosive and erosive properties. (chemical composition, acidic/basic nature, presence of sharp particles).
- Melting point of materials and tendency to form hard deposits on solidification.
- Heat transfer for the vessels from outside can be difficult when refractory bricks are provided internally. The materials inside the vessels may need internal heating by oil firing or electrical heating. However, it is necessary to provide impervious and non-conducting lining if there is possibility of formation of conducting slag during operation. Mica sheets/Ceramic paper sheets may be required as additional protection for the shell.
- Electrical resistance of the lining must be checked to prevent electrical short circuiting and damage to equipments due to electrical sparking.
- Consider carefully if there will be high operating vacuum in the equipments *which may peel off the lining.*
- Hardness of lining material shall be checked—for better erosion resistance. (for mixer and agitator blades, shafts) example—mixer shaft and paddles are made from Ni-hardened alloy for manufacture for single super phosphate by reaction of ground rock phosphate with sulphuric acid.

Test certificates

Vendor should provide test certificates for refractory bricks, rubber lining, glass lining etc. prior to application. Physical and chemical properties shall be confirmed for use as lining material. Small test pieces can be provided with the material and tested at actual operating conditions/by simulation in approved laboratories.

Some important tests for refractory bricks Draw the samples as per Sampling Plan agreed between vendor and purchaser:

- Alumina percentage,
- density and porosity of bricks,
- thermal conductivity,
- cold crushing strength,
- refractoriness under load (RUL) in Kg/cm^2 at specified temperature, Pyrometric Cone Equivalent (PCE) Values.
- coefficient of thermal expansion and variation in dimensions

Some important tests for Acid Resistant bricks Draw the samples as per Sampling Plan agreed between vendor and purchaser:

- Variation in dimensions (L,B,H) shall not be more than specified in purchase order
- Resistance to acid (% loss in weight in 24 hours when exposed to operating conditions. This shall not exceed 1.0–1.5%,
- Cold crushing strength as measured in kgs/cm^2,
- Glazing on surface to make it smooth and impervious
- Porosity (not to exceed 1.5%)

Installation of lining

The surface to be lined must be thoroughly cleaned by wire brush/grinding of the welding seams or by dilute inhibited acid before providing the protective lining.

Brick-lining of equipments shall be done at site since the lining can get disturbed during transportation.

Rubber lining, PTFE lining can be done at vendor's works.

Consult vendor when rubber lining is to be provided to ID fans with high tip speed. (At high tip speeds the rubber lining can come off exposing the internals to corrosive gases).

Glass lining shall be done at vendor's works and given heat treatment in special furnace by slow heating, maintaining at the elevated temperature and slow cooling thereafter. The vessel should not be subjected to vibrations or hammering from outside. Take special care during transportation to site.

The anti-vibration pads at supports must be checked.

Check the strainers in liquid inlet line so that hard sharp particles are filtered out before entering the glass lined vessel.

Spark Test

The linings shall be tested at 5000 volts by connecting one end of the supply to the shell and connecting the other to a smooth wire brush which is moved slowly all over the lined surface. Any sparking will indicate a fault. (**vendor should agree with the test voltage**)

Vendor to inform

- Resins and hardeners being used and their shelf life so far.
- Curing method which will be used by the contractor.
- Site facilities (steam, power, fuel etc.) required by vendor if the vessel is big and the job will be done at site only.
- When can the process vessel/equipment be put into service?
- Process vessels shall be carefully cured (after repairs/replacement of lining) by slowly heating, then holding to the higher temperature and finally cooled at a rate as advised by OEM.

Appendix VII: Some Welding Processes

Some common welding processes are as below. Consult welding specialist for such jobs. Fabricators should seek **approval from Statutory Authorities for fabrication drawings and the welding processes before fabrication** of the pressure vessel and those which are required to handle dangerous materials.

Some of the welding processes used are

1. SMAW—Shielded Metal Arc Welding (not for high pressure vessels)

Arc is struck between covered welding electrode and work piece. Shielding protects the molten metal from getting oxidized and form oxide.

This is a manual welding technique that uses a consumable electrode coated by a flux. It is then used to carry out the welding. This process is also called stick welding some times because it uses welding sticks or rods that are made up of filler material and flux. The flux protects the molten metal from oxidation and the filler inside the stick is used to join/fuse two pieces of metal together.

2. GTAW—Gas Tungsten Arc Welding (for higher pressure vessels after Statutory approval)

Inert gas cover is provided to protect the molten metal
Non-consumable Tungsten electrode is used to create arc

3. TIG—Tungsten Inert Gas process uses the heat generated by an electric arc between the metals to be joined and an infusible tungsten-based electrode, located in the welding torch. The arc area is enclosed in an inert gas shield that protects the molten weld pool and the tungsten electrode.

The filler metal as a rod is applied manually by the welder into the weld pool to join the two metals.

4. SAW—Submerged Arc Welding

Continuous feeding of wire for welding and feeding of flux to shield the welding zone from atmosphere

© Springer Nature Switzerland AG 2019
K. R. Golwalkar, *Integrated Maintenance and Energy Management in the Chemical Industries*, https://doi.org/10.1007/978-3-030-32526-8

5. GMAW—Gas Metal Arc Welding

May not be used for thicker plates because small wire diameter and low heat input can cause poor fusion.

Fabricator must possess valid certificates for fabrication of pressure vessels (as required by Statutory Authorities).

Post weld heat treatment PWHT *is mandatory* when vessel, pipe, plate, etc. are >38 mm thick, and when used in lethal service for handling dangerous chemicals such as H2S, concentrated acids, caustic soda etc.

It should be insisted upon.

- To relieve stresses introduced during fabrication
- To improve corrosion resistance
- To improve strength and toughness
- **No changes or corrections involving cutting/welding are permitted after PWHT**.

Vendor shall have proper facilities for PWHT as
- A furnace with heating systems with *calibrated thermocouples*
- *Automatic temperature control*
- *Proper Loading and Unloading facilities*
- *NDT facilities*
- Inspection after PWHT must be done once again for welded joints

Radiography of welded joints is a must when lethal fluid is handled, when the vessel will be used below minus 10°C as per Indian standards.

Follow equivalent international standards when in doubt.

Such precautionary steps are essential for better life of the vessel and minimizing breakdowns.

Index